AF738848

Springer Texts in Business and Economics

Springer Texts in Business and Economics (STBE) delivers high-quality instructional content for undergraduates and graduates in all areas of Business/Management Science and Economics. The series is comprised of self-contained books with a broad and comprehensive coverage that are suitable for class as well as for individual self-study. All texts are authored by established experts in their fields and offer a solid methodological background, often accompanied by problems and exercises.

More information about this series at http://www.springer.com/series/10099

Jiří Witzany

Derivatives

Theory and Practice of Trading, Valuation, and Risk Management

Springer

Jiří Witzany
Faculty of Finance & Accounting
University of Economics Prague
Prague, Czech Republic

ISSN 2192-4333 ISSN 2192-4341 (electronic)
Springer Texts in Business and Economics
ISBN 978-3-030-51750-2 ISBN 978-3-030-51751-9 (eBook)
https://doi.org/10.1007/978-3-030-51751-9

© The Editor(s) (if applicable) and The Author(s), under exclusive license to Springer Nature Switzerland AG 2020
This work is subject to copyright. All rights are solely and exclusively licensed by the Publisher, whether the whole or part of the material is concerned, specifically the rights of translation, reprinting, reuse of illustrations, recitation, broadcasting, reproduction on microfilms or in any other physical way, and transmission or information storage and retrieval, electronic adaptation, computer software, or by similar or dissimilar methodology now known or hereafter developed.
The use of general descriptive names, registered names, trademarks, service marks, etc. in this publication does not imply, even in the absence of a specific statement, that such names are exempt from the relevant protective laws and regulations and therefore free for general use.
The publisher, the authors, and the editors are safe to assume that the advice and information in this book are believed to be true and accurate at the date of publication. Neither the publisher nor the authors or the editors give a warranty, expressed or implied, with respect to the material contained herein or for any errors or omissions that may have been made. The publisher remains neutral with regard to jurisdictional claims in published maps and institutional affiliations.

This Springer imprint is published by the registered company Springer Nature Switzerland AG.
The registered company address is: Gewerbestrasse 11, 6330 Cham, Switzerland

Preface

The goal of this book is to cover basic and advanced topics in the valuation and modeling of (financial and commodity) derivatives, their institutional framework, and their risk management. It is dedicated mainly to graduate students and financial markets practitioners, but it may also be of interest to researchers in the fields of derivatives valuation, stochastic modeling, and risk management.

The book is based on its preceding version, Witzany (2013b), and on financial derivatives lecture notes, Witzany (2011), for the newly established financial engineering master's degree program students at the University of Economics in Prague. Compared to those publications, the text has been substantially revised and extended, to include new regulatory requirements such as Basel III, the Fundamental Review of the Trading Book (FRTB), the Interest Rate Risk of the Banking Book (IRRB), or the Internal Capital Assessment Process (ICAAP). There is also a detailed treatment of the counterparty credit risk, stochastic volatility estimation methods such as MCMC and particle filters, and the concept of model-free volatility, VIX index definition, and the related volatility trading.

The first four chapters can be used in an introductory derivatives course and cover the basic topics, i.e., forwards, futures, interest rate swaps, and options, including the Black–Scholes formula derivation in the context of binomial trees and the continuous time stochastic modeling approach. Regarding stochastic modeling, we systematically use an intuitive approach based on the concept of binomial trees extended to continuous time using the concept of infinitesimals. More details on the foundation of infinitesimals and of classical stochastic calculus are given in appendices A and B. The fifth chapter focuses on risk management issues. It deals not only with the elementary principles but also with more advanced topics such as value at risk estimation techniques, or credit valuation adjustment (CVA), which currently present a challenge to financial institutions and researchers. The chapter explains the principles of several recent important regulations that have a fundamental impact on all risk managers, namely the Fundamental Review of the Trading Book, which replaces the concept of value at risk with the conditional value at risk, operational risk, or ICAAP modeling requirements. The last three chapters deal with more advanced derivative topics. The valuation of interest rate options is not possible without deeper stochastic calculus foundations introduced in Chap. 6. This chapter also covers the classical Bachelier model, which has found new applications in times of negative interest rates. Besides the standard market model, in Chap. 7 we investigate the most well-known short-rate and term-structure models. The last chapter discusses a number of exotic derivatives, such as binary options, warrants, barrier or Asian options, and various alternative stochastic models that have become a necessity due to empirical phenomena such as volatility smile, skew, or surface. It turns out that the volatility of asset return is itself stochastic. Its modeling and estimation have been the subject of intensive research during recent years. Selected recent interesting results and advanced estimation approaches are covered in this chapter. The chapter concludes with the definition of the VIX volatility index and looks at volatility trading, which has become a relatively new phenomenon on the derivatives market.

The readers are certainly also encouraged to study other global derivatives textbooks that provide more focus and details on various other topics, such as Hull (2018), Wilmott (2006), Cipra (2010), and the mathematically more advanced Shreve (2004, 2005). Czech readers are also recommended Dvořák (2011) or Cipra (2008).

Although the materials presented here have been thoroughly checked, some mistakes may remain, and I will welcome any further remarks or recommendations sent to my e-mail address jiri.witzany@vse.cz.

Prague, Czech Republic Jiří Witzany
February 2020

Acknowledgments

This book has been supported by the the Czech Science Foundation Grant 18-05244S and VSE Institutional Grant IP 00040.

Contents

1 Introduction

Derivatives are financial instruments that are built on (derived from) more basic underlying assets. They are designed to transfer risk easily between different counterparties. Instruments such as forwards, futures, swaps, or options are nowadays normally used by banks, asset managers, or corporate treasurers for hedging or speculation. Trading with derivatives has become increasingly important in the last 30 years throughout the world. It has been made easier due to electronic communication and settlement systems, and has grown exponentially in recent years. On the other hand, derivatives are closely related to many bank failures and even many financial crises, including the global financial crisis of 2007–2008. The goal of this book is not to make derivatives more popular, but to understand their nature; in fact, our point of view will be rather critical. In order to understand modern financial markets, it has become necessary to know how derivatives work, how they can be used, and how they are priced. This text aims to give an overview of the basic (plain vanilla) derivatives as well of the more complex (exotic) ones. We will focus not only on the mechanics of trading and settlement, but also on the more difficult issues of valuation, hedging, and risk management in general. In order to understand the valuation and hedging techniques, we have to develop and apply the necessary mathematical and statistical tools.

Derivatives are financial instruments whose values depend on the market prices of one or more basic underlying instruments. The settlement of derivatives always takes time in the future, and their payoff can usually be calculated relatively easily at that time. However, it is generally more difficult to value derivatives before settlement, because the payoff, dependent on prices in the future, is not known. When valuing derivatives, we necessarily have to deal with the uncertainty of the future prices of the underlying assets. Derivatives also allow the elimination of physical settlement. This is, in particular, an advantage of commodity derivatives. Investors may invest, hedge, or speculate on oil, wheat or cows without physically dealing with any of these assets. The contracts can be, and, in fact, the majority of them are, settled financially, without the physical settlement of the underlying commodities.

© The Editor(s) (if applicable) and The Author(s), under exclusive license to Springer Nature Switzerland AG 2020

J. Witzany, *Derivatives*, Springer Texts in Business and Economics,
https://doi.org/10.1007/978-3-030-51751-9_1

This is why we may classify, in a broader sense, even commodity derivatives as financial ones. The first commodity derivative exchanges dealing with futures and options (The Chicago Board of Trade, CBOT, and later the Chicago Mercantile Exchange, CME) were established as early as the nineteenth and early twentieth centuries. The commodity derivative markets are, in fact, often more liquid than the spot markets, and the price relationship is partially reversed: the spot prices are derived from the derivative futures prices rather than vice versa. Derivatives are also typically used to increase leverage. For example, it is possible, using equity index futures, to "invest" 100 million USD in stocks while having just a fraction of the amount in cash and without holding the stocks at all.

1.1 Global Derivatives Markets and Derivatives Classification

The first derivative-like contracts existed in ancient times (Babylon, Ancient Greece, or Rome), but commodity derivatives have been actively traded on organized exchanges only since the nineteenth century. Equity derivatives can be traced back to the late nineteenth century. Trading in currency and interest rate derivatives came in the second half of the twentieth century, while later we see the spread of credit, energy, or weather derivatives. Figure 1.1 shows that the real boom in derivatives trading came in the late nineties and during the last decade, with the outstanding notional reaching 700 trillion USD, i.e., almost ten times global GDP. This exponential growth was, however, interrupted by the global financial crisis in 2008, followed by stagnating volumes. Alternatively, the overheated financial derivatives markets could, in fact, be partially blamed for the financial crisis.

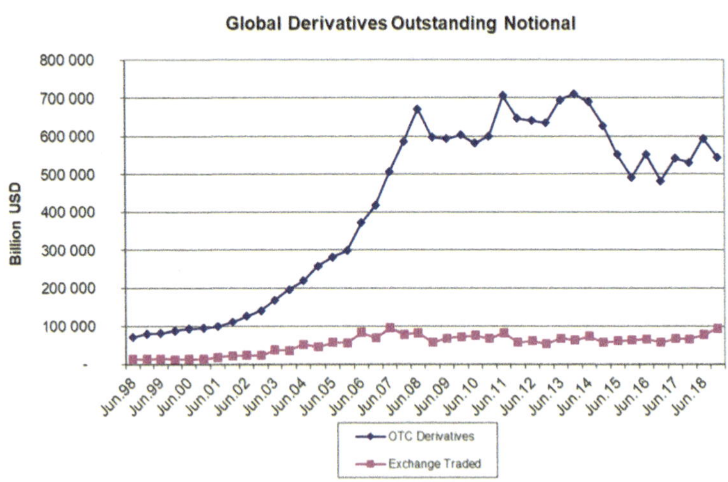

Fig. 1.1 OTC derivatives outstanding notional and exchange-traded derivatives open interest development in 1998–2018 (excluding commodity derivatives; source: www.bis.org)

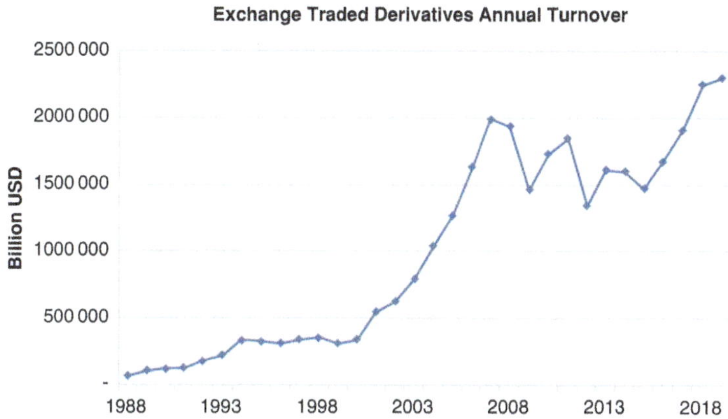

Fig. 1.2 Annual turnover of global exchange-traded derivatives in 1986–2018 (excluding equity and commodity derivatives; source: www.bis.org)

Figure 1.1 displays the aggregate outstanding notional of OTC traded derivatives and the open interest (which is a proxy for the outstanding notional) for exchange-traded derivatives over the period 1998–2018. The data are taken from the Bank for International Settlement (BIS), which collects statistics on international financial markets and coordinates global financial regulation.

OTC (*Over-the-Counter*) contracts are entered into directly by any two market participants with a large degree of flexibility. The contracts are usually settled by mutual payments or the contracted transfer of assets, and only exceptionally closed out (canceled with a profit/loss settlement) before maturity. The opposite is true for exchange-traded derivatives, where the majority of contracts are closed before maturity. The derivative contracts are entered into through a centralized counterparty (an organized exchange or its clearinghouse), and the opposite transaction can be easily canceled out (with a financial P/L settlement). The open interest represents the outstanding notional of all long, or equivalently short positions, with the centralized counterparty corresponding well to the outstanding notional of OTC derivatives. The numbers in Fig. 1.1 do not mean that the exchange-traded derivative markets are less important than the OTC markets, due to the effect of the netting and closing-out of transactions before maturity. Figure 1.2, with the development of annual exchange-traded turnover of FX and interest rate derivatives exceeding 2000 trillion USD, gives us quite a different picture.

Note that it is difficult to compare the outstanding notional (defined as the sum of the notional amounts of non-settled transactions at a given moment) and the turnover (defined as the sum of the notional amounts of transactions entered into during a given period). OTC derivatives with long maturity, typically interest rates swaps, stay on the books for many years and cumulate the aggregate statistics if the market is active. On the other hand, the turnover statistics may be magnified by derivatives with short maturity, or contracts that are closed out shortly after origination, as they are traded back and forth by the market participants. Consequently, the outstanding

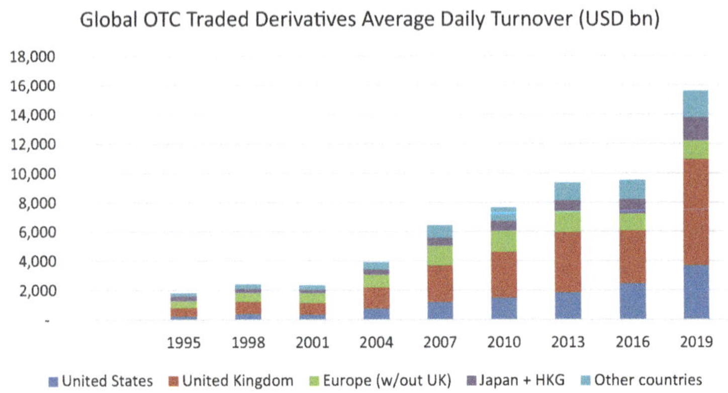

Fig. 1.3 Average daily turnover of global OTC derivatives trading in 1995–2019 by country and region (Triennial survey of FX and OTC interest rate derivatives trading; source: www.bis.org)

notional statistics are much more inertial and do not suddenly drop to low numbers, even if the activity on the derivatives market decreases to zero. The turnover, on the other hand, depends on the observed period and directly reflects the actual activity on the markets. This is illustrated when we compare Figs. 1.1 and 1.2.

In order to compare the regional activity on OTC and exchange-traded derivatives markets, let us look at the development of the average daily turnover in Figs. 1.3 and 1.5. The largest OTC market turnover goes through the United Kingdom, in particular, through the banks and brokers located in London that traditionally plays the role of an international financial center. The UK derivatives OTC market is, in terms of statistics, followed by the United States, Japanese, Hong Kong, and European markets. Figure 1.3 is based on a BIS Triennial Survey that covers only FX and interest rate derivatives and does not include other derivative types such as equity, commodity, or credit derivatives that are less important in terms of the turnover on the OTC market.

Regarding the Central and Eastern European countries the OTC derivatives market turnover (Fig. 1.4) has grown significantly since the beginning of the 1990s (mostly in Poland and the Czech Republic), however, remains still relatively negligible compared to overall numbers. On the other hand, a significant part of trades related to CEE currencies (especially PLN and CZK forwards and swaps), takes place through the UK (London) market.

The regional distribution of activity on exchange-traded derivative markets appears different according to Fig. 1.5. In this case, the leading role is traditionally played by the US derivative exchanges, namely the CME Group (composed of four exchanges—CME, CBOT, NYMEX, and COMEX), Cboe Global Markets, or ICE Futures US. According to the statistic the US market is followed by Europe (Eurex, Euronext, or ICE Futures Europe), and Asia-Pacific (Hong Kong, Australia, or Japan exchanges) with much smaller average daily turnover. However, it should be noted that the BIS statistics covers the FX and interest rate exchange-traded derivatives only.

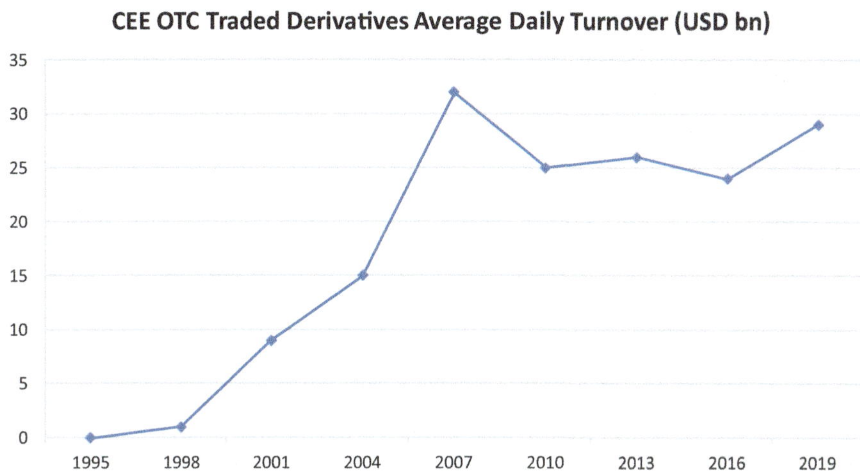

Fig. 1.4 Average daily OTC derivatives turnover in the CEE region (Triennial survey of FX and OTC interest rate derivatives trading; source: www.bis.org)

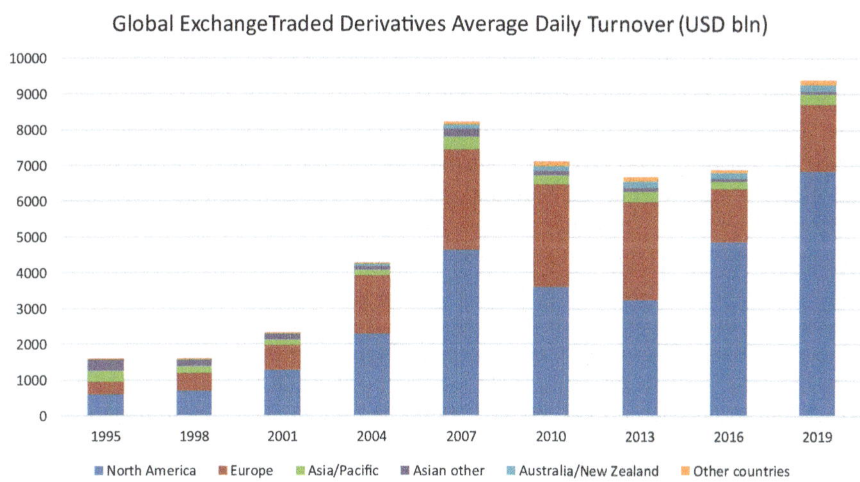

Fig. 1.5 Average daily turnover of global exchange-traded derivatives in 1995–2019 by location of exchange (excluding commodity and equity derivatives; source: www.bis.org)

The importance of Asia-Pacific exchanges vividly increases once we include equity and commodity derivatives (see Fig. 1.6). The statistics obtained from the World Federation of Exchanges 2018 Market Survey does not show the turnover in terms of notional amount but in terms of units of contracts. The disadvantage of this measure is that, as discussed in Sect. 2.2, the corresponding notional amount of one contract is variable depending on the underlying and the exchanging being usually around 100 to 200 K USD. For example, the total annual exchange-traded

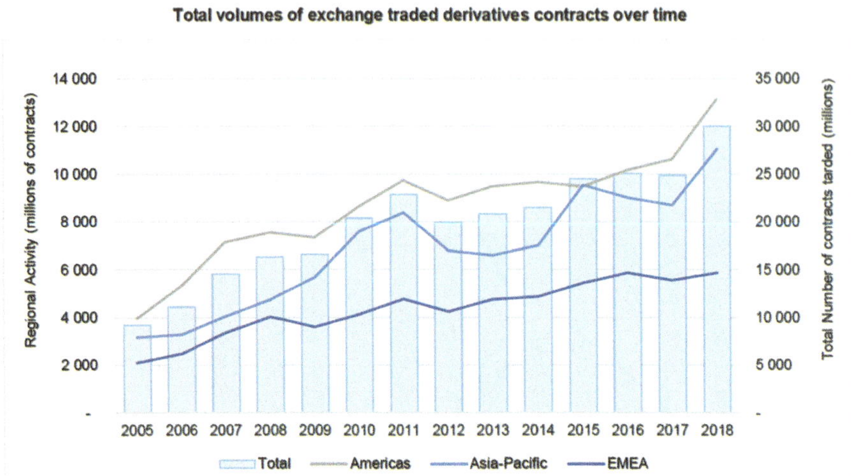

Fig. 1.6 Total annual number (millions) of derivative contracts traded on global organized exchanges (includes all underlying assets; source: WFE IOMA 2018 Derivatives Report, www.world-exchanges.org)

derivatives notional turnover in 2018 can be estimated only approximately from the WFE statistics to be around 45,000 tn USD assuming 150 K USD average contract size. The Americas region now includes also the South American largest exchange B3 SA Brasil Bolsa Balcao or the Canadian Montréal Exchange. On the other hand, the EMEA region (Europe, Middle East, and Africa) includes, for example, Johannesburg Stock Exchange or Tel Aviv Stock Exchange derivative markets. Regarding the CEE region, the only exchanges more active in derivatives products have been Warsaw Stock Exchange and Budapest Stock Exchange. For example, the Czech Republic Prague Stock Exchange has contemplated but never realized the introduction of CZK interest rate futures. The reason might be a low estimated turnover and a relatively liquid OTC market with CZK interest rate derivatives. Regarding PLN/EUR, CZK/USD or CZK/EUR foreign exchange futures, these are listed on CME and on ICE.

A more detailed view on the development of volumes (numbers of contracts) by split of the underlying asset and derivative product type is provided in Fig. 1.7. While the interest rate and equity derivatives have grown (in terms of volume) around 6% annually, the other two asset classes derivatives (currency and commodity) have grown substantially faster with CAGR (Constant Annual Growth Ratio) at 27% contributing to the overall average annual growth around 16%. The picture looks different compared to Fig. 1.2 base on the BIS statistics namely to exclusion of the equity and commodity derivatives.

Among the new OTC derivative products, the credit derivatives introduced in the late nineties, in particular Credit Default Swaps (CDS), experienced fast growth until the global financial crisis (Fig. 1.8).

The post-crisis decline in the CDS outstanding notional is much more significant compared to the other derivative products, since the credit derivatives were, in a sense, at the core of the financial crisis itself. It is worth noticing that the global derivative notional and turnover volumes are multiples of the GDP (estimated around 85 trillion USD globally or just around 20 trillion USD for the United States in 2017). For example, the 2007 CDS outstanding notional reached almost four times US GDP during that year. This looks dramatic, but it should be pointed out that the notional amounts are not (for most derivatives) real payment obligations between counterparties, but are only used to calculate certain fractional payments, e.g., interest payments. Hence, the settled cash flows are in practice much smaller compared to the notional amounts. This is also the case of credit derivatives under

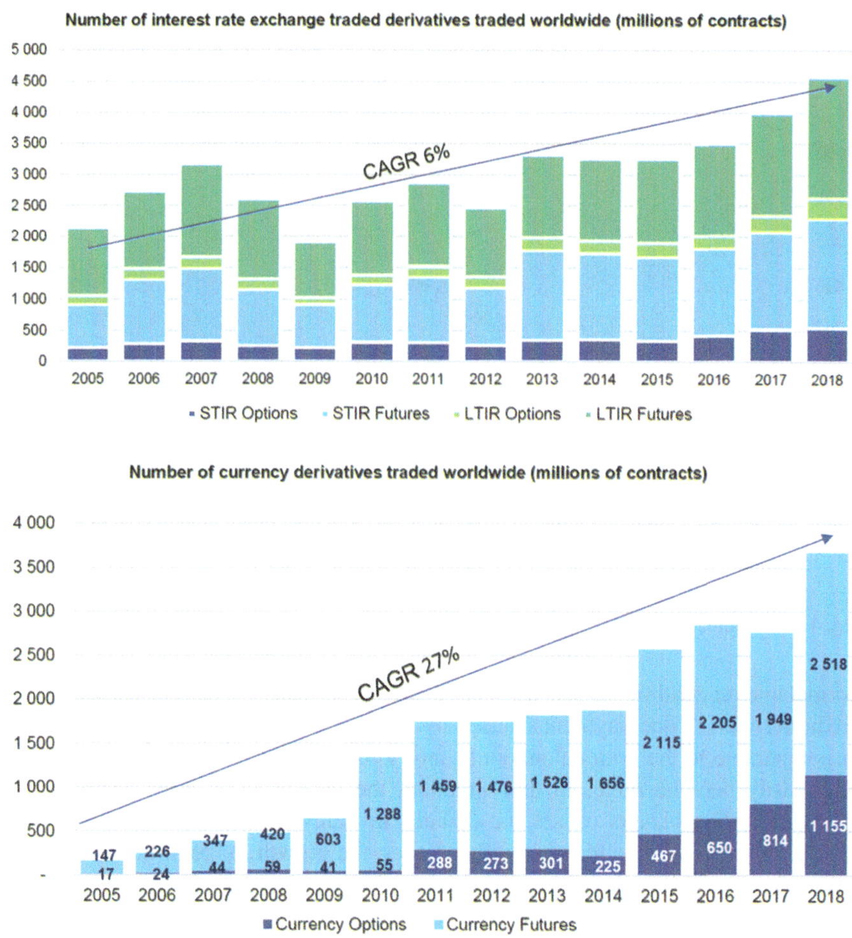

Fig. 1.7 Total numbers (millions) of exchange-traded derivative contracts by underlying assets and product type (source: WFE IOMA 2018 Derivatives Report, www.world-exchanges.org)

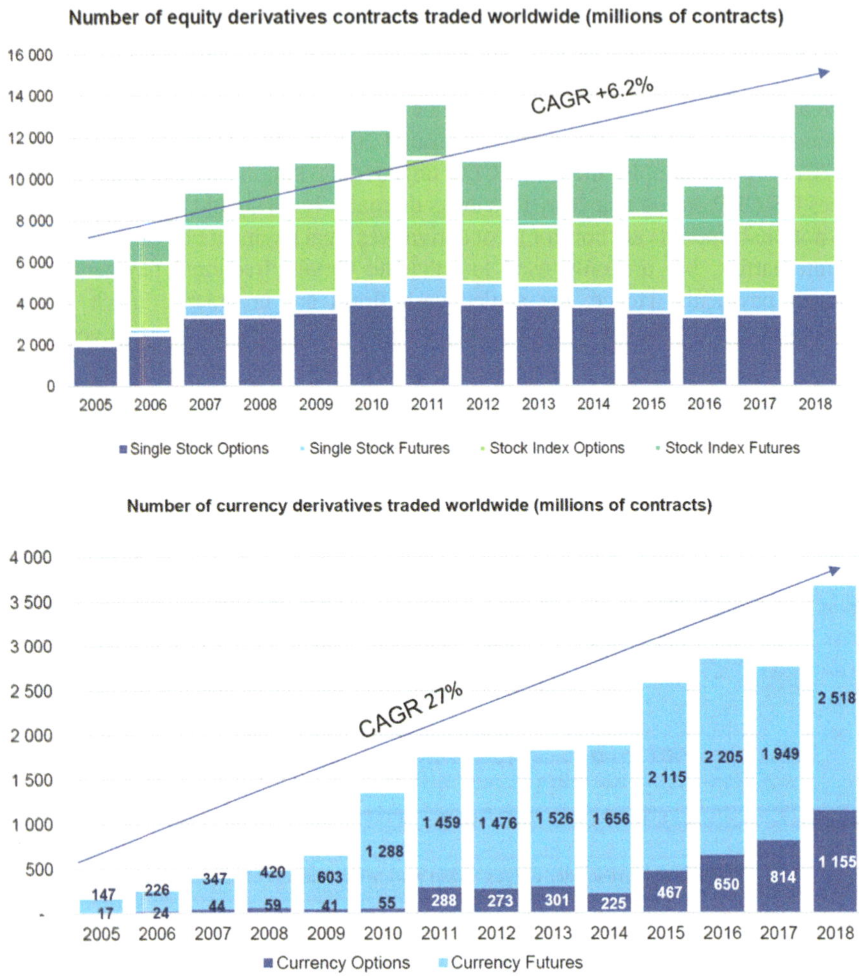

Fig. 1.7 (continued)

normal market conditions. But under stressed conditions, i.e., in a financial crisis, the credit derivatives due payments caused by many reference entity defaults are equal or comparable to the notional amounts, and so the mutual counterparty obligations may easily become huge. In that situation, the default of a group of important financial market players causes the defaults of many others in a kind of domino effect. That is why, in addition to other issues, the growth in credit derivatives has tremendously increased the systemic risk and interconnectedness of the financial markets, thus contributing to the depth of the crisis. Many disadvantages of OTC derivatives, in particular counterparty credit risk, are eliminated by exchange-traded derivatives, where settlement goes through a centralized counterparty. This may explain the relatively fast post-crisis recovery in exchange-traded derivatives activity

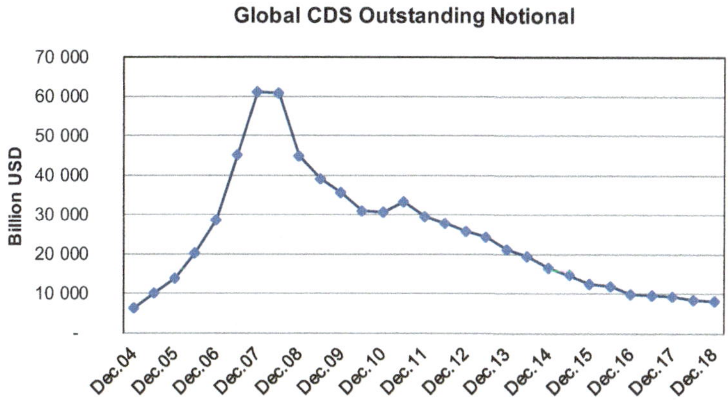

Fig. 1.8 Development of the CDS outstanding notional on the global OTC markets in 2004–2018 (source: www.bis.org)

that can be observed in Figs. 1.2 and 1.5. On the other hand, the stagnation of the global OTC derivatives outstanding notional (Fig. 1.1) and the sharp decline for credit derivatives after 2008 (Fig. 1.8) cannot be explained only by lower trading activity. In fact, we may notice in Fig. 1.3 that the average turnover on the FX and interest rate OTC market has significantly increased since 2007. The markets and the regulators have learned a lesson from the financial crisis and have introduced centralized settlement for some OTC products, in particular for credit default and interest rate swaps. Settlement through central counterparties (CCP) then allows the netting and closing of transactions before maturity, like on organized exchanges, significantly reducing the outstanding notional statistics in spite of increasing trading turnover.

1.2 Derivatives Classification

Derivatives can be classified according to different criteria: according to their market as OTC or as exchange traded, according to their underlying assets, or according to the derivative product type.

The structure of OTC markets can be seen in Table 1.1. The most important categories are foreign exchange (FX) and interest rate contracts, where the most frequently traded instruments are FX forwards, swaps, and options, currency swaps, forward rate agreements (FRA), interest rate swaps (IRS), and interest rate options. FX derivatives are predominantly traded on the OTC markets, while the volumes of FX futures and FX options on organized exchanges are relatively low, as indicated in Table 1.2. Interest rate derivatives are traded actively on both markets. On the other hand, equity and commodity derivatives are traded mostly in the organized markets. Finally, credit derivatives have, so far, been traded essentially only on the OTC

Table 1.1 Notional amounts outstanding and gross market values of over-the-counter (OTC) derivatives by risk category and instrument (in billions of US dollars, source: www.bis.org)

In billions of US dollars	Notional amount outstanding				Gross market values			
Risk category/investment	Jun 2017	Dec 2017	Jun 2018	Dec 2018	Jun 2017	Dec 2017	Jun 2018	Dec 2018
All contracts	542,439	531,911	594,833	544,386	12,683	10,956	10,326	9662
Foreign exchange contracts	88,429	87,117	95,798	90,662	2626	2293	2620	2257
Outright forwards and forex swaps	51,754	50,847	56,416	53,909	1259	1111	1249	1074
Currency swaps	24,532	25,535	26,012	24,858	1160	989	1155	990
Options	12,088	10,679	13,307	11,837	208	192	216	193
Other products	55	56	64	58				
Interest rate contracts	435,205	426,648	481,085	436,837	9045	7579	6644	6401
FRAs	75,414	68,334	84,131	67,636	129	112	107	134
Swaps	321,812	318,870	349,761	326,690	8131	6747	5914	5686
Options	37,641	39,112	46,833	42,154	786	719	623	581
Other products	338	332	361	357				
Equity-linked contracts	6964	6569	7071	6417	524	575	608	571
Forwards and swaps	2903	3210	3299	2938	184	197	228	248
Options	4061	3360	3772	3480	340	378	380	323
Commodity contracts	1762	1862	2133	1898	171	189	207	220
Forwards and swaps	1352	1414	1627	1450				
Total options	410	447	506	449				
Credit derivatives	9967	9578	8582	8373	307	313	238	191
Credit default swaps	9727	9354	8346	8143	300	304	232	187
Sigle-name CDS	5101	4570	4148	3954	152	130	118	105
Multi-name CDS	4626	4784	4199	4189	148	174	115	82

Table 1.2 Global turnover and open interest of derivatives traded on organized exchanges (in billions of US dollars, source: www.bis.org and www.world-exchanges.org)

In billions of US dollars	Open Interest		Turnover	
Risk category/instrument	Dec 2017	Dec 2018	2017	2018
All contracts	98,605	110,904	2,440,370	2,837,864
Foreign exchange contracts	413	396	32,508	40,572
Futures	289	257	28,980	36,792
Options	124	139	3528	3780
Interest rate contracts	80,572	94,368	1,875,888	2,215,332
Futures	33,381	38,783	1,452,024	1,758,456
Options	47,191	55,585	423,864	456,876
Equity-linked contracts	14,969	14,517	339,149	443,550
Futures	1983	1786	129,999	185,420
Options	11,803	11,145	209,150	258,130
Commodity contracts	2652	1623	192,825	138,410
Futures	2114	1170	185,295	129,632
Options	484	581	7530	8778

markets, but their trading mechanism is getting closer to organized exchanges in terms of standardization and centralized settlement in order to reduce systemic risk.

There are basically only two standard derivative instruments on the organized exchanges: futures and options. Of the exotic underlying assets traded on organized exchanges and not listed in Table 1.2, we should mention energy and, in particular, electricity, weather, or real estate.

The OTC markets, on the other hand, offer a much richer variety of derivative contracts. We will see that there are many types of options and swaps, starting from the most basic, typically called "*plain vanilla*," to extremely complex ones in terms of definition and valuation, which are often called "*exotic*."

1.2.1 Forward and Futures Contracts

The simplest derivative is a *forward contract* to buy or sell an underlying asset at a fixed (unit) *forward price* K at a future time (*maturity*) T. The forward settlement date normally goes beyond the ordinary spot settlement time, that is, typically, the trade date plus two or three business days, "T + 2," "T + 3," or more for currencies and equity trading due to technical reasons (note that the "T" in "T + 2" means the trade date, not the maturity date). The forward counterparty buying an asset is in a *long position* while the other counterparty selling the asset is in a *short position*. The forwards are usually *settled physically*, but can be also *settled in cash,* where the short position counterparty pays the difference between the asset spot price and the forward price $S_T - K$, calculated at time T, to the long position counterparty. If the difference is negative then, of course, the long position counterparty pays the amount to the short position counterparty. In the case of physically settled forwards, the

Fig. 1.9 Long forward payoff

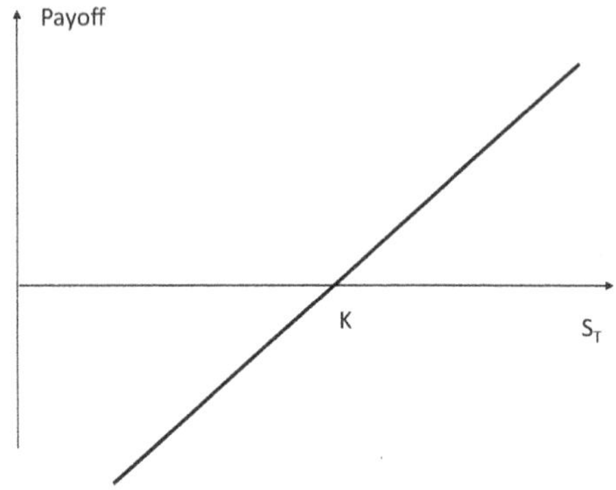

difference $S_T - K$ defines the forward *payoff*, since the long position counterparty can immediately sell the asset for S_T and receive the net profit $S_T - K$. Note that the payoff, i.e., the profit/loss at maturity, is not known at the time the contract is entered into, or at any time until maturity. We can only express the payoff as a function of the unknown price of the underlying asset at maturity (Fig. 1.9).

Futures are contracts traded on organized exchanges like forwards. However, there are a few differences that will be discussed in the next chapter.

1.2.2 Options

Forwards can be classified as *unconditional derivatives*, while options can be classified as *conditional*. An option is like a forward contract to buy or sell an asset at a specified price K in the future, but the settlement is conditional upon the decision of one of the counterparties. The counterparty that has the option has an advantage over the other counterparty, and thus pays an *option premium*. Note that there is no initial payment between forward counterparties. From the perspective of an option buyer, we distinguish a *call option* to buy the underlying asset and a *put option* to sell the asset. The fixed price K may now differ from the forward price, and so it is called the *exercise price* or *strike price* rather than the forward price. If the option holder decides to buy or sell the asset, we say that the option is *exercised*, or *realized*. Otherwise, the option *expires*. An option is of *European style* if it can be exercised only during the *expiration day* T. If an option can be exercised at any time until day T then it is called *American*. OTC options are usually of European style, while exchange-traded options are mostly of American style (originally introduced and traded on the US exchanges). As in the case of forwards, the contract value can be exactly defined by a *payoff function*. For example, in the case of a European call option the payoff function (Fig. 1.10) is given by:

$$\text{Payoff} = \max(S_T - K, 0). \tag{1.1}$$

The formula assumes a rational option holder who will exercise the option if, and only if, the actual value of the asset S_T is larger than the exercise price.

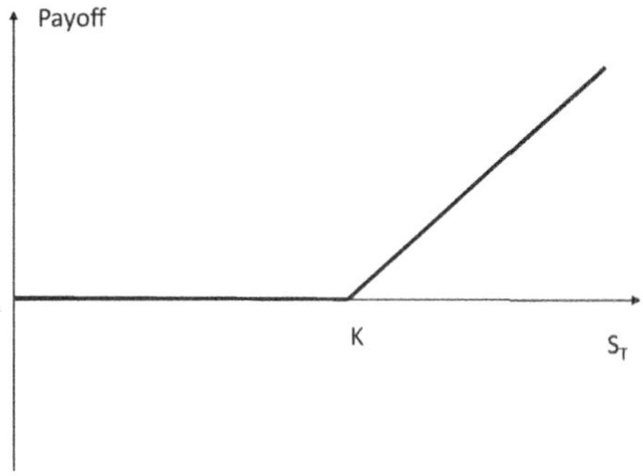

Fig. 1.10 European call option payoff

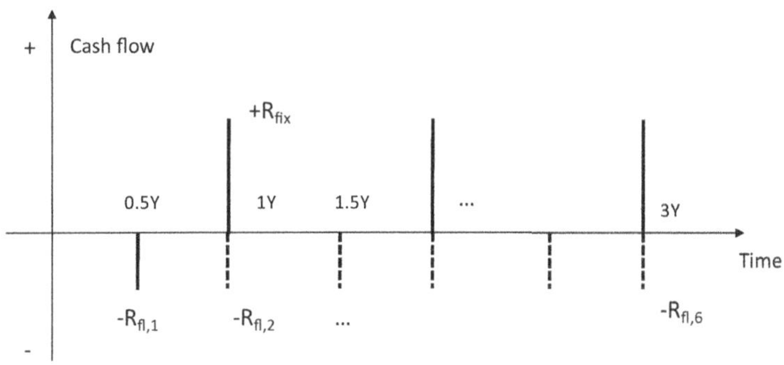

Fig. 1.11 Three-year interest rate swap cash flow

1.2.3 Swaps

Swaps are OTC contracts that take many different specific forms. Generally, swaps can be characterized as contracts to exchange a series of cash flows (or other assets) between two counterparties. The cash flows may be known in advance, but some of them are always contingent on certain future rates or prices. Table 1.1 shows the largest outstanding notional for the interest rate swaps that belong to the category of "*plain vanilla*" derivatives. Under an *interest rate swap* (IRS) contract, one counterparty periodically pays a *fixed interest rate* and the other counterparty pays a *floating interest rate* (defined as a reference rate, e.g., Libor and Euribor). The rates are calculated on a contractual *notional amount* and paid till the agreed maturity. The notional amounts are identical for both counterparties and thus are not exchanged at the start and maturity dates of the contract. Figure 1.11 gives an example of a three-year interest rate swap (3Y IRS) cash flow from the perspective of the float payer. The annual fixed interest rate payment corresponds to European OTC markets (while semiannual payments would be standard on the US market). The standard float payments' periodicity is 6 months. The float interest rate is always set at an appropriate *reference rate*, e.g., 6 M Libor (six-month London Interbank Offered Rate), for the next six months period. Note that the first float payment is known (the full line) while the subsequent floats (the dotted lines) are not known at the beginning of the contract.

1.3 Valuation of Derivatives

Notice that Table 1.1 shows the "Gross markets values" of the outstanding OTC instruments, while Table 1.2 does not contain anything like that. An OTC derivative contract, generally defined as a set of fixed or contingent cash flows and other mutual obligations, has, at any time, a value for each of the counterparties. The value is not just an arbitrary subjective value, but the *real value* that must be reflected in financial accounting. Sometimes, the real value can be defined as the market value directly or indirectly quoted on the financial markets. More often, the market with a particular derivative instrument is not liquid, yet the value can be calculated, or estimated from the values of other quoted instruments. Derivatives valuation based on other prices and information is, in fact, the most difficult part of the matter. While the prices of commodities or stocks reflect their fundamental value determined by the market supply and demand, the prices of derivatives need to be more or less exactly calculated (derived) from other market prices and factors. The valuation of derivatives is, to a large extent, an exact mathematical science.

Let us consider the trade date of an OTC derivative contract when it is negotiated and entered into between two counterparties. If there is no initial payment, which is usually the case of forwards and swaps, and if the contract is entered into under market conditions, then there should be an equilibrium between the two parties, i.e., the market value should be close to zero for both counterparties. In the case of forwards, the price K that makes the initial market value equal to zero is called the theoretical market (or *fundamental*) *forward price*. It should be close to the quoted market forward prices but does not have to be the same due to the existence of *bid-ask spreads* and other inefficiencies on the market. The forward price should not be confused with the forward contract market value, initially zero and later positive or negative depending on the market development (Fig. 1.12). In the case of interest rate swaps, the fixed rate (besides notional and maturity) is the only negotiable parameter that makes the initial market value zero. Again, it should be close to the quoted IRS rates.

In the case of OTC options, there is an initial option premium payment that makes the overall cash flow value equal to zero, i.e., the premium should be equal to the market value of the option payoff. If there is no outright premium quotation, the question is how the market premium should be determined. While the valuation of forwards and plain vanilla swaps remains relatively elementary, based on the principle of discounted cash flows, the valuation of options requires the introduction of a stochastic model for the underlying price dynamics.

The statistics from organized exchanges (Table 1.2) do not show any gross market values. The exchange-traded derivative contracts certainly also have market

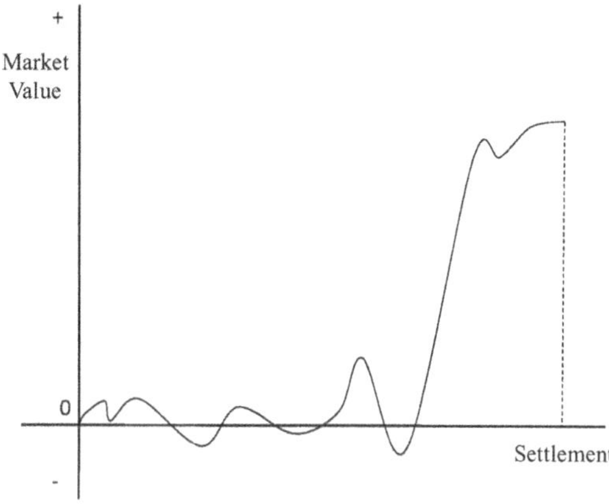

Fig. 1.12 Possible development of an FX forward market value

values, and the market participants need to know the fundamental prices in order to price the contracts correctly, but the daily profits/losses are settled through a *daily settlement and margin mechanism*, and so there is essentially no market value at the end of each business day. This mechanism will be explained in more detail in Sect. 2.2. There is, therefore, no need to show the aggregate gross market value in the statistics.

1.4 Hedging, Speculation, and Arbitrage with Derivatives

As for other basic assets, traders on the financial markets can be classified as *market-makers* and *market-users*. Market-makers provide price quotations and are prepared to buy and sell the instruments to make a profit on the differences between the buying (bid) and selling (ask/offer) prices. Their existence is important for the liquidity of the markets. Market users, on the other hand, from time to time just use the market to hedge, speculate, or perform arbitrage. By the term hedging, we mean entering into a new contract that will reduce our risk in one or more underlying assets. On the other hand, a speculative transaction will create or increase the risk, while arbitrage would be a combination of two or more transactions that generate a profit without any risk. Let us illustrate these concepts with a few examples.

Example 1.1 A CZK based company will receive one million EUR in 1 year, it will need to exchange the amount into CZK, and would like to hedge against the possible depreciation of EUR. Let us assume that the current 1Y EUR/CZK quoted forward price is 25. The exchange rate risk could be simply hedged by entering into the 1Y forward to sell one million EUR for 25 million CZK. One year later, the company will exchange the EUR income at the fixed exchange rate independently of the spot exchange rate. For example, if the exchange rate goes down to 23.50, the forward can be viewed as profitable. On the other hand, if EUR appreciated to 27 CZK, then the result of the hedging operation would appear negative. The forward hedging was nevertheless correct as the future spot exchange rate is not known in advance.

Example 1.2 Let us consider the same situation as above, i.e., the company needs to sell one million EUR for CZK in 1 year, and assume that the financial manager wants to hedge the downside risk, but at the same time wants to keep the upside potential. These two goals can be easily achieved using an option. Assume that the prices of *at-the-money* 1Y EUR/CZK call and put options (i.e., with the exercise price equal to the current forward exchange rate of 25) are 0.50 CZK per option on 1 EUR. The simple solution is to buy the 1Y EUR/CZK put option on one million EUR. The total premium of 0.5 million CZK is an initial cost that must be paid by the company. If the exchange rate S_1 in one year is lower than 25, then the put option will have a positive payoff and will be exercised. Otherwise, it does not make sense to use it. Consequently, taking the initial hedging cost into account, the effective selling price will be max $(S_1–0.5, 24.5)$ per one EUR. In this approach, the financial manager keeps the upside potential (appreciation of EUR), but the minimum exchange rate is 24.5, not 25 as in the forward hedging approach with no initial cost.

Example 1.3 A trader expects EUR to appreciate against CZK during the next month. She/He would like to speculate on the appreciation taking a long position in ten million EUR, but she is not allowed to take cash positions (i.e., buying EUR on the spot market and keeping an amount for a time) due to liquidity restrictions. The same result could, however, be achieved by a long 1M (one-month maturity) EUR/CZK forward on ten million of EUR. Entering into the position does not usually require any cash. Sometimes, depending on the institution's credibility, the counterparty might require a margin deposit that would, nevertheless, be just a small fraction of the full notional. The position would be normally closed by a spot transaction selling the ten million EUR settled the same day as the forward contract. If the fixed forward rate is K and the settlement spot exchange rate is S_1 then the final gain/loss is indeed $(S_1 - K) \times$ ten million CZK.

Example 1.4 Although options seem to be designed purely for hedging purposes, they can be also used for wild speculations. Let us assume that the trader from Example 1.3 is allowed to invest up to 20 million CZK. If one 1 M EUR/CZK call option with a strike of 25 costs 0.25 CZK then the trader can speculatively buy the call on 80 million EUR. If S_1 is the rate in one month then the total net gain/loss will be

$$\max\left(-20 \text{ million CZK}, (S_1 - 25) \times 80 \text{ million} - 20 \text{ million CZK}\right).$$

The gain/loss profile is shown in Fig. 1.13. Thus, the trader may easily lose the full invested amount of 20 million CZK. On the other hand, the potential gains are high. For example, if the EUR appreciated to 26, then the net gain would be 60 million CZK.

Finally, let us give an arbitrage example. Generally, *arbitrage* is a (static or dynamic) combination of transactions that leads to a profit (with a positive

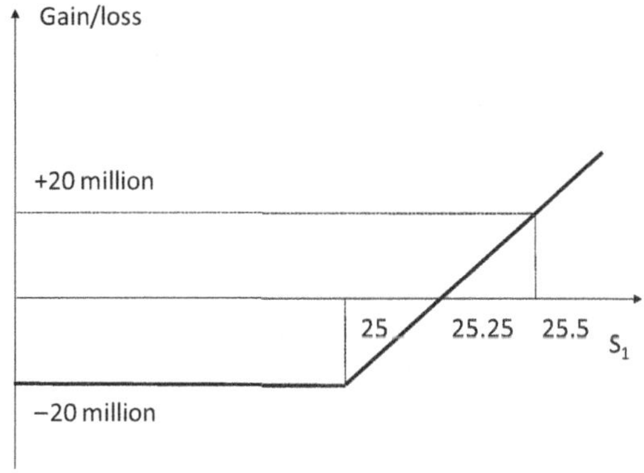

Fig. 1.13 The gain/loss on a long option position as a function of the settlement spot price

probability) without any possibility of loss. The simplest example of an arbitrage is buying an asset on one market and immediately selling the asset for a higher price on another market. Arbitrage is like a free lunch. If there is an arbitrage opportunity, everybody tries to use it, and so it cannot last long, as the supply and demand forces change the prices almost immediately. That is why the prices of identical assets on different markets should be (almost) equal. Certain differences might exist due to bid/ask spreads, different transaction costs, taxes, etc.

The pricing of derivatives is, generally, based on arbitrage arguments. Possible arbitrage strategies between the underlying asset market and the derivatives market force the derivative prices to be in line with the underlying prices. We shall now give a basic derivative arbitrage example related to FX spot and forward quoted prices.

Example 1.5 Let us assume that the quoted EUR/CZK spot exchange rate is 24.5 and that the 1-year forward is 25. It seems that the forward price is relatively high compared to the spot price. An arbitrageur can try to borrow CZK, buy EUR on the spot market, deposit EUR, and sell them on the forward market. At this point, we need to take the actual interest rates into account. Assuming that the CZK 1Y interest rate is 1.5% and that the EUR 1Y interest rate is 1%, she can borrow 24.5 million CZK at 1.5%, buy one million EUR on the spot market, and deposit the EUR amount for one year at 1%. At the same time, she can enter into a forward contract selling 1.01 million EUR in one year for $1.01 \times 25 = 25.25$ million CZK. She/He also has to repay the CZK loan, i.e., $24.5 \times 1.015 = 24.8675$ million of CZK. Finally, the remaining arbitrage profit is a nice 382,500 CZK.

Notice that the possibility to take a speculative position using derivatives relatively easily without any cash creates, at the same time, a new significant operational risk. Since the speculative forward or option positions require (almost) no initial cash payment, they can be easily overlooked or even intentionally hidden from the trading room manager, financial control unit, or risk management department. The trader might take too large speculative positions, being tempted by moral hazard considerations: losses will be paid for by the institution, but profits will bring fat bonuses to the trader. Unfortunately, this scenario occurred in the past in many different variations with serious consequences for the financial institutions.

For example, Nick Leeson, an employee of Barings Bank, lost over $1 billion in 1995 speculating on Nikkei 225 futures. Originally, he was supposed to carry out arbitrage operations between different markets, but later he became a speculator without the bank's authorization. He was based at the bank's Singapore office. Moreover, he was responsible not only for trading, but also for back-office operations, i.e., settlement and accounting, thus it was easier for him to hide the money-losing operations from his supervisors, who did not fully understand the dangers of derivatives. When the losses were discovered, it was too late, and the bank had to be closed down after 200 years of existence.

Hedge funds have become major users of derivatives. Similarly to mutual funds, hedge funds invest funds on behalf of their clients. Contrary to their name, hedge funds do not hedge, but rather speculate, or seek arbitrage opportunities using derivatives. Long-Term Capital Management (LTCM) was a successful and popular hedge fund in the early nineties. Its investment strategy was known as convergence arbitrage and was based on the idea that bonds issued by the same issuer but traded on different markets would eventually converge to the same value. However, the fund managers, including the Nobel prize laureates R. C. Merton and M. S. Scholes, underestimated the liquidity risk. During the Russian crisis in 1998, the fund was forced to unwind its huge positions and suffered losses of over $4 billion. The fund was considered *too-big-to-fail* (*TBTF*), and most of the losses were covered by the Federal Reserve, i.e., paid for by the taxpayers at the end of the day.

In a more recent case, a Société Générale trader, Jérôme Kerviel, lost over 5 billion EUR speculating on the future direction of equity indices in 2008. He was able to hide his money-losing operations in equity futures due to his nonstandard access rights to the information system.

Most recently, in September 2011, Swiss bank UBS trader Kweku Adoboli lost $2.3 billion in unauthorized trading. The rogue trader placed bets on EuroStoxx, DAX, and S&P 500 index futures. To cover the loss-making positions, the trader created fictitious hedging operations that hid the actual loss. The trader was arrested under a suspicion of fraud, and the scandal led to the resignation of the UBS CEO.

It appears that large losses on derivatives are often closely related not only to the pure market risk but even more to fraud or operational risk. We will discuss the risk management issues in more detail in Chap. 5. Derivatives are useful and extremely successful tools for hedging, speculation, or arbitrage, but they can also be compared to electricity, which can cause great damage if not used properly. Consequently, it is important to understand derivatives' mechanics, valuation, hedging, and risk management!

Forwards and Futures

2

2.1 Pricing of Forwards

Forwards are, in general, OTC contracts to buy or sell a specified asset at a specified price K, at a future time T, and settled later than for normal spot operations. Futures are similar contracts traded on organized exchanges. The arbitrage idea applied in the previous chapter (Example 1.5) can be generalized to obtain a precise relationship between the spot and the forward (Futures) prices of a financial asset that must hold on an arbitrage-free and perfectly liquid market.

2.1.1 FX Forwards

Let us firstly analyze *FX forwards*, i.e., forward contracts to exchange one currency for another, in more detail. The arbitrage strategy can generally be performed as indicated in Fig. 2.1—borrow a certain amount $N \times S_0$ of the domestic currency, exchange it on the FX spot market at the rate S_0, deposit the corresponding foreign currency amount N, and sell the amount plus accrued interest on the forward market at the rate F_0 negotiated today.

The arbitrage yields a positive profit if

$$N \times S_0 \times \left(1 + r_{DC}\frac{d}{360}\right) < N \times F_0 \times \left(1 + r_{FC}\frac{d}{360}\right),$$

where r_{DC} and r_{FC} are the domestic and foreign currency interest rates, respectively, in the standard money market convention (Act/360). If the market is arbitrage-free, i.e., market participants take advantage of arbitrage opportunities as they occur, then the opposite inequality must hold. This gives as an upper bound on the forward price given the spot price and the two interest rates:

© The Editor(s) (if applicable) and The Author(s), under exclusive license to Springer Nature Switzerland AG 2020

J. Witzany, *Derivatives*, Springer Texts in Business and Economics,
https://doi.org/10.1007/978-3-030-51751-9_2

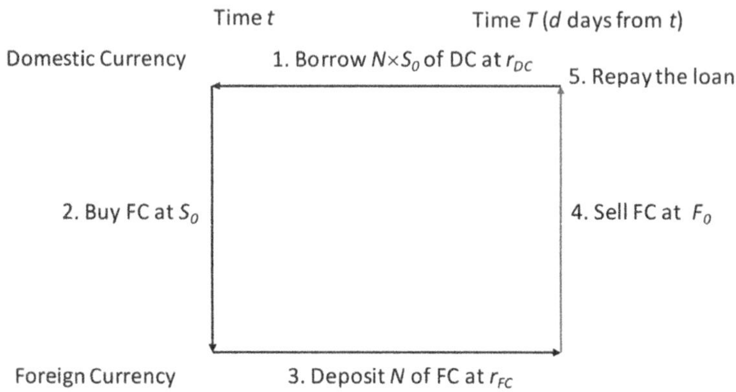

Fig. 2.1 A possible arbitrage strategy between the FX spot and FX forward markets

$$F_0 \leq S_0 \times \frac{1 + r_{DC}d/360}{1 + r_{FC}\,d/360}.$$

In order to get the opposite inequality, we need to reverse the order of the arbitrage operations: borrow a certain amount N of the foreign currency, exchange it at the FX spot market at the rate S_0, deposit the corresponding domestic currency amount $N \times S_0$, and, finally use the amount plus the accrued interest on the forward market to buy the foreign currency at the rate F_0 and to repay the loan. If this can be done, then we conclude that

$$F_0 = S_0 \times \frac{1 + r_{DC}\,d/360}{1 + r_{FC}\,d/360} \approx S_0(1 + (r_{DC} - r_{FC})d/360). \tag{2.1}$$

The combination of the two deposits and the spot operation in fact *replicates* the forward operation, and the replication works in both directions. Hence, the replication price must be equal to the quoted forward price. It is useful to summarize the implicit assumptions used in this argument:

1. There are no transaction costs and taxes.
2. Market participants can borrow and lend money at the same risk-free interest rate (for both domestic and foreign currencies).
3. There are no arbitrage opportunities.

In practice, the assumptions above hold only partially. There are transaction costs and taxes, in particular, there are bid-ask spreads, there is a difference between the borrowing and lending interest rate, and arbitrage opportunities may temporarily exist. The arbitrage can still be realized in both directions, but there is a different buy/sell spot price, different bid/ask interest rates, and so the Eq. (2.1) is, in practice, rather an inequality giving an upper and lower bound for the forward price F_0. Figure 2.2 shows an example of USD/CZK forward quotes. While the spot exchange

0#CZKF= USD/CZK FORWARD ENHANCED GUIDE

CZK= 22.744 / 758

Term	Bid	Offer	Source	TIME	Start	Maturity	Outright	DayCount
CZKON=	-0.60	-0.52	KOBA	12:42	10JUL19	11JUL19	22.74506/5922	1
CZKTN=	-0.62	-0.54	KOBA	12:42	11JUL19	12JUL19	22.74454/5862	1
CZKSN=	-1.51	-1.03	SANM	12:41	12JUL19	15JUL19	22.74249/5697	3
CZKSW=	-4.75	-3.25	BAFX	12:42	12JUL19	19JUL19	22.73925/5475	7
CZK2W=	-8.25	-6.25	BOAF	12:42	12JUL19	26JUL19	22.73575/5175	14
CZK3W=	-11.72	-9.14	INGY	12:42	12JUL19	02AUG19	22.73228/4886	21
CZK1M=	-14.70	-12.60	KOBA	12:42	12JUL19	12AUG19	22.72930/4540	31
CZK2M=	-26.40	-22.30	KOBA	12:42	12JUL19	12SEP19	22.71760/3570	62
CZK3M=	-38.30	-32.30	KOBA	12:42	12JUL19	15OCT19	22.70570/2570	95
CZK4M=	-50.92	-41.94	INGI	12:42	12JUL19	12NOV19	22.69308/1606	123
CZK5M=	-64.28	-49.97	INGI	12:42	12JUL19	12DEC19	22.67972/0803	153
CZK6M=	-105.00	-95.00	BAFX	12:42	12JUL19	13JAN20	22.63900/6620	185
CZK7M=	-115.58	-89.09	CKLU	12:42	12JUL19	12FEB20	22.62842/6891	215
CZK8M=	-129.80	-92.10	INGY	12:42	12JUL19	12MAR20	22.61420/6590	244
CZK9M=	-123.50	-111.50	BAFX	12:42	12JUL19	14APR20	22.62050/4650	277
CZK10M=	-130.49	-112.64	CKLU	12:42	12JUL19	12MAY20	22.61351/4536	305
CZK11M=	-142.58	-104.88	INGY	12:42	12JUL19	12JUN20	22.60142/5312	336
CZK1Y=	-140.66	-126.66	BAFX	12:42	12JUL19	13JUL20	22.60334/3134	367
CZK15M=	-206.59	-131.59	CBFW	12:41	12JUL19	13OCT20	22.53741/2641	459
CZK18M=	-260.50	-147.50	SANM	12:39	12JUL19	12JAN21	22.48350/1050	550
CZK21M=	-321.33	-167.33	INGY	12:35	12JUL19	12APR21	22.422670/0670	640
CZK2Y=	-356.00	-206.00	SANM	12:42	12JUL19	12JUL21	22.38800/5200	731
CZK3Y=	-610.70	-410.30	SANM	12:40	12JUL19	12JUL22	22.13330/4770	1096
CZK4Y=	-734.4	-663.81	INGI	12:42	12JUL19	12JUL23	22.00960/9419	1461

Fig. 2.2 An example of USD/CZK forward quotes (Source: Reuters, 10.7.2019)

rate is fixed (22.744/22.754) the forward bid/offer quotes for different maturities are in fact given as the differences between the forward and the spot exchange rates (forward points), e.g., 1Y bid outright forward implied by the forward point quotation (−140.66) is 22.744–0.14066 = 22.60334. Notice that the forward points are negative since the CZK interest rate is lower than the USD interest rate at the time of the quotation.

The arbitrage argument can be generalized for other underlying assets as well, but we need to distinguish assets that can and cannot be borrowed (shorted), investment and consumption assets, and between storable and non-storable assets. In order to simplify the formulas we are going to use the continuous compounding (see Sect. 3.1) where the interest factor from the spot settlement time t to maturity T has the exponential form $e^{r(T-t)}$. Thus, the Eq. (2.1) above can be simply written as

$$F_0 = S_0 \times e^{(r_{DC}-r_{FC})(T-t)}. \tag{2.2}$$

2.1.2 Investment Assets

First, let us consider an *investment asset* held for investment purposes by a significant number of investors. We will also assume that the asset pays a known income. Even if the investors are not willing to lend the asset to someone else, they can use it for short-term speculations or arbitrages, or alternatively, allow their asset managers to perform short-term speculations, or arbitrages with the assets that have to be returned back to the managed portfolio. Let I denote the present value of the known income (on one unit of the asset) over the period from t to T. Thus, if we hold the asset over the period, we collect I. On the other hand, if we borrow it over the period, we have to pay back I as the cost of borrowing. Under those assumptions, the arbitrage strategy outlined in Fig. 2.1 can be easily generalized to obtain the following relationship:

$$F_0 = (S_0 - I) \times e^{r(T-t)}. \tag{2.3}$$

Let us illustrate the argument with a stock that pays a known dividend.

Example 2.1 Consider a stock that is sold for 50 EUR on the spot market. The six-month forward contract on the stock is quoted at 48 EUR. It is known that the stock will pay 2 EUR dividend in 3 months. Let us assume that the interest rate for 3 as well as for 6 months is 4% p.a. in continuous compounding. One possible arbitrage strategy is to borrow 50 EUR, buy one stock, after 3 months collect the dividend, invest it for the remaining 3 months, sell the stock on the forward market for the price of 48 EUR negotiated today, and finally repay the loan. The final balance of this operation $48 + 2e^{0.01} - 50e^{0.02} = -0.99$ EUR is, unfortunately, negative. But it can be reversed with the opposite result: borrow 1 stock, sell it for 50 EUR on the spot market, deposit $50 - 2e^{-0.01}$ EUR for 6 months and $2e^{-0.01}$ for 3 months, repay the 2 EUR dividend to the stock owner after 3 months, then buy back one stock at the forward price 48 EUR negotiated today, and return it to the owner. In this case, our result will be $(50 - 2e^{-0.01})e^{0.02} - 48 = 0.99$ EUR, i.e., a positive arbitrage profit that can be arbitrarily multiplied by implementing the strategy with a larger number of stocks. Note that the arbitrage opportunity indeed disappears, if and only if, the forward price equals

$$F_0 = (S_0 - I) \times e^{r_{DC}(T-t)} = \left(50 - 2e^{-0.01}\right) \times e^{0.02} = 48.99.$$

Other examples of investment assets are bonds, stock index portfolios, or precious metals (gold, silver, etc.). For bonds, the income I would be the present value of coupons paid over the period, S_0 and F_0 would be the spot and the forward cash prices, respectively. In the case of a large stock index portfolio (usually traded through index futures without physical settlement), we often accept a simplifying assumption of continuously paid dividends at an average continuously compounded rate q per annum. That is, if S_0 were the current value of the portfolio then qS_0dt

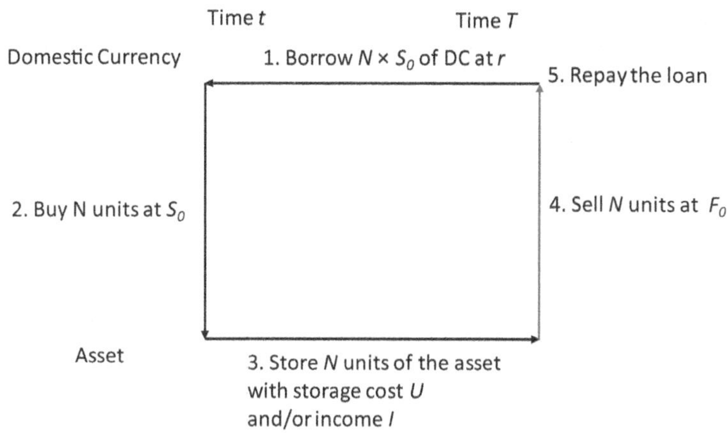

Fig. 2.3 General arbitrage scheme between the spot and forward markets

would be paid over the short time interval of length *dt*. The dividend payment can be immediately reinvested to buy an additional fraction *qdt* of the portfolio (assuming the arbitrary divisibility of the stocks). It can be shown that over the time period from *t* to *T* the initial portfolio will nominally grow $e^q(T-t)$ times. Hence, if we borrow S_0, buy 1 index portfolio, reinvest dividends from *t* to *T*, and sell $e^{q(T-t)}$ units of the portfolio at F_0 negotiated today to repay the loan, or vice versa, then the arbitrage-less market condition is

$$S_0 e^{r(T-t)} = F_0 e^{q(T-t)}, \quad \text{i.e.}$$

$$F_0 = S_0 \times e^{(r-q)(T-t)}. \tag{2.4}$$

A similar formula holds for investment in precious metals like gold or silver. The rate *q* is, in this case, called the gold or silver *lease rate*. Since precious metal producers need to hedge against future movements of prices, and financial institutions providing the hedging contracts need to hedge their long forward positions by shorting the metals on the spot market, there is a demand to lease the metals, in particular from central banks and from investors holding large amounts of the metals. Gold is usually borrowed for a lease rate, however, in the case of silver there may also be a storage cost just to pay for the safekeeping of the asset.

If *U* is the present value of the storage cost (paid, for example, at the beginning of the period) then, considering the general arbitrage scheme shown in Fig. 2.3 and assuming a zero lease rate, we get the modified formula:

$$F_0 = (S_0 + U) \times e^{r(T-t)}. \tag{2.5}$$

Alternatively, if u were an average continuously paid storage cost (this would be a rather theoretical assumption) and q continuously paid income[1] then we get a nice formula that allows us to analyze easily the relationship between the spot and the forward prices:

$$F_0 = S_0 \times e^{(r+u-q)(T-t)}. \tag{2.6}$$

If the *cost of carry* $r + u - q > 0$ is positive, then the forward prices should be higher than the spot price and increase with maturity (the *market is normal*), and if $r + u - q < 0$ then the forward prices are below the spot price decreasing with maturity (the *market is inverted*).

Example 2.2 The spot price of 1 ounce of gold is \$1530.35 and the 1-year forward (or futures) price is quoted at \$1540.50. Assume that the 1-year interest rate (in continuous compounding) is 3% and the gold lease rate is 2%. Find out if there is an arbitrage opportunity.

First, let us calculate the arbitrage-free price according to (2.4):

$$F_0 = 1530.35 \times e^{0.03-0.02} = 1545.73. \tag{2.7}$$

Since the theoretical arbitrage-free forward price is higher than the quoted forward price, there is an arbitrage opportunity, and the potential arbitrage profit is \$1545.73 – \$1540.50 = \$5.23 per one ounce. Recall that the price (2.7) is achieved by a replication using the spot market price of gold according to the general scheme shown in Fig. 2.3. In this case, we need to buy gold on the forward market and sell it on the spot market. In detail: lease N ounces of gold at 2% for 1 year (we have to return $Ne^{0.02}$ ounces of gold), sell the gold on the spot market and deposit $N \times 1530.35$ at 3%, finally buy $Ne^{0.02}$ ounces of gold on the forward market at the price \$1540.50/oz. and return the gold. The remaining profit is indeed positive

$$Ne^{0.02}\left(1530.35e^{0.01} - 1540.50\right) = Ne^{0.02} \times 5.23,$$

i.e., \$5.23 per one ounce of gold settled in 1 year. The absolute arbitrage profit depends only on our capacity to borrow gold.

2.1.3 Consumption Assets

Consumption assets are commodities that, by definition, are held predominantly for consumption. In the context of pricing, we need to distinguish storable assets from the other consumption assets that cannot be stored or are difficult to store. For

[1] The storage cost and income are usually mutually exclusive—if gold or silver is leased then there is no storage cost and if it is stored in a safe then we do not collect any lease income.

example, oil, gas, raw materials, and certain agricultural products are storable, at least for a limited time. On the other hand, electricity, live cattle, fresh oranges, etc. are difficult to store. For storable assets the arbitrage strategy can be implemented only in one direction (buy the consumption asset on the spot market and store it) but not in the other direction (the consumption assets cannot be borrowed and/or shorted at all, or only with difficulty). We can assert only that

$$F_0 \leq (S_0 + U) \times e^{r(T-t)} \text{ or } F_0 \leq S_0 \times e^{(r+u)(T-t)}. \tag{2.8}$$

If the inequality is strict, then there is a unique positive rate y (just solving the equation) so that

$$F_0 = S_0 \times e^{(r+u-y)(T-t)}. \tag{2.9}$$

The rate y is called the *convenience yield* and it has, in fact, a natural economic interpretation. Producers prefer to keep some consumption assets physically in stock rather than through a forward contract to be delivered in the future. For example, an oil refiner may use its stock of crude oil to increase production in periods of gasoline shortage and higher prices. This would not be possible if the refiner was simply long in crude oil through a forward contract. In fact, the oil forward market is quite often inverted due to a high convenience yield $y > r + u$.

In the case of non-storable consumption assets, such as electricity or live cattle, the arbitrage argument cannot be applied in any direction. Forward prices cannot be mechanically derived from spot prices and may, in general, be based on the seasonally expected supply and demand equilibrium and other factors.

2.1.4 Normal Backwardation and Contango

According to (2.6) or (2.9), the forward prices may be, starting from the spot price, increasing (*Normal market*) or decreasing (*Inverted market*) with maturity (Fig. 2.4).

The simple concept of normal and inverted forward prices should not be confused with the notions of *Contango* and *Normal Backwardation*. One could naively argue that at the maturity T the forward price F_0 should be equal to the expected future spot price $\mathrm{E}[S_T]$. If $F_0 < \mathrm{E}[S_T]$ then speculators would get long in the forward contract expecting a positive profit $E[S_T] - F_0$. If $\mathrm{E}[S_T] < F_0$ then speculators could get short expecting the profit $F_0 - E[S_T]$. The point is that the speculation involves the risk of a negative payoff if $S_T < F_0$ (or $S_T > F_0$ taking the short position) and investors usually require a premium for taking a risk. The premium could be modeled in line with the *Capital Asset Pricing Model* (CAPM). According to the CAPM, the expected return of a stock (including dividends) can be decomposed into the risk-free return and a positive premium depending on the stock's systematic risk measured by the beta index and on the market risk premium:

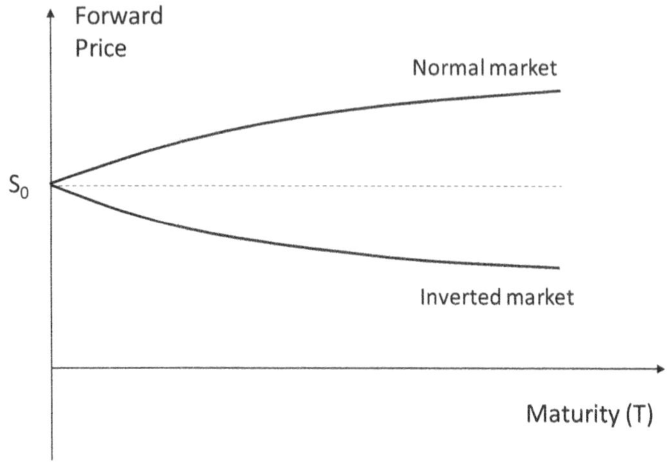

Fig. 2.4 Normal and inverted term structure of forward prices

$$E[R_i] = R_0 + \beta \times RP_M \tag{2.10}$$

Therefore, if S denotes the price of a dividend paying stock at a rate q then, in the exponential notation, we can write:

$$E[S_T] = S_0 e^{(r-q+p)(T-t)}, \tag{2.11}$$

where the annualized risk premium p depends on the systematic risk measure beta and on the market risk premium RP_M. If the spot price S_0 is expressed from (2.11) and substituted into (2.4) we obtain:

$$F_0 = E[S_T]e^{-p(T-t)}.$$

Therefore, if the factor beta is positive, then p is positive, and the forward price lies below the expected spot price. This relationship is called the *Normal Backwardation* since systematic risk is positive for most investment assets. There are a few exceptions such as, for example, gold or oil. The opposite relationship, when $p < 0$ and forward prices are larger than the expected spot prices, is called the *Contango* (Fig. 2.5). Note that in Fig. 2.4 we fix the current time t and look at forward prices for varying maturities T, while in Fig. 2.5 we fix the maturity T and let the time t go to T.

2.1.5 Valuation of Forward Contracts

So far, we have discussed the determination of equilibrium forward prices. At the trade date (when $t = 0$) the value of such a forward contract should be zero—there should be a market equilibrium. As time goes on, the spot price, interest rates, and

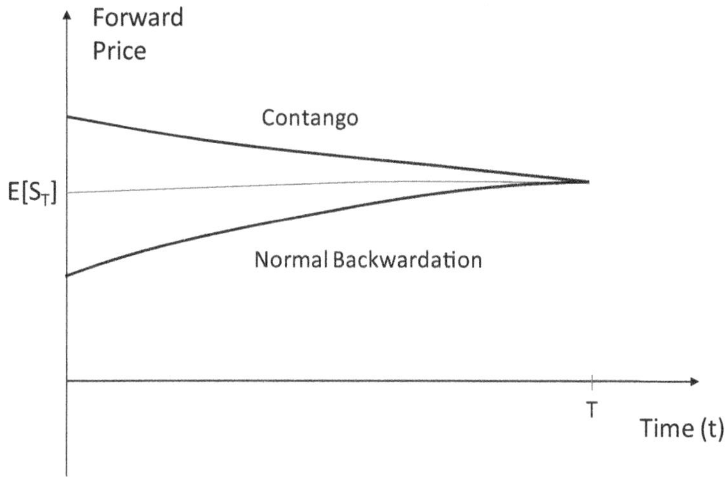

Fig. 2.5 Contango and Normal Backwardation

other factors change, and the value becomes positive or negative as illustrated in Fig. 1.12. Let us assume that we are in a long position buying one unit of the underlying asset for K with the forward maturity T. If F_t is the current forward price, then the position can be closed entering into a short contract on the same amount of the asset and with the same maturity. It means that in this case, at maturity, we buy and sell the asset and end up with the difference $F_t - K$ per one unit of the asset. Note that the closing transaction has the value zero, since it is entered into under the actual market conditions, and therefore, in order to value the original position, we need only to value the combined position. But this is a fixed cash flow, which can be valued by discounting to the time t, i.e., the market value of the long forward contract on one unit of the underlying asset is

$$f_t = (F_t - K)e^{-r(T-t)}. \tag{2.12}$$

The forward price F_t can be replaced by an appropriate forward price formula obtained above. For example, for FX forwards applying (2.2) we get

$$f_t = S_t e^{-r_{FC}(T-t)} - K e^{-r_{DC}(T-t)}. \tag{2.13}$$

2.2 Futures

Futures are the financial equivalents of forward contracts traded on organized exchanges all over the world. The Chicago Mercantile Exchange (CME), which merged with the Chicago Board of Trade (CBOT) and the New York Mercantile Exchange (NYMEX) into the CME Group (www.cmegroup.com), is the largest

Gold Futures Quotes

Globex

View Another Product

Quotes | Settlements | Volume | Time & Sales | Contract Specs | Margins | Calendar

Month	Options	Charts	Last	Change	Prior Settle	Open	High	Low	Volume	Hi / Low Limit	Updated
JUL 2019			-	-	1397.5	-	-	-	0	No Limit / No Limit	08:19:07 CT 10 Jul 2019
AUG 2019			**1411.2**	+10.7	1400.5	1400.0	1413.7	1391.8	237,561	No Limit / No Limit	08:46:17 CT 10 Jul 2019
SEP 2019			**1414.9**	+11.0	1403.9	1401.2	1416.3	1397.5	64	No Limit / No Limit	08:28:00 CT 10 Jul 2019
OCT 2019			**1417.9**	+11.1	1406.8	1406.2	1419.7	1398.2	1,141	No Limit / No Limit	08:40:04 CT 10 Jul 2019
DEC 2019			**1423.6**	+10.8	1412.8	1412.2	1426.0	1404.4	25,967	No Limit / No Limit	08:46:17 CT 10 Jul 2019
FEB 2020			**1429.5**	+10.8	1418.7	1414.3	1431.8	1411.0	3,029	No Limit / No Limit	08:45:40 CT 10 Jul 2019
APR 2020			**1435.2**	+11.9	1423.3	1418.2	1435.2	1416.9	954	No Limit / No Limit	08:19:08 CT 10 Jul 2019
JUN 2020			**1439.6**	+12.0	1427.6	1421.2	1439.9	1421.2	1,444	No Limit / No Limit	08:19:08 CT 10 Jul 2019

Fig. 2.6 Gold Futures quotations (Source: www.cmegroup.com, 10.7.2019)

derivatives exchange in the United States and in the world. Trading with futures in the United States has a long tradition going back to the nineteenth century and is well developed. The largest exchange in Europe is Euronext (www.euronext.com), which merged with the London International Financial Futures and Options Exchange (LIFFE) and with the New York Stock Exchange (NYSE) Group to form the Euro-American NYSE Euronext Group. The world's third largest derivatives exchange is the Eurex (www.eurexchange.com) belonging to the Deutsche Börse Group. Other large derivative exchanges are the Tokyo Financial Exchange (www.tfx.or.jp), Singapore Exchange (www.sgx.com), and the Australian Securities Exchange (www.asx.com). In less developed markets, like the Czech Republic, trading with derivatives takes place mostly OTC and there is almost no trading with futures or options on the local organized markets.

The main differences between forwards and futures are standardization, the existence of a centralized counterparty, daily settlement and the margin mechanism. Figure 2.6 shows an example of CME gold futures quotes. The exchange must necessarily specify a limited set of maturities for which the futures are listed and traded. A futures maturity is denoted by a month, but the exchange must exactly specify during which period the settlement takes place and what the rules are. In the case of financial futures, the delivery takes place during one specific day, for example, the third Friday of the month, while for commodity futures the delivery can often take place during the whole month, and the counterparty in the short position has the option to decide when the asset is delivered. It files a notice of intention to deliver with the exchange. The notice also specifies the grade of the asset to be delivered and a delivery location selected from a list given by the exchange. In the case of financial assets, such as foreign currencies or stocks, there is little ambiguity regarding the asset to be delivered, but in the case of commodities, the

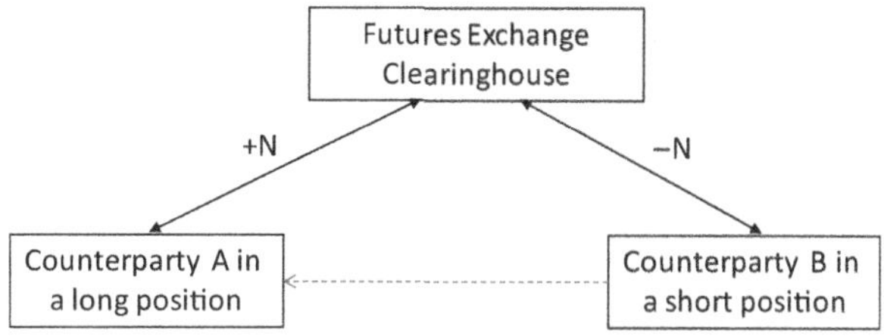

Fig. 2.7 The futures exchange clearinghouse stands between the market participants as a centralized counterparty

quality and grade must be precisely specified. For example, in the case of the Gold Futures, the contract specification says that the gold "shall assay to a minimum of 995 fineness, . . ." A futures contract always has a specified size, for example, for the gold futures, it is 100 troy ounces. Usually, the size of one futures contract corresponds to an equivalent of 100–500 K USD. It is not possible to buy or sell a smaller amount through futures unless there is a *mini-futures* contract where the underlying volume could be smaller. The contract specification also stipulates the quoting convention, for example, for the Gold Futures it is in US dollars per one troy ounce, but for Corn futures it is in cents per bushel.

Investors place orders to buy and sell futures to brokers, who execute the trades on the exchange. When two orders are matched, there are, in fact, legally two contracts with the exchange clearinghouse stepping in between the two counterparties as an intermediary (Fig. 2.7). The two contracts with the clearinghouse are on the same number of contracts and with the same price. Even if the counterparties A and B do not close the positions before maturity, the settlement does not necessarily take place directly between them. At maturity, the clearinghouse will randomly match counterparties in long and short positions. The counterparties in short positions are then obliged to deliver to the assigned counterparties in long positions. The clearinghouse guarantees that the settlement takes place. One of the advantages of this scheme is that positions can be easily closed-out before maturity. For example, if the counterparty A later decides to close the long position by selling N futures (the same asset and maturity) to another counterparty C, the clearinghouse will net out the long and short position with respect to A, settling, of course, the price differences, and A will not have a position anymore, i.e., A will not have any obligation to deliver or accept delivery of the asset. This is why the outstanding number of short or long futures contracts with respect to the clearinghouse—*the open interest* or outstanding notional—may go up and down as new positions are opened and some outstanding positions closed.

In order to minimize the counterparty risk, i.e., the possibility of a counterparty defaulting and not settling its position, the clearinghouse requires a collateral in the form of a margin account deposit. The required margin balance is relatively low, usually around 5% of the underlying value, in order to cover potential daily, not cumulative, losses as the gains/losses are settled daily. This is a significant difference compared to forwards, where the gains and losses are settled only at maturity. Moreover, the daily futures gain/loss is calculated only based on price differences, and disregards the time value of money, i.e., it is without discounting. Specifically,

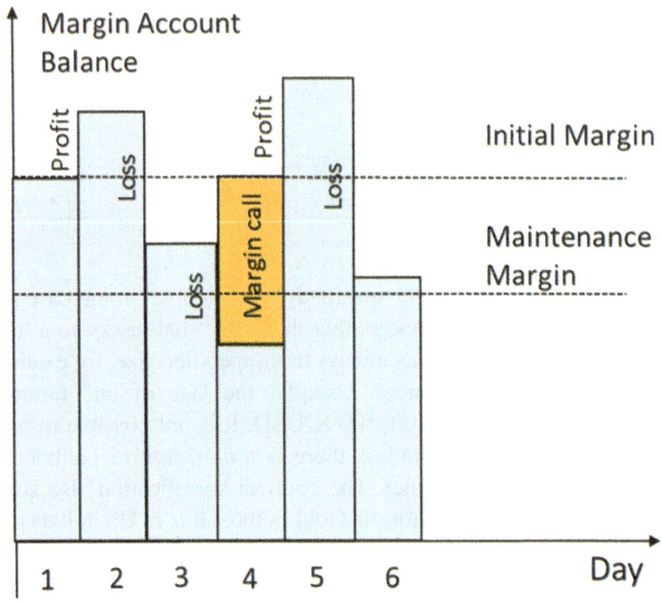

Fig. 2.8 Margin account development example

considering a long position, if F_0 is the previous day (initial contracted) futures price and F_1 today's closing settlement price, then $F_1 - F_0$ is the gain/loss per one unit of the asset that is settled against the margin account. Hence, if the difference is positive, it is credited to the margin account; if it is negative, then it is debited. By settling the differences, the previous day futures price F_0 is effectively reset to the new settlement F_1 price (consequently, if physical delivery takes place, then the last settlement price, not the initial contracted price, is paid). In order to keep the collateral amount sufficient, the clearinghouse sets not only an *initial margin* that has to be deposited when the position is opened, but also a *maintenance margin,* usually around 75% of the initial margin. If the balance drops below the maintenance margin, then there is a *margin call*, and the investor must deposit additional funds in the margin account in order to be at least at the initial margin again (see Fig. 2.8 for an illustration). If the investor fails to top-up the margin account, then the clearinghouse automatically closes out the position at the prevailing market price. The remaining margin account balance should be sufficient to cover the losses, if there are any. On the other hand, if the futures position is profitable, the investor may withdraw any amount above the initial margin from the margin account.

Example 2.3 Prices of gold went up dramatically after the financial crisis. Let us assume that we expect the prices to go down during the next few weeks and are ready to risk up to $30,000 of our own cash funds in a speculation. The initial margin for Gold Futures is $2000 and the maintenance margin is $1500 per one contract. We decide to short 10 gold futures corresponding to a short position in 1000 oz. of gold

Table 2.1 Margin account development for a short position in 10 gold futures

Day	Futures price ($)	Daily gain/loss ($)	Cum. gain/loss ($)	Margin acc.bal. ($)	Margin call ($)
0	1423.60			20,000.00	
1	1414.80	8800.00	8800.00	28,800.00	
2	1409.40	5400.00	14,200.00	34,200.00	
3	1388.00	21,400.00	35,600.0	55,600.00	
4	1413.20	−25,200.00	10,400.00	30,400.00	
5	1426.60	−13,400.00	−3000.00	17,000.00	
6	1430.70	−4100.00	−7100.00	12,900.00	7100.00
7	1446.30	−15,600.00	−22,700.00	4400.00	15,600.00
8	1433.00	**13,300.00**	**−9400.00**	Pos. Closed	
9	1399.50	**33,500.00**	**24,100.00**		
10	1383.00	**16,500.00**	**40,600.00**		

with current value around $1,500,000. The total initial margin is $20,000 and we still have $10,000 as a reserve in case of margin calls. It is July 2019, and we decide to use December 2019 contracts entered at $1423.60. We plan to close out the position by November. The futures price development during the first 10 trading days is shown in Table 2.1. On day one, the closing settlement price (officially set by the exchange—see column "Prior settle" in Fig. 2.6) moves down to $1414.80. The first day looks good since the gain of $10 \times 100 \times (1423.60–1414.80) = \$8800.00$ is credited to our margin account. The amount could be withdrawn, but we keep it as a reserve (variation) margin. The positive development continues until the day three, when the cumulative gain exceeds $35,000. We could close the position, but since we expect the gold to go down much more (being greedy speculators) we hold on to the position. Unfortunately, during the following three days we lose over $42,000 and there is a margin call to top-up $7100. When the amount is deposited, we hope to recover the lost profits. Unfortunately, the next day an additional amount of $15,600 is lost. There is another margin call of $15,600 that we cannot meet, since our remaining cash is only $2900. The position will be automatically closed, and we end up with a total loss of over $27,000.

Unfortunately, we were not able to hold the position for a long time. From the very beginning, we could have realized that an increase in the price by $1 causes a loss of $1000. Hence if the price of gold goes up more than $20, we do not have enough cash to cover the margin call and are out. Such a swing would be quite possible even if our medium-term expectation of a significant price of gold decrease was correct. The lesson learned might be that speculative position was too aggressive; we should have taken a short position in fewer gold futures.

2.2.1 Stock Index Futures

A popular and easy way to speculate on the stock market or to hedge an equity portfolio is to use stock index futures. A stock index can, in principle, be defined as the value of an underlying stock index portfolio.

For example, the traditional Dow Jones Industrial Average (DJIA) is defined as the sum of prices of 30 US blue-chip stocks divided by a divisor, i.e., its value equals the value of a portfolio of 30 stocks with equal weights. The divisor was originally set at 30 (it used to be an "average"), but since 1928 the divisor has been adjusted any time there is a stock split or large dividend payout, in order to avoid the discontinuity of the index, and currently it is significantly less than one (around 0.15 in 2019). It follows that the relative return of DJIA equals to the price-weighted average of the individual returns of the 30 stocks. That explains why the index is called *price-weighted.*

On the other hand, Standard & Poor's 500 Index (S&P 500) is based on the market capitalization of 500 selected US stocks. Its value is defined as the sum of market capitalizations of the stocks divided by a divisor, i.e.,

$$I_t = \sum_{i=1}^{500} w_i p_i(t).$$

The weights $w_i = n_i/d$ (the number of stocks issued divided by a divisor) do not change over time, unless a stock is removed and a new stock added to the portfolio, or there is an issue of new stocks, split of stocks, etc. The weights are fractional numbers, mostly less than 1 (the divisor value was 8302.34 billion in June 2019) due to a large number of stocks in the portfolio (the initial value of S&P 500 was initially set equal to 1000). The return of the index turns out to be the *market capitalization weighted* average of the individual stock returns. In its standard form it is published as the *price return* index, but it exists also in the *total return* and *net total return* version when dividends and/or taxes are accounted for. Most of the world's stock indices used by the markets are price return and market capitalization weighted. Other well-known indices are the DJ Euro Stoxx 50, Nikkei 225, British FTSE 100, or German DAX.

A stock index futures contract is, in principle, a forward contract to buy or sell a multiple of the underlying stock index portfolio. It is, in general, very difficult or impossible to achieve, all weights being integers. Moreover, the physical settlement of a broad index (like S&P 500) would bear relatively high transaction costs. The index futures contracts are, therefore, settled only in cash: on the settlement day there is an official index fixing $I_{closing}$, and the difference between the fixed futures price K and the index value, times the futures *multiplier* M, is paid; i.e.,

$$M \times \left(I_{\text{closing}} - K\right)$$

Nikkei/USD Futures Quotes
Globex

View Another Product

Quotes | Settlements | Volume | Time & Sales | Contract Specs | Margins | Calendar

Month	Charts	Last	Change	Prior Settle	Open	High	Low	Volume	Hi / Low Limit	Updated
SEP 2019		**21615**	+70	21545	21535	21635	21510	2,543	23250 / 19810	04:32:41 CT 11 Jul 2019
DEC 2019		-	-	21425	-	-	-	0	23080 / 19640	20:26:39 CT 10 Jul 2019

E-mini S&P 500 Futures Quotes
Globex

View Another Product

Quotes | Settlements | Volume | Time & Sales | Contract Specs | Margins | Calendar

Month	Options	Charts	Last	Change	Prior Settle	Open	High	Low	Volume	Hi / Low Limit	Updated
SEP 2019			**3006.00**	+8.50	2997.50	2996.00	3006.75	2995.25	110,486	3146.00 / 2847.00	04:38:15 CT 11 Jul 2019
DEC 2019			**3008.25**	+8.75	2999.50	2998.00	3008.50	2998.00	113	3148.00 / 2849.00	04:36:28 CT 11 Jul 2019
MAR 2020			**3001.00**	-1.25	3002.25	3002.00	3002.00	3001.00	2	3150.50 / 2851.50	20:26:36 CT 10 Jul 2019
JUN 2020			-	-	3002.75	-	-	-	1	3151.00 / 2852.00	04:25:51 CT 11 Jul 2019

Fig. 2.9 Nikkei 225 (Dollar) and E-mini S&P 500 Futures quotes (source: www.cmegroup.com, 11.7.2019)

would be the payoff for the counterparty in the long position. The multiplier is an integer of a magnitude set in order to get the desired futures contract basic volume. The DJIA Futures multiplier is $10, the S&P 500 standard futures multiplier is $250, the S&P 500 mini futures is $50, and the Euro Stoxx 50 Eurex traded futures multiplier is €10. Normally, the settlement amount (and the multiplier) is denominated in the index market domestic currency, but the calculation also allows the use of a different currency. For example, Fig. 2.9 shows Nikkei 225 (dollar) Futures quoted on CME where the multiplier is in Dollars, although the Nikkei index stocks are traded in Yen. This is convenient for US investors, but we will see later that this feature brings a complication to value the contracts (which belong to the class of *quanto derivatives*) precisely.

Stock indices can be used as alternatives to classical stock investments. The gain/loss on a long position combined with a corresponding cash position (invested for example in risk-free bonds) is (almost) equivalent to the gain/loss on the corresponding stock index portfolio investment.

Example 2.4 We have up to $600,000 to invest in US stocks. We expect the market to grow over the following six or seven months and do not want to pick any particular stocks or pay unnecessary fees to a professional asset manager. The solution would be to invest in a representative stock index like the S&P 500. This could be easily done by entering into long Mar 2020 E-mini S&P Futures contracts currently quoted at $F_0 = 3001$ (see Fig. 2.9) and keeping the long position until the maturity. The number of contracts can be calculated as $600\,000/(50 \times 3001) \approx 4$. Only a fraction of our funds needs to be deposited in the margin account (where it normally accrues the market interest), and the remaining part could be invested in a money market account. In March 2020, we will collect the accrued interest and the difference between the closing index value $I_{closing}$ and the initial futures price F_0 (multiplied by $4 \times \$50$). If we invested the same amount directly into the S&P 500 index portfolio ($4 \times \$50$ multiple), quoted today at I_0, then in March 2020 we would collect the stock dividends paid over the year plus the difference between $I_{closing}$ and the initial index value I_0. It turns out, as the initial futures price F_0 is not equal to the index value, that the difference between the prices offsets the difference between the expected interest and dividends, see (2.4), and so the two investment strategies are virtually equivalent. However, the direct investment into the index portfolio entails much larger transaction costs compared to the index futures strategy.

2.2.2 Pricing of Futures

Since futures are essentially equivalent to forwards, it can, in general, be assumed that the futures and forward prices are equal (for contracts with the same underlying assets and maturities). Consequently, the relationship between the spot and forward prices for investment assets given by (2.6) and for consumption assets given by (2.9) holds for futures as well. However, it should be noted that the equivalence between the forward and futures prices is only approximate. We will prove below that this is so, provided interest rates are constant. The argument could be generalized if the interest rates were deterministic, but futures prices start to depart from the forward prices if the interest rates are stochastic and correlated with the underlying asset prices. This is, in particular, the case of long-term interest rate futures, where the traders must calculate with the so-called *convexity adjustment* (see Sect. 6.4).

Proof of futures and forward price equivalence provided the interest rates are constant: The key difference between futures and forwards lies in the daily settlement mechanism. The price differences are not settled at the maturity of the contract, as in the case of forwards, but daily during the life of contracts without any discounting, i.e., in a sense prematurely. Assuming constant interest rates, the following strategy has been proposed by Cox et al. (1981). Suppose that a futures contract lasts n days and F_i is the closing price at the end of day $i = 0, \ldots, n$. Let G_0 be the market forward price for the same asset and maturity. We want to prove that $F_0 = G_0$. Let $\delta = r/360$ be the daily interest rate, assuming no holidays for simplicity. Consider the following strategy:

1. Take a long futures position in e^{δ} units of the asset (assuming the perfect divisibility of the futures contract) at the end of day 0.
2. Increase the position to $e^{2\delta}$ at the end of day 1.
3. Increase the long position to $e^{3\delta}$ at the end of day 2, and so on.

At the beginning of day i the position is $e^{i\delta}$ and so the day i gain/loss will be

$$e^{i\delta}(F_i - F_{i-1}).$$

This amount will accrue interest (negative or positive) for the following $(n-i)$ days and so at the end of day n it will be

$$e^{(n-i)\delta}e^{i\delta}(F_i - F_{i-1}) = e^{n\delta}(F_i - F_{i-1}).$$

Finally, the cumulative gain/loss plus the interest accrued on the margin account will be

$$\sum_{i=1}^{n} e^{n\delta}(F_i - F_{i-1}) = e^{n\delta}(F_n - F_0) = e^{n\delta}(S_T - F_0), \tag{2.14}$$

where $F_n = S_T$ is the futures closing price that is equal to the asset spot price at maturity. On the other hand, the payoff of the long forward on $e^{n\delta}$ units of the asset maturing after n days and with the market forward price G_0, entered into at the end of day 0, is

$$e^{n\delta}(S_T - G_0). \tag{2.15}$$

There is no initial cost in taking the futures and forward positions, so we can take the long futures position and short forward and achieve, deducting (2.15) from (2.14), the fixed result $e^{n\delta}(G_0 - F_0)$, or we can take the opposite short futures position and long forward position obtaining $e^{n\delta}(F_0 - G_0)$. Thus, assuming that there are no arbitrage opportunities, the two prices must necessarily be equal, and we have proved that $F_0 = G_0$, as needed.

2.2.3 Hedging with Futures and Forwards

Hedging the uncertain price of an asset with forwards or futures contracts is straightforward, if the contracts are available exactly for the asset we need to hedge and for the maturity when the asset is to be bought or sold. However, this is not always the case—futures are available only for a limited set of assets and maturities. Forwards are more flexible, but the OTC forward markets are not usually as liquid as the futures markets, or do not exist at all.

Often, we can only use a hedging futures contract with maturity T_2 that goes substantially beyond the desired hedging date, $T_1 < T_2$. Then, there is *the time basis risk* illustrated by Fig. 2.10. Assume that we need to sell an asset at a fixed price at time T_1 and enter at the time 0 into a short futures contract with the initial price F_0 and maturity T_2. We plan to close out the futures position at time T_1. When we sell the asset at the spot price S_1 the total income will be

Fig. 2.10 Variation of the basis (spot and futures price difference) over time

Corn Futures Quotes
Globex

View Another Product

Quotes | Settlements | Volume | Time & Sales | Contract Specs | Margins | Calendar

Month	Options	Charts	Last	Change	Prior Settle	Open	High	Low	Volume	Hi / Low Limit	Updated
JUL 2019			**434'4**	+0'2	434'2	433'2	434'4	433'2	28	472'4 / 422'4	03:53:53 CT 11 Jul 2019
SEP 2019			**433'6**	-1'2	435'0	434'2	435'0	432'4	8,549	460'0 / 410'0	06:26:33 CT 11 Jul 2019
DEC 2019			**437'4**	-2'0	439'4	438'6	439'2	436'2	11,789	464'4 / 414'4	06:25:29 CT 11 Jul 2019
MAR 2020			**444'2**	-2'0	446'2	445'2	446'0	443'0	2,309	471'2 / 421'2	06:24:02 CT 11 Jul 2019
MAY 2020			**447'2**	-2'2	449'4	448'4	449'0	446'2	693	474'4 / 424'4	06:23:49 CT 11 Jul 2019

Fig. 2.11 Corn Futures quotes (source: www.cmegroup.com, 11.7.2019)

$$S_1 + (F_0 - F_1) = F_0 + (S_1 - F_1) = F_0 + b_1.$$

Unfortunately, the *basis* value $b_1 = S_1 - F_1$ does not necessarily equal to zero. Since at T_1 there is still some time to maturity, the value b_1 is uncertain, see (2.9), although the risk is usually quite negligible.

Example 2.5 A farmer plans to sell his corn production (approximately 500,000 bushels; 1 bushel = 35.24 l) on a local market six months from now (July 2019), let us say in February 2020. The future market price of live cattle is rather uncertain, so the farmer decides to use ten Corn futures contracts to fix his selling price. Figure 2.11 shows an example of quoted Corn futures (one futures trade unit is 5000 bushels and the price is in US cents per pound). Let us propose an effective hedging strategy for the farmer.

The trader can simply enter into $100 = 500{,}000/5000$ short Mar 2020 contracts at the quoted price $444'2 = \$4.445$. The position should be closed in February 2020 when the trader sells his corn on the local market for a price S_1 per pound. The closing price of the short futures position will be F_1 and the total income from the sale and the hedging operation

$$500000 \times (S_1 + (\$4.445 - F_1)) = \$2222500 + 500000 \times (S_1 - F_1).$$

The March 2020 futures prices quoted in February 2020 are expected to be close to the spot price, but there could be a small difference caused by, for example, seasonal effects, temporary shortage or excess supply, etc. Recall that corn is neither an investment asset nor a perfectly storable asset, so the price does not follow the ordinary futures pricing rule (2.9). The basis risk can hardly be eliminated, unless the farmer finds a buyer who will enter into a direct forward contract with maturity in February 2020.

The basis risk becomes more serious when the underlying asset of the futures contract is not identical, but only similar to the hedged asset. This could be the case even in the example above—the farmer's corn could be of some different kind and quality compared to the standardized "CME corn." In this case, it is also the basis value $b_2 = S_2 - S_2^*$, the difference between spot prices of the two assets at the futures maturity time T_2, that generally differs from zero and is uncertain. If the difference between the two spot prices cannot be neglected, and we can only assume a positive correlation, then we should use, rather, the *minimum variance* or *cross-hedging approach*.

Let us assume that we have a long position in N units of an asset. Today's value of the portfolio is $V_0 = N \times S_0$ and we would like to fix its value at a future time T. Without hedging, the value would be just $V_T = N \times S_T$ and we might be afraid of the spot price going down. If exact hedging is not available, i.e., if there are no futures or forward contracts matching the asset and the maturity T, our goal should be at least to minimize the uncertainty (risk) of the hedged portfolio value at time T. Since there is a correlation between the spot price change $\Delta S = S_T - S_0$ and the futures price change $\Delta F = F_T - F_0$, we consider a short futures position corresponding to $h \times N$ units of the underlying asset with an unknown coefficient h called the *hedging ratio*. The change of the value of the hedged portfolio loss between today and the value at time T is

$$\Delta V = V_T^H - V_0 = N(\Delta S - h\Delta F).$$

The risk can be, as usual, measured by the variance σ_V^2 of ΔV viewed as a random variable. The variance can be expressed in terms of the variance σ_S^2 of ΔS, the variance σ_F^2 of ΔF, and their correlation ρ:

$$\sigma_V^2 = N^2\left(\sigma_S^2 - 2h\rho\sigma_S\sigma_F + h^2\sigma_F^2\right). \tag{2.16}$$

The variance of the hedged portfolio depends only on the unknown hedging ratio h. Since (2.16) is a quadratic function in h with a positive coefficient of h^2, it is sufficient to take the first derivative and find the coefficient h that makes it equal to zero, i.e.,

$$\frac{\partial}{\partial h}\left(\sigma_V^2\right) = N^2\left(-2\rho\sigma_S\sigma_F + 2h\sigma_F^2\right) = 0,$$

$$h = \rho\frac{\sigma_S}{\sigma_F}. \tag{2.17}$$

The coefficient given by (2.17) minimizing the variance of the hedged position is called *the optimal hedging ratio*. It can be seen that it is also, by definition, equal to the slope coefficient of the OLS (ordinary least squares) regression equation $\Delta S = \alpha + h\Delta F + \varepsilon$.

The variances and the correlation can be estimated from historical data by different statistical methods (see Chap. 5). Let $x_0, \ldots, x_n$ be a series of prices over certain regular periods (business days, weeks, months, etc.) and $r_i = \frac{x_i - x_{i-1}}{x_{i-1}}, i = 1, \ldots, n$ the relative returns with the sample mean $\widehat{\mu} = \sum_{i=1}^{n} r_i/n$. The simplest volatility estimate is then given by the sample standard deviation

$$\widehat{\sigma} = \sqrt{\frac{1}{n-1}\sum_{i=1}^{n}\left(r_i - \widehat{\mu}\right)^2}.$$

That is, we implicitly assume that the returns in the future have the same standard deviation as the returns observed in the past estimated by $\widehat{\sigma}$. If we consider that the hedging horizon consists of K elementary periods used for the past returns calculation (e.g., K days or weeks) then the simple *square root of time rule* could be applied, i.e., we estimate the cumulative K period return standard deviation as $\widehat{\sigma}\sqrt{K}$. This follows (approximately) from the assumption of independence between returns over nonoverlapping time periods (which means that the variances over K periods can be added up). Finally, we have to keep in mind that the standard deviations in (2.17) are based on absolute price changes. Therefore, the relative return standard deviations need to be adjusted accordingly. If $\widehat{\sigma}_S\sqrt{K}$ is our estimate of the relative return $\Delta S/S_0$ standard deviation, then $\widehat{\sigma}_S\sqrt{K}S_0$ is an estimate of the standard deviation of the absolute changes ΔS (since S_0 is known); similarly $\widehat{\sigma}_F\sqrt{K}F_0$ estimates the standard deviation of ΔF; and the sample correlation ρ estimated from elementary period returns[2] could also be applied for the K periods as the correlation between absolute price differences. Finally, plugging in the estimates to (2.17) we obtain

---

[2]If $x_0, \ldots, x_n$ and $y_0, \ldots, y_n$ are two observed price series then the estimated return correlation is $\widehat{\rho} = \sum_{i=1}^{n}\left(r_{x,i} - \widehat{\mu}\right)\left(r_{y,i} - \widehat{\mu}_y\right)/\left((n-1)\widehat{\sigma}_x\widehat{\sigma}_y\right)$.

$$h = \widehat{\rho}\frac{\widehat{\sigma}_S\sqrt{K}S_0}{\widehat{\sigma}_F\sqrt{K}F_0} = \widehat{\rho}\frac{\widehat{\sigma}_S}{\widehat{\sigma}_F}\frac{S_0}{F_0} \tag{2.18}$$

Example 2.6 Let us reconsider the hedging strategy from Example 2.5 in the case that the farmer's corn type differs significantly from the standard "CME corn." The Latest 2-year data can be used to estimate weekly return sample standard deviations and the sample correlation of the farmer's corn local spot prices and CME corn futures priccs with maturity around 6 months. The estimates are $\widehat{\sigma}_S - 1.9\%$, $\widehat{\sigma}_F - 1.3\%$, and $\widehat{\rho} = 0.8$. The initial spot price of the farmer's corn is $S_0 = \$4.85$ per pound, while the futures corn price is $F_0 = \$4.445$. According to (2.18) the optimal hedging ratio is

$$h = 0.8 \times \frac{0.019}{0.013} \times \frac{4.85}{4.445} \approx 1.276.$$

Thus, the optimal number of short Corn Futures to hedge the farmer's position

$$1.276 \times 500\,000/5\,000 = 127.6 \approx 128$$

differs significantly from the naive approach, where one would use only 100 short futures contracts. The difference is caused by the higher volatility and price level of the farmer's corn type compared to the CME corn. The difference is only partially offset by the 80% correlation, which reduces the hedging ratio.

2.2.4 Stock Portfolio Hedging

The cross-hedging technique can be easily applied to hedge a stock portfolio against the systematic market risk. It is enough to know the beta of the portfolio. According to the CAPM (2.10) the portfolio returns satisfy the regression equation

$$R_P = R_0(1 - \beta) + \beta \times R_M + \varepsilon_P, \tag{2.19}$$

where R_0 is the risk-free return, R_M is the efficient market (or a market index) portfolio return, R_P is our portfolio return over a given time period, and ε_P is the residual risk term. Since the (percentage) return can be viewed as the price change of one (currency) unit of the initial portfolio investment, the beta is exactly the optimal hedging ratio. Thus, if V_0 is the initial stock portfolio value, F_0 is the initial index futures value, and M is the index multiplier, then the optimum number of short futures positions is calculated as

$$N = \frac{\beta V_0}{F_0 M} \tag{2.20}$$

rounded to the nearest integer.

Example 2.7 An asset manager has put together an aggressive portfolio of US stocks with a high 1.6 beta and current market value \$7.5 million. There is market turmoil, and the manager is afraid of large losses on the portfolio. He does not want to liquidate the portfolio, but only to hedge it against the potential systematic risk over the next 3 months. It is July 2019 and he can use, for example, the Dec 2019 E-mini S&P 500 Futures quoted at 3008.25 in Fig. 2.9. The optimum number of short futures contracts according to (2.20) would be

$$\frac{1.6 \times 7.5 \times 10^6}{3008.25 \times 50} = 79.78 \doteq 80.$$

Let us assume that the market index indeed goes down to 10%. Then, according to (2.19) the loss on the portfolio without hedging would be approximately 16%, i.e., $0.16 \times \$7.5$ million $= \$1.2$ million. The short futures position, on the other hand, corresponds to 1.6 times the portfolio value $80 \times 3008.25 \times 50 \doteq \$12$ million, and so the 10% drop means a profit of \$1.2 million almost exactly offsetting the loss. The results would be similar for all other scenarios of the stock index development. The calculation above is a little bit approximate, to be more precise, we have to take into account the risk-free interest rate and dividends paid by the stocks in the index portfolio. For example, if $R_0 = 2\%$ and the dividend rate $q = 1\%$, then the expected return of the portfolio would be

$$E[R_P] = 0.02\frac{1}{4}(1 - 1.6) + 1.6\left(-0.1 + 0.01\frac{1}{4}\right) = -0.159.$$

More importantly, we must not forget that (2.19) is just a statistical relationship with mean zero random error ε_P representing the *specific risk* of the portfolio. The specific risk depends on the correlation ρ between the portfolio returns and the stock index returns.[3] If the correlation is low, then the actual portfolio return might deviate significantly from the expected return given by (2.19).

The asset manager from the example above might only want to reduce beta to a certain lower value β^*. Generalizing the argument above slightly, we are looking for a number of contracts N^* so that $\beta V_0 - N^* F_0 M = \beta^* V_0$, hence.

$$N^* = \frac{(\beta - \beta^*)V_0}{F_0 M}.$$

The strategy can be used not only to hedge a stock portfolio temporarily, but also to bet on a stock, or a portfolio of stocks, against the market. If we believe that a stock with certain beta will perform better than the market, we can hedge the beta, and effectively speculate on the specific risk term ε_P from (2.19) being positive. If we are right, then the strategy will be profitable even if the market declines.

[3]In fact, ρ^2 equals exactly to R-squared of the regression Eq. (2.19).

2.2.5 Rolling the Hedge Forward

Another problem hedgers often face is that the available forward and futures maturities are too short compared to the desired hedging horizon. International stock or bond mutual fund portfolio managers also typically need to hedge against FX risk in an indefinite horizon. The portfolio needs to be hedged, but the horizon depends on the investors' decision to sell back the shares. In addition, the value of the portfolio to be hedged may change over time. The solution is to roll the hedge forward.

Let us hedge a long position in one unit of an asset with an initial spot price S_0 over a horizon that needs to be divided into shorter periods $T_0, T_1, \ldots, T_k$ for which the futures contracts are available. We want to offset the difference $S_k - S_0$ by an opposite gain/loss from the hedging strategy. By entering into a short futures contract from T_0 to T_1 we obtain a profit loss approximately offsetting the difference $S_1 - S_0$. The subsequent short position will offset the difference $S_2 - S_1$, etc. Since

$$S_k - S_0 = \sum_{i=1}^{k} (S_i - S_{i-1}) \tag{2.21}$$

the total rolling forward hedging strategy gain/loss will approximately offset the difference $S_k - S_0$. Nevertheless, the hedge gain/loss will not be exactly equal to the price difference (2.21) due to the cost of carry that accumulates from T_0 to T_k. If the futures prices follow the Eq. (2.9) then the first hedging result $F_0 - F_1$ can be (approximately) expressed as $(S_0 - S_1) + (r + u - y)(T_1 - T_0)S_0$, the second as $(S_1 - S_2) + (r + u - y)(T_2 - T_1)S_0$, etc. The total hedging result could be approximately written as $(S_k - S_0) + (r + u - y)(T_k - T_0)S_0$. Since the interest rate r, storage cost u, and convenience yield y may change over time, there is still a residual risk.

Example 2.8 Let us consider a €10 million portfolio of French government (EUR denominated) bonds paying a 4% coupon managed on behalf of US based investors. The portfolio manager decides to roll over 1-year EUR/USD futures to hedge the FX risk. The advantage of futures is that the position can be closed at any time when investors decide to liquidate the portfolio. This would be difficult in the case of OTC EUR/USD forwards. Assume that today's spot price is $S_0 = 1.20$, the 1-year interest rates in USD and EUR are $r_{USD} = 3\%$ and $r_{EUR} = 1\%$. The standard EUR/USD futures underlying amount is €125,000, and so the manager should initially enter into $83 \approx 10{,}400{,}000/125{,}000$ short 1-year futures (i.e., selling EUR) at a price that should be around $1.20 \times (1 + (0.03 - 0.01)) = 1.224$. The first-year hedge is closed at the spot price S_1, and then the rolled over short futures position is entered approximately at $S_1(1 + (r_{USD} - r_{EUR}))$, recalculating the number of contracts, and so on.

Table 2.2 shows two possible scenarios of the EUR/USD exchange rate development over the 3 years, with the exchange rate going up or down. The bond portfolio values in USD including the hedging gain/loss are almost the same in the two scenarios. The 4% portfolio yield calculated in EUR is enhanced by the positive interest rate differential 2% = 3%–1%. Indeed, the annualized 3-year USD return of the portfolio is in both cases around 6% = 4% + 2%. Possible changes of EUR and

Table 2.2 An example of a rolling 1-year EUR/CZK futures hedge—fixed bond price, two FX scenarios

Month	0	12	24	36
Value in EUR	10,000,000.00	10,400,000.00	1,081,000.00	11,248,640.00
Value in USD	12,000,000.00	11,960,000.00	13,520,000.00	12,373,504.00
EUR/USD Spot	1.200	**1.150**	**1.250**	**1.100**
EUR/USD 12 MFut	1.224	1.173	1.275	1.122
# Futures	83.000	87.000	90.000	
Cum.Hedg. P/L (mil USD)		765284.65	−51,608.29	1,912,808.81
Hedged port. In USD	12,000,000.0	12,725,284.65	13,468,391.71	14,286,312.81

Month	0	12	24	36
Value in EUR	10,000,000.00	10,400,000.00	10,816,000.00	11,248,640.00
Value in USD	12,000,000.00	13,000,000.00	14,060,800.00	15,185,664.00
EUR/USD Spot	1.200	**1.250**	**1.300**	**1.350**
EUR/USD 12 MFut	1.224	1.275	1.326	1.377
# Futures	83.000	87.000	90.000	
Cum.Hedg. P/L (mil USD)		−272,215.35	−554,948.64	−844,493.14
Hedged port. In USD	12,000,000.00	12,727,784.65	13,505,851.36	14,241,170.86

Table 2.3 An example of a rolling 1-year EUR/CZK futures hedge with variable bond prices

Month	0	12	24	36
Bond price	100%	**95%**	**97%**	**100%**
Coupons		400,000.00	804,000.00	1,212,040.00
Value in EUR	1,000,000.00	9,900,000.00	10,504,000.00	11,212,040.00
Value in USD	12,000,000.00	11,385,000.00	13,130,000.00	12,333,244.00
EUR/USD Spot	1.200	**1.150**	**1.250**	**1.100**
EUR/USD 12 MFut	1.224	1.173	1.275	1.122
# Futures	83.000	82.00	87.000	
Cum.Hedg. P/L (mil USD)		765,284.65	−3340.97	1,896,991.97
Hedge port. In USD	12,000,000.00	12,150,284.65	13,126,659.03	14,230,235.97

USD interest rates during the next 3 years, however, mean that the total cost of carry cannot be exactly predicted. Moreover, the simulation neglects the changing bond portfolio value due to the changing market value of the bonds. The rolling hedge strategy allows the regular adjustment of the number of futures contracts accordingly at the end of every year. Table 2.3 shows a scenario where the bond price goes down to 95% and then back up to 100%. Thanks to the rolling FX hedge, the 3-year USD return equals approximately 6% p.a. again.

Interest Rate Derivatives

3

3.1 Interest Rates

The *time value of money* is a key concept for the valuation of all financial instruments. The value of \$1 received 1 year from now is not the same as the value of \$1 received today; \$1 deposited today earns the interest received in 1 year and is, therefore, financially equivalent to \$1 plus the accrued interest received in 1 year. Therefore, the present value of \$1 received in a year is (normally) less than \$1.

Zero-coupon bonds are bonds that pay no coupons, only their face value at maturity T. The bonds are traded at time t at a discounted market value quoted as a percentage of the face value. It is denoted as $P(t,T)$ and used as the *discount factor* from T to t. The discounted zero-coupon bond value certainly depends on currency, and so we will also sometimes use the notation $P^{\mathrm{XYZ}}(t,T)$ for currency XYZ discount factors to avoid ambiguity when we are working with more than one currency.

3.1.1 Present Value

If a financial instrument is defined as a fixed cash flow $C_1,\ldots,C_n$ paid at times $T_1,\ldots,T_n$, then its market value must be equal, by a straightforward arbitrage argument, to the value of the portfolio of a C_1 face value zero-coupon bond maturing at T_1, a C_2 face value zero-coupon bond maturing at T_2, …, and a C_n face value zero-coupon bonds maturing at T_n. Consequently, the instrument's *market (or present) value* from the perspective of time t must be

$$PV = \sum_{i=1}^{n} C_i \times P(t,T_i). \tag{3.1}$$

© The Editor(s) (if applicable) and The Author(s), under exclusive license to Springer Nature Switzerland AG 2020

J. Witzany, *Derivatives*, Springer Texts in Business and Economics,
https://doi.org/10.1007/978-3-030-51751-9_3

In this elementary valuation formula, we either assume no counterparty credit risk or we admit that there is a more significant credit risk and the value must, in addition, be explicitly adjusted by a Credit Valuation Adjustment (CVA). In both cases, we need to know the discount factors $P(t,T)$ corresponding to risk-free zero-coupon bonds, i.e., to bonds issued by issuers with no possibility of default. This sounds simple, but in reality, it presents a difficult problem. Government or top-rated banks' bonds have been traditionally considered (almost) risk-free. However, in today's financial markets even bonds issued by financially relatively sound governments have a measurable credit risk perceived by the markets. Nevertheless, as a starting point we assume that government bonds and interbank loans have no credit risk; later we will look at other possibilities and refinements, for example, the discount curve defined on the basis of interest rate swap (IRS) prices or other instruments (such as over-night index swaps—OIS).

Zero-coupon bond prices are obviously available only for a limited set of maturities, but in order to price general instruments we need the discount factors $P(t,T)$ for all maturities $T > t$ up to a reasonable time horizon (sometimes up to 30 or even 50 years). Fortunately, the discounted value can be calculated or extrapolated from other market instruments' prices. For example, if R_{3M} is the quote of an interbank money market 90-day deposit then the corresponding discount factor will be

$$P(t_S, t_S + 90/365) = \left(1 + R_{3M}\frac{90}{360}\right)^{-1}. \tag{3.2}$$

Formula (3.2) uses t_S for the settlement day, usually "$T + 2$", i.e., two business days from the trade date, and the standard money market day convention Act/360. The unit of time in $P(t,T)$ is 1 year, and so 90 calendar days correspond to the fraction 90/360 of a year (disregarding leap years). These details may seem tedious, but we have to keep in mind that even a *basis point* (0.01%) difference may cause large absolute differences in valuation results when the rates are applied to notional amounts in billions of Dollars or Euros.

3.1.2 Interest Compounding

Interest rates for different maturities are usually expressed on an annual basis (p.a.—*per annum*), but the calculation of the total interest amount accrued by maturity might differ. Bank deposits have different compounding frequencies; the interest may be accrued, for example, monthly, quarterly, or just annually. If the interest rate R is accrued $m > 1$ times a year then the end-of-year value of \$1 investment will be higher than in the case of the same rate R in the *simple annual* compounding:

$$\left(1+\frac{R}{m}\right)^m > 1+R, \text{assuming positive } R.$$

Therefore, a monthly accrued deposit account with the rate R is more valuable than an account with the same annually accrued interest rate R. In order to have a canonical and mathematically well-behaved interest rate convention, financial engineers, and mathematicians use the *continuous compounding* convention. In this convention interest is accrued over every infinitesimal time period. The infinite frequency compounding is certainly not used in practice, but it allows us to translate different interest conventions into a technically convenient common basis. In fact, we have already used this convention in Sect. 2.1. It follows from elementary calculus that

$$\lim_{m\to\infty}\left(1+\frac{R}{m}\right)^m = e^R,$$

and so the value of \$1 continuously accrued with interest rate R over a period from t to T can be expressed simply as $e^{R(T-t)}$. Alternatively, the time t value financially equivalent to \$1 received at time T, i.e., the discount factor, is $P(t,T) = e^{-R(T-t)}$. The time here is calculated on the actual basis, i.e., the actual number of calendar days divided by the actual number of days in a year, or alternatively as the actual number of seconds divided by the total number of seconds in a year, etc.

3.1.3 Day Count Conventions

Money market deposits of different maturities use a simple compounding, where the interest is calculated and paid only at maturity, but we must take into account the day-count convention. The standard convention is Act/360. In general, the interest paid for a d calendar day rate R deposit is calculated as $R \times d/360$. Note that the time adjusted interest paid on a 1-year deposit $R \times 365/360$ is larger than the nominal interest R.

On the other hand, fixed-coupon bond markets use rather the 30/360 or Act/Act day count conventions to calculate the *accrued interest* (AI) over a certain period. The accrued interest is settled by counterparties when bonds are traded between the coupon payment days. The full (cash) bond price $Q = P + AI$ is calculated as the sum of the quoted net price P and the accrued interest AI. In the 30/360 convention each month has 30 days regardless of the actual number of days. For example, if 6% is the coupon paid annually, then the accrued interest for the first 3 months would simply be $6\% \times 90/360 = 1.5\%$. The 30/360 day-count convention is used for US corporate and municipal bonds and generally on European markets; US Treasury bonds use the Act/Act day convention.

3.1.4 Zero-Coupon Curve Construction

Let us assume that the current time is $t = 0$. Our goal is to calculate the discount factors $P(0,T)$, and the corresponding interest rates

$$r(T) = -\frac{1}{T}\ln P(0.T) \tag{3.3}$$

in continuous compounding convention. The function $r = r(T)$ assigning zero-coupon interest rates to different maturities is called the *zero-coupon curve*. It shows the term structure of interest rates which can be flat, increasing, decreasing, or may have any other shape.

Since money market deposits are usually settled two business days from the trade date, we should first look at the Over-Night (O/N) interest rate $R_{\mathrm{O/N}}$ on deposits between Today and Tomorrow (Today +1 business day) and Tomorrow-Next (T/N) interest rate $R_{\mathrm{T/N}}$ on deposits between Tomorrow and the day after Tomorrow (Next = Tomorrow +1 business day = Spot maturity). We also need to count the calendar days d_1 between Today and Tomorrow, and d_2 between Tomorrow and the Next. If R is the d-day deposit rate between the Spot and Spot + d calendar days (i.e., time T) then the precise discount factor should be calculated as

$$P(0,T) = \left(1 + R_{O/N}\frac{d_1}{360}\right)^{-1}\left(1 + R_{T/N}\frac{d_2}{360}\right)^{-1}\left(1 + R\frac{d}{360}\right)^{-1}. \tag{3.4}$$

The money market rates might be the actual publicly quoted (bid or mid) interest rates, or *reference rates* like Libor (London Inter-bank Offered Rate), Euribor, Pribor, etc. The reference rates are used not only for valuation but also for the settlement of FRN (floating rate notes) coupons and interest rate derivatives payments. Therefore, the objectivity of the reference rates is of utmost importance.

The reference rates are published daily by a financial authority as certain averages of quotations provided by contributing banks (usually eliminating outliers). For example, the most well-known Libor used to be published by the British Bankers' Association (and is currently by the Intercontinental Exchange) for ON, SW, 1M, 2M, 3M, 6M, and 12M maturities and five currencies: USD, EUR, GBP, JPY, CHF. Libor's reliability was shaken after the financial crisis by the Libor scandal, when it was discovered that some reference banks had manipulated their quotations to profit from their positions. Despite the EUR Libor fixings, EUR denominated instruments normally use the Euribor reference rates, which have been published by the European Banking Federation and the European Money Markets Institute since 2015. Figure 3.1 gives an example of USD Libor fixings. CZK denominated instruments use Pribor (the Prague Inter-bank Offered Rate) published currently by the Czech Financial Benchmark Facility (CFBF) authorized under the European Benchmark Regulation (BMR), etc. Notice that the set of maturities of quoted USD deposit rates (Fig. 3.2) is much finer, and that the rates are given as bid/ask quotes

USDLIBOR= ICE LIBOR LINKED DISPLAYS

	USD	12/07/19			DEALING
ON	2.35475		IBA	LON	12:55
SW	2.38425		IBA	LON	12:55
1M	2.36913		IBA	LON	12:55
2M	2.36363		IBA	LON	12:55
3M	2.33950		IBA	LON	12:55
6M	2.26238		IBA	LON	12:55
1Y	2.28575		IBA	LON	12:55

Fig. 3.1 USD Libor fixings (Source: Thomson Reuters, 12.7.2019)

USDDEPO= USD DEPO RATES

RIC		Bid/Ask	Contributor	Loc	Srce	Deal	Time	High	Low	Date
USDOND=	↑	2.42/2.52	KLIEM	FFT	KLMM	KLMM	12:31	2.42	2.35	12JUL19
USDTND=	↑	2.30/2.60	PREMEX	MOW		PRMX	12:35	2.30	2.36	12JUL19
USDSND=	↓	2.30/2.50	LLOYDS BANK	LON	LOYL	LOYQ	12:18	2.40	2.37	12JUL19
USDSWD=	↓	2.32/2.52	LLOYDS BANK	LON	LOYL	LOYQ	12:18	2.52	2.38	12JUL19
USD2WD=	↑	2.32/2.52	LLOYDS BANK	LON	LOYL	LOYQ	12:18	2.36	2.40	12JUL19
USD3WD=	↓	2.29/2.49	O.H.V	AMS	OHVA	OOLD	12:19	2.40	2.39	12JUL19
USD1MD=	↓	2.28/2.44	LLOYDS BANK	LON	LOYM	LOYQ	12:18	2.33	2.37	12JUL19
USD2MD=	↑	2.28/2.42	LLOYDS BANK	LON	LOYL	LOYQ	12:18	2.30	2.34	12JUL19
USD3MD=	↑	2.24/2.36	LLOYDS BANK	LON	LOYL	LOYQ	12:18	2.28	2.34	12JUL19
USD4MD=	↑	2.22/2.34	LLOYDS BANK	LON	LOYL	LOYQ	12:18	2.26	2.31	12JUL19
USD5MD=	↑	2.21/2.33	LLOYDS BANK	LON	LOYL	LOYQ	12:18	2.22	2.27	12JUL19
USD6MD=	↑	2.20/2.32	LLOYDS BANK	LON	LOYL	LOYQ	12:18	2.20	2.24	12JUL19
USD7MD=	↑	2.19/2.31	LLOYDS BANK	LON	LOYL	LOYQ	12:18	2.19	2.27	12JUL19
USD8MD=	↑	2.18/2.30	LLOYDS BANK	LON	LOYL	LOYQ	12:18	2.18	2.27	12JUL19
USD9MD=	↑	2.18/2.30	LLOYDS BANK	LON	LOYL	LOYQ	12:18	2.18	2.22	12JUL19
USD10MD=	↑	2.18/2.30	LLOYDS BANK	LON	LOYL	LOYQ	12:18	2.18	2.27	12JUL19
USD11MD=	↑	2.18/2.30	LLOYDS BANK	LON	LOYL	LOYQ	12:18	2.18	2.27	12JUL19
USD1YD=	↑	2.18/2.30	O.H.V	AMS	OHVA	OOLD	12:19	2.20	2.19	12JUL19

Fig. 3.2 USD deposit interest rates quotations (Source: Thomson Reuters, 12.7.2019)

differ slightly from the respective maturity Libor rates, and (unlike the fixings) change more or less continuously during the day.

Discount factors can be calculated according to Eq. (3.4) for a set of maturities up to 1 year. To extend the zero-coupon curve to longer maturities we need to use a capital market instrument where government bonds are the first choice. Before we continue constructing the curve, it is useful to interpolate the interest rates and discount factors between any two points where the rates have already been calculated. The discount factors are first translated to interest rates in continuous compounding, and then the interest rates can be interpolated linearly (see Fig. 3.3), or using more advanced interpolation techniques (e.g., spline interpolation). Finally, for any maturity T where the interest rate $r(T)$ has been obtained let us set

$$P(0,T) = e^{-r(T)T}. \tag{3.5}$$

Note that it would not be correct to interpolate the discount factors directly since the function shape should be exponential in line with Eq. (3.5).

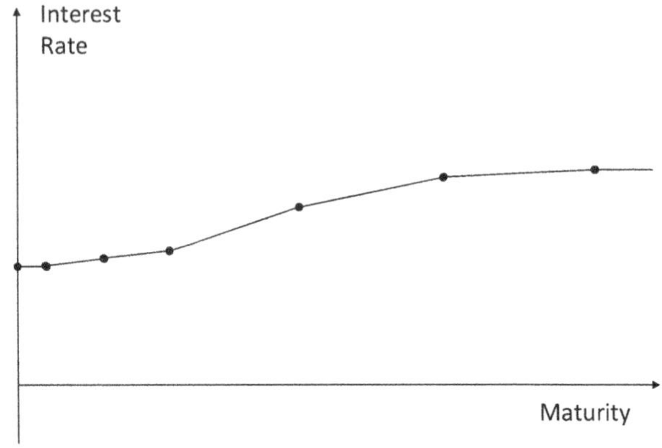

Fig. 3.3 Linearly interpolated zero-coupon curve

Given government bond market prices, the method of *bootstrapping* can be used to obtain the zero-coupon rates. Let Q be the full market price (settled at t_S, typically Today +3 or 4 business days) of a government bond with a fixed coupon C payable at $T_1, \ldots, T_{n-1}$, and at maturity T_n together with the principal amount A. Then according to Eq. (3.1) we must have

$$Q = \sum_{i-1}^{n-1} C \cdot P(t_S, T_i) + (C + A)P(t_S, T_n). \tag{3.6}$$

The idea of bootstrapping is that once we know the discount factors up to the maturity T_{n-1} we can use Eq. (3.6) to express the discount factor with maturity T_n as

$$P(t_S, T_n) = \frac{Q - \sum_{i=1}^{n-1} C \cdot P(t_S, T_i)}{C + A}. \tag{3.7}$$

Thus, starting from the money market zero-coupon curve constructed up to 1 year we can use a 2-year government annual frequency coupon bond with a price quotation to get the 2-year rate by Eqs. (3.7) and (3.3). The rates between one and two-year maturity are interpolated and we go on calculating (approximately) the three-year maturity rate, and so on. The shape of the curve may then look like in Fig. 3.3.

The zero-coupon curve obtained according to the described procedure certainly depends on the benchmark instruments used in the calculation. For short maturities (up to 1 year), we may use not only money market deposits and government zero-coupon bonds but also the repo rates. *Repo* or *repurchase agreements* are basically deposits collateralized by high-quality securities (technically the sale and repurchase of a security). Since repo operations have a very small credit risk, the zero rates obtained from the repo rates would be more "risk-free" than the zero rates based on the interbank rates. Similarly, forward rate agreements, interest rate futures, interest

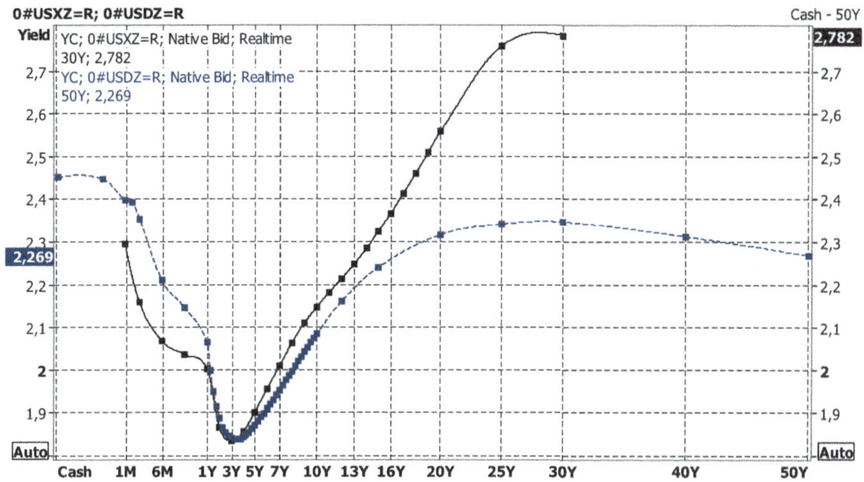

Fig. 3.4 USD bond (full black line) and swap (dashed blue line) zero-coupon curves generated by Thomson Reuters (Date: 12.7.2019)

rate swaps (IRS), or interest rate index swaps described in Sects. 3.2 and 3.3 bear significantly lower counterparty credit risk, and so they might be preferred in the zero-coupon curve construction. Figure 3.4 shows two USD zero-coupon curves automatically generated by the Thomson Reuters system from two different sets of benchmark instruments with cubic spline interpolation on continuously compounded rates. The first curve is obtained from the outstanding US Treasury notes and bonds while the second is built from the money market deposit and interest rate swap rates. The discrepancy between the two zero curves can be explained by the different credit and liquidity risk of the two types of benchmarks. It appears lower for short-term Treasury notes compared to inter-bank deposits but greater for long-term bonds compared to the swaps. The differences between the two curves illustrate the importance of careful benchmark set selection and zero-curve construction.

3.1.5 Forward Interest Rates

Forward interest rates are the future short-term rates implied by the current term structure of interest rates. If the markets were arbitrage free (with no transaction costs and bid/ask spreads) then the forward interest rate from the time T_1 to T_2 (also denoted as $T_1 \times T_2$) would be the market rate on a loan or a deposit that could be contracted today, starting at T_1, and maturing at time T_2. The forward interest rate $r(T_1,T_2)$ in continuous compounding can be easily calculated from the zero rates $r(T_1)$ and $r(T_2)$. If there is a market for forward deposits from T_1 to T_2, then the ordinary money market loan or deposit maturing at T_2 and negotiated today at $r(T_2)$ can be replicated by a combination of a loan/deposit maturing at T_1 with the rate $r(T_1)$ and a forward loan/deposit from T_1 to T_2 at the rate $r(T_1,T_2)$, see Fig. 3.5.

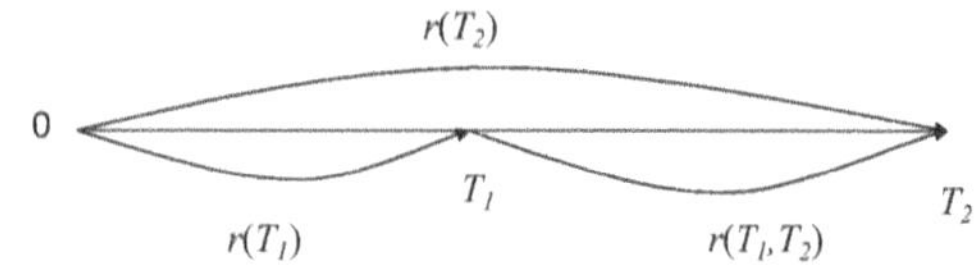

Fig. 3.5 Potential arbitrage scheme using regular and forward money market loans and deposits

In other words, the regular deposit maturing at T_1 can be financed by the compounded loan and, vice versa, the compounded deposit can be financed by the regular loan. Consequently, if there are no arbitrage opportunities, the following identity must hold:

$$e^{r(T_2)T_2} = e^{r(T_1)T_1} \cdot e^{r(T_1,T_2)(T_2-T_1)} = e^{r(T_1)T_1+r(T_1,T_2)(T_2-T_1)}.$$

Since the exponents on the left-hand side and right-hand side must be equal, we can easily solve the equation for $r(T_1,T_2)$ and get

$$r(T_1,T_2) = \frac{r(T_2)T_2 - r(T_1)T_1}{T_2 - T_1}. \tag{3.8}$$

In the following section, we shall discuss interest rate derivatives where forward rates are indeed contracted. The interest rate convention for the market rates is the ordinary money-market convention. The forward rates calculated according to Eq. (3.8) can be transformed back from the continuous compounding convention to money market simple compounding, but it would be more straightforward to use the market convention only. If R_1 and R_2 are the market rates for d_1 calendar days and d_2 calendar days deposits, then the d_2–d_1 forward rate R_F starting d_1 calendar days from the spot maturity date repeating the argument above is equal to:

$$R_F = \frac{1}{d_2 - d_1} \frac{R_2 d_2 - R_1 d_1}{1 + R_1 d_1/360}. \tag{3.9}$$

Example 3.1 Let us assume that 6-month (183 days) deposits are currently quoted at 2%, 1-year (365 days) deposits are 2.5%, and 2-year government bonds with 3% coupon are quoted at 99% of their face value. The O/N (1 day) and T/N (1 day) rates are 1%. The zero-coupon rates for the maturities $T_1 = 185/365$ and $T_2 = 367/365$ can be calculated according to Eqs. (3.4) and (3.3) as follows:

$$r(T_1) = \frac{365}{185} \ln\left(\left(1 + 0.01\frac{1}{360}\right)^2 \left(1 + 0.02\frac{183}{360}\right)\right) = 1.978\% \quad \text{and}$$

$$r(T_2) = \frac{365}{367} \ln\left(\left(1 + 0.01\frac{1}{360}\right)^2 \left(1 + 0.025\frac{365}{360}\right)\right) = 2.461\%.$$

The 2-year ($T_3 = 1 + T_2$) interest rate can be then obtained by applying the bootstrapping formulas (3.7) and (3.3):

$$r(T_3) = \frac{-1}{T_2} \ln \left(\left(1 + 0.01 \frac{1}{360}\right)^{-2} \left(\frac{0.99 - 0.03 e^{r(T_1)T_1}}{1.03} \right) \right) = 3.474\%.$$

The forward interest rates in continuous compounding are calculated according to Eq. (3.8) are $r(T_1,T_2) = 2.950\%$ and $r(T_2,T_3) = 4.493\%$. The 6 × 12 months forward interest rate can be recalculated to the money market convention by

$$R_F = \frac{360}{d_2 - d_1} \left(e^{r(T_1,T_2)(T_2 - T_1)} - 1 \right) = 2.936\%.$$

The rate coincides with the result of Eq. (3.9), but the direct computation would be generally more precise due to possible numerical rounding in the indirect calculation approach.

3.1.6 Expectation and Liquidity Preference Theory

Forward interest rates allow for a better analysis of the term structure of interest rates. It is usually upward sloping, sometimes flat, from time-to-time downward sloping, and exceptionally bumpy. The 6 × 12 months forward rate calculated in Example 3.1 is approximately 1% higher than the 6-month interest rate, and the 12-month rate is approximately an average between the two rates. The *expectation theory* says that the long-term interest rates are determined by the expected short-term interest rates. It means that the implied forward rates would be, according to the theory, just the expected rates. So, if a 6-month rate is considered to be the short-term rate, then the 1-year rate would be determined by the 6-month rate and the expected 6 × 12 rate. Similarly, the 2-year interest rate would be determined by the 6-month rate, the expected 6 × 12, 12 × 18, and 18 × 24 rates, etc.

The empirical fact that the term structure of interest rates is more often upward than downward sloping contradicts the pure expectation theory. An upward sloping curve implies that the forward interest rates are higher than the current short-term interest rates. However, the short-term interest rates fluctuate around a long-term average. It means that the forward interest rates are definitely biased estimates of the "real" expected future interest rates (at least based on the historical experience). The phenomenon is well explained by the *liquidity preference theory* according to which investors prefer to invest funds for shorter periods of time in order to preserve their liquidity. Borrowers, on the other hand, prefer to borrow at fixed rates for longer maturities. The standard argument of supply and demand implies that borrowers must pay a liquidity premium over the expected short interest rates if they want to attract funds over longer periods of time. Hence, according to the liquidity preference theory, the forward rates implied by the long-term market interest rates should be above the expected short-term rates.

3.2 Interest Rate Forwards and Futures

Uncertain future interest rates present a serious financial risk for banks, corporations, and investors. The main purpose of interest rate derivatives like OTC Forward Rate Agreements (FRA) or exchange-traded interest rate futures is to hedge or speculate on the interest rate risk. The contracts allow for the fixing of an interest rate paid or received in the future (FRA and short-term interest rate futures) or to fixing of the price of a long-term interest rate instrument (bond futures). It can be seen from Tables 1.1 and 1.2 that the market with the interest rate derivatives is very active. Note, however, that most of the contracts do not settle the notional amounts, but only interest rate differences.

3.2.1 Forward Rate Agreements

Forward rate agreements are popular Money Market OTC derivatives, which make it possible to fix an interest rate R_K on a deposit or loan starting at a future time T_1 and maturing at time T_2, denoted $T_1 \times T_2$. The contract at time T_1 does not realize the forward deposit or loan, but only settles the difference $R_K - R_M$ between the contracted interest rate and the market interest rate determined at time T_1. The market rate R_M is defined as an appropriate reference rate (Libor, Euribor, Pribor, etc.) valid for the period from T_1 to T_2. Technically, the reference rates are published two business days before T_1. The idea is that the fixed interest rate deposit could be financed with the current market interest rate loan, or vice versa the fixed interest rate loan could be offset (i.e., closed) by a market interest rate deposit. The difference R_K–R_M must be certainly adjusted for the length of the interest accrual period (number of calendar days d) and multiplied by the contracted notional amount A. Moreover, if the forward deposit or loan were realized and closed by an opposite contract with the rate R_M, then the difference would be paid at T_2, and so it should be discounted to T_1 (by the market rate valid at T_1). Therefore, the settlement amount for the *fixed interest rate receiver* (the counterparty contracting the virtual forward deposit) payable at time T_1 is

$$\text{Payoff} = A\frac{(R_K - R_M)\frac{d}{360}}{1 + R_M\frac{d}{360}}. \tag{3.10}$$

The payoff can be positive or negative depending on the sign of the difference R_K–R_M. If it is positive, then the *fixed rate receiver* gets a positive compensation. If it is negative, then the fix rate receiver pays the compensation to the *fixed rate payer*. Regarding the terminology, the fixed rate receiver is also called *FRA seller* (selling the loan) while the fixed rate payer is called *FRA buyer*. In this logic, an FRA buyer is in a long FRA position, while the seller is in a short position. Note that this differs from all other money market instruments. In the cash market, the party buying a

CZKFRA CZK FRA FOCUS LINKED DISPLAYS MONEY

<CZKVIEW> <CZKIRS> <CZKOIS> <CZKFRA> <CZK/1>

	CZK				DEALING	
1X4	2.15	2.19	42 FIN SRV	PRG	FTFS	15:05
1X7	2.180	2.230	KOMERCNI	PRG	KOBA	15:00
3X6	2.13	2.17	42 FIN SRV	PRG	FTFS	15:05
3X9	2.15	2.19	42 FIN SRV	PRG	FTFS	15:05
6X9	2.03	2.07	42 FIN SRV	PRG	FTFS	15:05
6X12	2.06	2.10	42 FIN SRV	PRG	FTFS	15:05
9X12	1.91	1.95	42 FIN SRV	PRG	FTFS	15:05
12X15	1.76	1.80	42 FIN SRV	PRG	FTFS	15:05
12X18	1.75	1.79	42 FIN SRV	PRG	FTFS	15:05
18X24	1.710	1.760	KOMERCNI	PRG	KOBA	15:00

Fig. 3.6 CZK FRA market quotes (Source Thomson Reuters, 12.7.2019)

treasury note is the lender of funds (loan seller), and so it is preferable to use the fixed rate payer/receiver terminology in order to avoid confusion.

The interest rates in Eq. (3.10) are in the regular money-market convention and the length of the forward period does not exceed 1 year. Figure 3.6 shows an example of FRA quotes. The prices are available for a variety of start and end dates in months. Although money market instruments should have maturity, by definition, of up to 1 year, there are 18 $\times$ 24 months FRA where the virtual forward deposit matures in 24 months, but the length of the forward deposit is still formally shorter than 12 months.

FRA contracts can be used like other derivatives to hedge, speculate, or to perform arbitrage. Speculation with the contracts is straightforward. For example, if a speculator expects short-term interest rates to go up in 6 months then (s)he enters a long (fix interest rate payer) 6 $\times$ 9 or similar FRA position. If the interest rates indeed go up, then the increase is collected by the speculator, but if the rates go down then the speculator suffers a loss. An arbitrage can be based, for example, on a mismatch between the ordinary money-market rate and FRA rates by applying the scheme outlined in Fig. 3.5. The following example illustrates how to use FRA for interest rate hedging.

Example 3.2 A company needs to draw a 3-month (90 days) 100 million CZK short-term loan in 9 months. The treasurer expects to pay 1% margin over the 3M Pribor. (S)he is, however, afraid of the Central Bank hiking of the key repo rate and a subsequent increase in the Pribor relative to the current rates. The treasurer can use the 9 $\times$ 12 FRA contract to hedge against the risk. The contract does not ensure the loan itself—it has to be combined with a loan taken in 9 months at prevailing market conditions. The company enters into a 100 million CZK 9 $\times$ 12 FRA contract in the position of a fixed interest rate payer as it needs to draw a loan and pay an interest rate fixed today. The negotiated FRA rate would be around the ask rate (Fig. 3.6), e.g., 1.95% p.a. There is no initial payment for the FRA, and in 9 months the company will borrow 100 million and close the FRA contract. Let R_M denote the 3 M Pribor published in 9 months and assume that the interest on the loan is $R_M + 1\%$ p.a. The FRA payoff will be given by Eq. (3.10), but with the opposite sign, $R_K = 1.95\%$, $A = 100$ million, and $d = 90$. There is a time mismatch between the payoff settled in 9 months and the loan interest paid in 12 months. However, the payoff (3.10) is equivalent to $A(R_M - R_K) \times d/360$ paid in 12 months, perfectly offsetting the loan interest payment $-A(R_M + 1\%) \times d/360$. Indeed, the sum of the two cash flows is $-A(R_K + 1\%) \times d/360$ corresponding to the fixed interest rate of 2.95% p.a. on the 9 × 12 forward loan.

3.2.2 FRA Market Value

As in the case of other forwards, the initial market value of an FRA contract entered into under market conditions should be close to zero. Later, when the interest rates go up or down, the FRA market value changes accordingly. Since the contract (e.g., from the fixed rate receiver perspective) is equivalent to a deposit of the notional amount A at time T_1, repaid at T_2 by $A + I$, where $I = AR_K d/360$, the time t market value can be calculated simply by discounting the fixed cash flow by the zero-coupon rates

$$f = -e^{-r(T_1-t)\times(T_1-t)}A + e^{-r(T_2-t)\times(T_2-t)}(A + I). \tag{3.11}$$

A more straightforward calculation would be to consider the time t market FRA rate $\widetilde{R}_K$ for the period from T_1 to T_2 and the possibility of closing the fix receiving position by the opposite fix paying position entered into at the current market rate. The payoff of the closed position would be the fixed amount $A\left(R_K - \widetilde{R}_K\right)d/360$ payable at T_2 or discounted to T_1. Since the (theoretical) closing FRA's market value is zero, the outstanding FRA market value must be

$$f = e^{-r(T_2-t)\times(T_2-t)}A\left(R_K - \widetilde{R}_K\right)d/360. \tag{3.12}$$

The interest rate in continuous compounding can be replaced by the rate in an appropriate money market quotation. The second approach directly reflects the difference between the contracted and actual market forward rates, and so it should be used rather than the first approach if an appropriate forward quote is available.

Example 3.3 A CZK 100 million 6 × 9 FRA contract was entered into 3 months ago, paying the fix $R_K = 2.3\%$. Calculate the market value of the outstanding position using the rates given in Fig. 3.6 and the 6 months interest rate quoted at 1.9%. The short FRA position could be closed by the currently quoted 3 × 6. Since the mid quote is 2.15 = (2.13 + 2.17)/2 the difference would be—0.15% and it is obvious that the position is in a loss. If the number of days corresponding to 6 months is 183, and the FRA contract is on 90 days, then the market value can be precisely calculated as

$$f = -\frac{100 \times 10^6 \times 0.15\% \times 90/360}{1 + 1.9\% \times 183/360} = -37141.28 \text{ CZK.}$$

The calculation neglects the (almost negligible) impact of the O/N and T/N rates. In fact, it is the present value with respect to the spot settlement date. Alternatively, the formula (3.12) with exactly constructed continuously compounded zero rates could be used.

3.2.3 Short-Term Interest Rate Futures

The exchange traded equivalents of forward rate agreements are *short-term interest rate* (STIR) futures. The most popular ones are the Eurodollar, Euribor, Euroswiss, or the 3-month Sterling contracts. While FRAs offer different lengths of the underlying loan/deposit, the standard length of the STIR futures contracts is 90 days. The futures are, on the other hand, traded for a wide range of maturities going up to 10 years. The standardized notional amount of a Eurodollar futures contract is \$1 million and, similarly, €1 million for the Euribor futures.

Eurodollar interest rate futures contracts should not be confused with EUR/USD currency futures. Eurodollars are dollars deposited and traded outside the United States. Since the USD Libor reference rate used to settle the contracts is based on the interest rates of deposits traded in London, the contracts are called "Eurodollar." Euribor, on the other hand, is an "ordinary" reference rate published by Thomson Reuters and based on the European money market quotes.

Like other futures, the STIR futures settle their profit/loss daily. As in the case of FRA, the daily settlement is based on the difference R_K–R_M between the futures contracted (previously settled) interest rate R_K and the actually quoted (settlement) futures rate R_M. However, there is an important distinction: the futures settlement formula does not take into account the time value of money—there is no discount factor like in Eq. (3.10). Hence, the daily or cumulative payoff can be simply calculated (from the perspective of the fixed rate receiver) by the formula:

$$\text{Payoff} = 10^6 \times (R_K - R_M)\frac{90}{360} = (R_K - R_M) \times 250\,000 \tag{3.13}$$

The advantage of the simplification is that it is easy to remember that one basis point (0.01%) of futures interest rate movement is equivalent to 25 (USD or EUR). The final settlement takes place at the maturity of the futures contract, typically the third Wednesday of the contract expiry month. The closing market rate R_M is defined as the 3-month USD Libor (3M Euribor, etc.). Neglecting the time value of money and the daily settlement mechanism cause a small difference between futures and FRA prices. While FRA rates should be theoretically equal to the forward

Eurodollar Futures Quotes
Globex

View Another Product

Quotes | Settlements | Volume | Time & Sales | Contract Specs | Margins | Calendar

Month	Options	Charts	Last	Change	Prior Settle	Open	High	Low	Volume	Updated
JUL 2019			97.69	-0.01	97.70	97.6975	97.705	97.6825	36,690	08:36:10 CT 12 Jul 2019
AUG 2019			97.845	+0.005	97.84	97.84	97.845	97.825	47,715	08:32:32 CT 12 Jul 2019
SEP 2019			97.92	+0.01	97.91	97.915	97.925	97.90	170,226	08:37:03 CT 12 Jul 2019
OCT 2019			97.945	+0.005	97.94	97.945	97.95	97.935	2,936	08:23:15 CT 12 Jul 2019
NOV 2019			-	-	97.975	-	-	-	367	08:23:14 CT 12 Jul 2019
DEC 2019			98.00	+0.01	97.99	97.99	98.005	97.975	114,129	08:37:03 CT 12 Jul 2019
JAN 2020			-	-	0.00	-	-	-	0	16:45:00 CT 11 Jul 2019
MAR 2020			98.165	UNCH	98.165	98.155	98.18	98.14	85,036	08:37:03 CT 12 Jul 2019
JUN 2020			98.25	UNCH	98.25	98.25	98.265	98.225	108,796	08:37:04 CT 12 Jul 2019
SEP 2020			98.31	UNCH	98.31	98.30	98.325	98.285	96,043	08:37:05 CT 12 Jul 2019
DEC 2020			98.30	UNCH	98.30	98.295	98.315	98.275	76,153	08:37:03 CT 12 Jul 2019
MAR 2021			98.325	+0.005	98.32	98.31	98.34	98.30	58,717	08:37:07 CT 12 Jul 2019
JUN 2021			98.30	UNCH	98.30	98.285	98.31	98.275	55,623	08:37:04 CT 12 Jul 2019

Fig. 3.7 Eurodollar Futures quotations (Source: www.cmegroup.com, 12.7.2019)

interest rates, the futures rates slightly deviate from the forward rates, and the difference may become significant for longer maturities.

Figure 3.7 shows only a fraction of Eurodollar futures quotations. In fact, the contracts are quoted until the June 2029 maturity and the market is relatively liquid up to the 2025 maturities. The futures prices are quoted conventionally as 100–R_K [%] and appear analogously to the prices of discounted (1 year) zero-coupon bonds (without the percentage sign). The fixed rate receiver futures position can in this case be called long ("buying" the bonds) and fixed rate payer position can be called short ("selling" the bonds). Nevertheless, the settlement is still based on the formula (3.13). For example, the December 2019 last price of 98.00 is equivalent to the contractual interest rate $R_K = 100–98.00 = 2.00\%$.

The contracts can be used to speculate or to lock short-term interest rates in the future as in the in case of FRA.

Example 3.4 A company took a $10 million bank 2-year loan with interest payable quarterly and defined as 3M USD Libor +2%. The treasurer is afraid that the interest rates may go up in the next 2 years and would like to lock in the current relatively low rates. The first interest payable in 3 months is already known, but the rates payable in 6, 9, . . ., and 24 months will be set in 3, 6,. . ., and 18 months. It would be difficult, or impossible, to use FRA contracts for the periods where the settlement (i.e., Libor fixing) date goes beyond 12 months. But the treasurer can easily use Eurodollar futures contracts, specifically 10 fixed rate paying futures contracts with 3-month maturity, then 10 contracts with 6-month maturity, and so on until 18-month maturity. The contracts maturing in 3 months (e.g., in October 2019) will pay a 90-day interest rate differential $R_M–R_K$ on \$10 million (with $R_K = 2.055\%$ according to the Oct 2019 quote in Fig. 3.7) and the company will pay on the loan $R_M + 2\%$ where R_M is the same 3M Libor fixed in 3 months. The net cash flow in 6 months is fixed at $R_K + 2\%$. In this way, the treasurer will lock the rates at around 4% (The Eurodollar futures rates + 2% with locked rates varying for different payment dates). There is, however, a time mismatch in the cash flows; for example, the second loan interest payment takes place in 6 months, while the corresponding futures gain/loss settlement is realized in 3 months. The difference is fortunately negligible. Another issue is that the fixing dates of the loan rates and of the futures Libor rates will not usually exactly coincide, since one must choose from a given list of standardized futures maturities. The time discrepancy might be from 1 day up to several weeks. Hence, there is a residual basis risk, which can be quite significant, focusing on a single interest payment. Over a longer horizon, hedging with a series of futures contracts, a sudden move of interest rates up or down should be relatively well offset. In any case, it needs to be kept in mind that the hedging is only approximate.

3.2.4 FRA Versus STIR Futures Interest Rates

If future interest rates were deterministic there would be no point in interest rate derivatives. Necessarily, we must work with stochastic (random) future interest rates. In order to analyze the difference between FRA and STIR Futures we have to choose a specific interest rate model. In the following example, we will use a very simple model to illustrate the issue.

Example 3.5 Let us consider short futures and FRA contracts both on \$1 million and maturing in 12 months. The FRA contract is 12×15. The fixed rate to be paid by both contracts is 4%, and this is also the current level of interest rates for all maturities. Let us assume the following simplistic interest rate model and compare the two contracts: There are just two scenarios, each has 50% probability: in the first scenario the interest rates tomorrow go up by 1% and remain constant, and in the second scenario the rates go down by 1% and remain constant. The futures contract gain/loss is realized (i.e., credited or debited to the margin account) immediately and accrues the market interest until maturity. The payoff in 12 months in the two scenarios will be:

(1) $\$2500 \times 1.05 = \$2625$,
(2) $-\$2500 \times 1.03 = -\$2575$.

In the case of the FRA contract, the cash settlement takes place in 12 months and, moreover, it takes into account the discount factor between 15 and 12 months' time. The payoffs in the two scenarios will be:

(1) \$2500/1.0125 = \$2469.14
(2) –\$2500/1.0075 = –\$2481.39.

The expected, i.e., probability-weighted futures contract payoff value is positive \$25 while in the case of FRA it is negative \$6.13. Consequently, the futures contract will be preferable to the FRA contract with the same fixed interest rate. The law of supply and demand will cause the futures rate to go up, or the FRA rate to go down, or both to move in the opposite directions (since the FRA rate should be equal to the forward interest rate as we have shown, it is rather the futures rate that needs to go up).

In practice, analysts use the following *convexity adjustment* between the futures and FRA $T_1 \times T_2$ rates:

$$\text{Futures rate} = \text{FRA rate} + \frac{1}{2}\sigma^2 T_1 T_2. \tag{3.14}$$

Both rates are expressed in continuous compounding, and σ is the standard deviation of the 90-day interest rate changes over a 1-year period. The normal value of σ would be around 1% to 1.5%, and so the adjustment becomes more significant only for longer-term maturities. For example, if $\sigma = 1\%$, $T_1 = 10$, $T_2 = T_1 + 0.25 = 10.25$ then the adjustment is more than 50 basis points! Neglecting this fact may become quite costly and, in fact, this has happened to some traders in the past. The convexity adjustment (3.14) is based on the Ho and Lee (1986) model. Alternative interest rate models (see Chap. 7) might lead to slightly different adjustments.

3.2.5 Long-Term Interest Futures

The settlement of *long-term interest rate futures* (LTIR) is not based on the long-term interest rates, but on bond prices that reflect the long-term interest rates. The most popular ones are the US Treasury Bond and Treasury Note futures, UK Gilt futures, and German (Italian and Swiss) Euro Bund Futures. The contracts are quoted according to the same convention as the underlying bonds and are settled physically. A bond futures contract normally has not only one underlying bond, but includes a list of eligible bonds that can be delivered. The counterparty in the short position has the option of choosing the bond to be delivered. Since different bonds with different coupons will have different market values, the contract specification involves a *conversion factor* (CF), which is used to recalculate the futures price to the cash price depending on the bond to be delivered. Figure 3.8 shows, as an example, a partial list of Treasury bond futures conversion factors. The conversion

U.S. TREASURY BOND FUTURES CONTRACT

	Coupon	Issue Date	Maturity Date	Cusip Number	Issuance (Billions)	IX. 2019	XII. 2019	III. 2020
1.)	4 1/2	02.15.06	02.15.36	912810FT0	$24.0	0.8456	0.8471	0.8484
2.)	4 3/4	02.15.07	02.15.37	912810PT9	$14.0	0.8667	0.8679	0.8690
3.)	5	08.15.07	05.15.37	912810PU6	$14.0	0.8926	0.8934	0.8943
4.)	4 3/8	02.15.08	02.15.38	912810PW2	$15.0	0.8212	0.8226	0.8239
5.)	4 1/2	08.15.08	05.15.38	912810PX0	$20.0	0.8337	0.8349	0.8363
6.)	3 1/2	02.17.09	02.15.39	912810QA9	$25.0	0.7168	0.7188	0.7208
7.)	4 1/4	05.15.09	05.15.39	912810QB7	$36.0	0.8004	0.8017	0.8032
8.)	4 1/2	08.17.09	08.15.39	912810QC5	$39.0	0.8277	0.8289	0.8300
9.)	4 3/8	11.16.09	11.15.39	912810QD3	$42.0	0.8122	0.8133	0.8147
10.)	4 5/8	02.16.10	02.15.40	912810QE1	$42.0	0.8400	0.8411	0.8420
11.)	4 3/8	05.17.10	05.15.40	912810QH4	$42.0	0.8098	0.8109	0.8122
12.)	3 7/8	08.16.10	08.15.40	912810QK7	$42.0	0.7496	0.7512	0.7527
13.)	4 1/4	11.15.10	11.15.40	912810QL5	$42.0	0.7926	0.7938	0.7951

Fig. 3.8 Treasury bond futures conversion factors (www.cmegroup.com, 12.7.2019)

factors are defined in quarterly increments, while the column "IX.2019" gives CFs valid for bond futures maturing until September 2019; the column "XII.2019" shows CFs for maturities from October to December 2019, etc. The conversion factors are based on a 6% notional coupon bond, i.e., they are calculated as percentage prices on the listed bonds' cash flows discounted with 6% yield-to-maturity. Thus, the factors of bonds with a coupon higher than 6% would be larger than 1 and the factors of bonds with coupons less than 6% are below 1. In the case of Treasury Bond Futures, the underlying bonds must have remaining maturity of at least 15 years, but less than 25 years, and so the list of eligible bonds changes when we go on to longer futures maturities. Treasury Note Futures contracts (2-, 5-, and 10-years) work the same way, but have shorter maturity bonds (notes) as their underlying ones.

When a bond futures contract is settled with settlement futures price F, the counterparty in the short position must choose a bond i out of the list of eligible bonds. The full cash price to be paid by the counterparty in the long position is then

$$F \times CF_i + AI_i$$

where CF_i and AI_i are the conversion factor and accrued interest of the bond i, respectively. On the other hand, the cost of delivering the bond i is $P_i + AI_i$ where P_i is the quoted net bond price. For the counterparty in the short position, it is natural to choose bond i minimizing the cost of delivery, i.e., the difference

$$P_i - F \times CF_i. \tag{3.15}$$

Although the conversion factors are designed to make the delivery of various eligible bonds more or less equal, there will be always a single bond that is the *cheapest to deliver*, which is denoted as CTD. For this bond, the difference (3.15) must be equal to zero, otherwise, there would be an arbitrage opportunity. If it were negative, then it would be profitable to enter into the short position immediately before settlement, and if it were positive, it would be profitable to enter into the long position. Therefore,

$$P_{CTD} = F \times CF_{CTD}. \tag{3.16}$$

The CTD can be also defined at any time before maturity by replacing the spot by the forward price. It turns out that the defined CTD remains relatively stable even though it can change from time to time. The concept of CTD helps to analyze the properties of the bond futures and propose hedging (or speculation) strategies.[1]

Before we look at bond futures applications to hedge interest rate risk, it will be useful to recall some basic concepts of bond mathematics. If $Q = P + AI$ is the full market price (including the accrued interest) of a fixed coupon bond paying cash flows C_i at times t_i (coupons and principal), then the *yield-to-maturity* (YTM) y is defined as the unique solution of the equation:

$$Q = \sum_{i=1}^{n} \frac{C_i}{(1+y)^{t_i}}. \tag{3.17}$$

The yield to maturity characterizes the level of interest rates until the bond's maturity, but it is not the same as the zero-coupon rate. In a sense, it is a mix of rates with different maturities depending on the coupon rate. That is why the zero-coupon rates are constructed as an unambiguous characterization of the interest rate structure.

The bond price $Q = Q(y)$ can be also quoted as a function of the YTM, and it is useful to analyze its sensitivity with respect to y. Mathematically, sensitivity is measured by the derivative Q with respect to y, i.e., $\frac{dQ}{dy}$. In finance, we rather use the related concept of the *(modified) duration*

$$D_{\text{mod}} = -\frac{1}{Q}\frac{dQ}{dy}$$

that can be interpreted as an estimated percentage increase in the bond market price when the yield goes down by 1%. The duration can be expressed analytically by differentiating the Eq. (3.17), that is,

$$D_{\text{mod}} = \sum_{i=1}^{n} t_i \frac{1}{Q} \frac{C_i}{(1+y)^{t_i+1}}. \tag{3.18}$$

[1]The hedging arguments usually assume CTD determined using forward prices remains unchanged. The fact that the CTD determined at the futures maturity might in fact change means that there is a residual hedging risk.

If the modified duration Eq. (3.18) is multiplied by the discount factor $1 + y$ we obtain more traditional *Macaulay duration*, which can be interpreted as the average time to maturity of the cash flow weighted by the discounted payments:

$$D_{\mathrm{Mac}} = D_{\mathrm{mod}}(1 + y) = \sum_{i=1}^{n} t_i \frac{1}{Q} \frac{C_i}{(1 + y)^{t_i}}.$$

Note that many financial calculators or spreadsheets like Excel by default calculate the Macaulay duration rather than the modified one.

The dependence of bond prices on interest rate movements can be further analyzed by looking at the second-order derivative $\frac{d^2Q}{dy^2}$ and the related concept of *convexity*:

$$C = \frac{1}{Q} \frac{d^2Q}{dy^2}.$$

The convexity can be used to estimate the change of duration if there is a change in interest rates, but it is more useful to apply the Taylor's second-order expansion according to which

$$\begin{aligned} \Delta Q = Q(y + \Delta y) - Q(y) &\approx \Delta y \frac{dQ}{dy} + \frac{1}{2} \Delta y^2 \frac{d^2Q}{dy^2} \\ &= -\Delta y \times Q \times D_{\mathrm{mod}} + \frac{1}{2} \Delta y^2 \times Q \times C. \end{aligned}$$

Hence, convexity provides an improvement of the first-order (linear) approximation $\Delta Q \approx -\Delta y \times Q \times D_{\mathrm{mod}}$ given by the duration.

The basic interest rate management strategies are based on the concept of duration or sensitivity—the goal is to make the interest rate sensitivity of a portfolio of assets and liabilities equal or close to zero. More advanced approaches take convexity into account as well. To use the long-term interest rate futures for hedging we need, first, to estimate their duration. The Eq. (3.16) holds at the futures' maturity; if we assume that the CTD can be determined now and does not change until maturity, then Eq. (3.16) must hold as well, but with P_{CTD} replaced by the bond's forward price (the forward price can be expressed by Eq. (2.3)). Differentiating the equation with respect to the CTD bond yield to maturity and dividing by $P_{\mathrm{CTD}} = F \times CF_{\mathrm{CTD}}$ we obtain

$$D_{CTD} = -\frac{1}{P_{CTD}} \frac{dP_{CTD}}{dy} = -\frac{1}{CF_{CTD} \times F} \frac{CF_{CTD} \times dF}{dy} = -\frac{1}{F} \frac{dF}{dy} = D_F.$$

Hence, we conclude that futures price duration equals to the CTD forward price duration.

Long-term interest futures contracts can be used to hedge a portfolio of bonds and other rate sensitive instruments. The goal of the strategy called *duration matching* or *portfolio immunization* is to minimize the sensitivity of the portfolio value with respect to parallel interest rate shifts. The strategy has a hedging horizon (maturity), and we want to minimize the sensitivity to interest rates and, by extension, the variability of the portfolio value at that maturity. Let F be the actual price of an appropriate bond futures contract and D_F its duration equal to the CTD modified duration. We assume that the portfolio is in one currency, hence, US Treasury bonds would be used if the currency were USD, Euro Bund Futures if the currency were EUR, etc. The maturity of the futures must be equal to or longer than the hedging maturity. If it is longer, then we close out the futures position before maturity. Let P be the forward price of the portfolio at the hedging maturity and D_P the modified duration of the portfolio forward price. The sensitivity of the portfolio forward price with respect to a parallel yield curve shift Δy can be expressed by the (approximate) equation:

$$\Delta P = -D_P P \Delta y.$$

Similarly, the futures price sensitivity is

$$\Delta F = -D_F F \Delta y.$$

Since the goal is to minimize the sensitivity of the given portfolio, we need to enter N short futures contracts, where N is the nearest integer satisfying the equation $\Delta P \approx N \times \Delta F$, i.e.,

$$N \approx \frac{D_P P \Delta y}{D_F F \Delta y} = \frac{D_P P}{D_F F}. \tag{3.19}$$

Practitioners often calculate *the basis point value* (BPV) of the portfolio and the futures contract as the price change corresponding to one basis point decrease $\Delta y = -0.01\%$. Solving the equation $BPV_P = N \times BPV_F$ for N then gives the same result as Eq. (3.19). The result is called the *duration-based* or *price-sensitivity based hedge ratio*.

Example 3.6 A bond portfolio manager holds 100 Treasury bonds maturing in 2035 and 200 Treasury bonds maturing in 2044. (S)he expects the long-term interest rates to rise and would like to hedge the portfolio against the corresponding price decline over the next 2 months. The 2035 bonds 3M forward price is 104% of the \$100,000 nominal and the 2044 price is 96%. The total forward value of the portfolio is \$29.6 million. The forward price duration of the former bonds is 13.5 years and of the latter it is 21 years. The fund manager would like to use the T-bond futures quoted in Fig. 3.9.

U.S. Treasury Bond Futures Quotes
Globex

View Another Product

Quotes | Settlements | Volume | Time & Sales | Contract Specs | Margins | Calendar

Month	Options	Charts	Last	Change	Prior Settle	Open	High	Low	Volume	Updated
SEP 2019			153'11	-0'06	153'17	153'09	153'24	152'28	258,510	16:40:41 CT 12 Jul 2019
DEC 2019			152'15	-0'08	152'23	152'19	152'19	152'07	7	16:40:41 CT 12 Jul 2019
MAR 2020			-	-	151'23	-	-	-	0	16:40:40 CT 12 Jul 2019

Fig. 3.9 US Treasury bond futures quotations (Source: www.cme.com, 12.7.2019)

The first task is to calculate the whole portfolio's forward price duration. It follows from the definition of (modified) duration that the portfolio duration equals to the average bond duration weighted by the bonds' market value, that is,

$$D_P = \frac{13.5 \times 100 \times \$104\,000 + 21 \times 200 \times \$96\,000}{100 \times \$104\,000 + 200 \times \$96\,000} = 18.36.$$

The basis point value of the portfolio is

$$BPV_P = 18.36 \times \$29.6 \times 10^6 \times 10^{-4} = \$54345.6,$$

which means that an increase of interest rates by a single basis point causes a loss in the portfolio value of over \$54,000. It is July 2019, and to hedge over the next 2 months the manager uses Sep 2019 futures currently quoted at 153′11. In the US market convention, this quote means $\$153\frac{11}{32}\% = 153.344\%$. Let us assume that the actual CTD duration is 18. The basis point value of one futures contract can be estimated as

$$BPV_F = 18 \times \$1.53344 \times 10^5 \times 10^{-4} = \$276.0192.$$

In order to offset BPV_P the manager needs to enter into N short contracts, so that $BPV_P = N \times BPV_F$, so. $N = \frac{\$54\,345.6}{\$276.0192} = 196.89 \approx 197$.

Therefore, the portfolio manager should short 197 September 2019 contracts that will be closed out in September 2019 before the futures settlement date. The realized gain/loss on the futures position will approximately offset the change in the bond portfolio value due to an increase or decrease in interest rates.

The duration-based strategy has several weak points. The most important weakness to mention consists in the futures duration that is estimated assuming that the CTD bond does not change. But if there is a change of CTD, then the duration may significantly jump up or down and the hedge has to be adjusted. The strategy also does not take into account convexity and assumes only parallel shifts of the yield curve. Short-term interest rates are more volatile than long-term interest rates and sometimes the shifts may go in opposite directions. Generally, if a portfolio of assets and liabilities is sensitive to interest rates with different maturities, then a more advanced strategy immunizing the sensitivity with respect to short-term, medium-term, and long-term interest rates should be used. The hedging then may use STIR Futures, Treasury Note Futures (with maturity at most 10 years), bond futures, or interest rate swaps with different maturities, which will be discussed in the next section.

3.3 Swaps

Hedging of a series of float interest payments, as in Example 3.4, could be quite cumbersome, or impossible, if the series becomes too long. STIR futures with longer maturity might have low liquidity or might not be listed at all. In addition, the futures contracts lock the rates at the short-term forward rates that depend on maturity. If the zero-coupon curve is increasing, then the forward rates increase and vice versa. There could be quite significant jumps in the forward rates from period to period as shown in Example 3.1. A company treasurer would normally prefer fixed interest payments that are constant over all the periods. This requirement can be easily solved with an interest rate swap, or cross-currency swap, if the treasurer also wants to exchange cash flows in different currencies.

3.3.1 Interest Rate Swaps

A *plain vanilla* (i.e., basic) *interest rate swap* (IRS) is an OTC contract where counterparties exchange fixed and floating interest rates calculated on a notional amount and until an agreed maturity (Fig. 3.10).

There is no initial payment and no exchange of notional amounts. The swap can be interpreted as an exchange of two loans between the two counterparties with the same principals and maturities, but different types of interest rates. Since the principals (notional amounts) are identical, their exchange can be netted out and is not needed. The floating rate is defined as the currency standard reference rate (6M or 3M Libor, Euribor, Pribor, etc.) set at the beginning (two business days before) and paid at the end of each 6-months interest period (on the European markets and quarterly on the US markets). The fixed rate is paid annually (semiannually in the United States) for a standard IRS (respecting the bond interest rate conventions). Hence, besides the notional amount and maturity, the key parameter negotiated between counterparties is the fixed swap rate. Figure 3.11 shows an example of CZK IRS quotations provided by CSOB IRS desk. For example, the 10 years Bid/Ask quote 1.57/1.62 means that the quoting bank is prepared to pay/receive the fixed percentage rate against the standard semiannual float (6 M

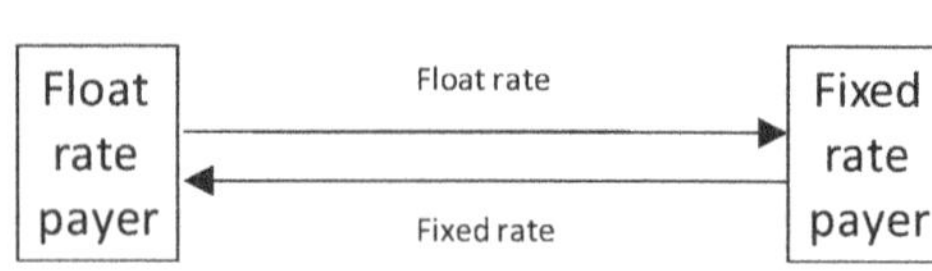

Fig. 3.10 A plain vanilla interest rate swap cash flow

CZKIRS=CSOB CZK IRS LINKED DISPLAYS CSOB

CESKOSLOVENSKA OBCHODNI BANKA - IRS DESK +420-261.353.512

	CZK	AM-A360/3MPRIBOR			DEALING	
1Y/3	2.09	2.14	CSOB	PRG	CSOM	12:16
2Y/6	1.99	2.04	CSOB	PRG	CSOM	12:28
3Y/6	1.84	1.89	CSOB	PRG	CSOM	14:54
4Y/6	1.74	1.79	CSOB	PRG	CSOM	15:43
5Y/6	1.67	1.72	CSOB	PRG	CSOM	15:43
6Y/6	1.62	1.67	CSOB	PRG	CSOM	12:28
7Y/6	1.58	1.63	CSOB	PRG	CSOM	15:43
8Y/6	1.56	1.61	CSOB	PRG	CSOM	15:43
9Y/6	1.57	1.62	CSOB	PRG	CSOM	15:02
10Y/6	1.57	1.62	CSOB	PRG	CSOM	15:43
12Y/6	1.61	1.66	CSOB	PRG	CSOM	14:51
15Y/6	1.68	1.78	CSOB	PRG	CSOM	13:33

Fig. 3.11 CZK interest rate swap rate quotations (Source: Thomson Reuters, 13.12.2019)

Pribor) for 10 years. The main users of IRS contracts for hedging are companies and financial institutions managing their assets and liabilities. A swap contract could be theoretically entered into between two companies with opposite hedging needs, but, as for other products, there is a wholesale interbank market with market-making banks and brokers providing liquidity to the market users.

As indicated by Table 1.1 the global swap markets have become very active. In order to simplify trading and settlement procedures, the International Swap and Derivatives Association (ISDA) introduced standard framework documentation (the *ISDA Master Agreement*), which is usually signed between large swap counterparties. A specific swap contract is then entered into simply by negotiating a few key swap parameters (dates, rates, notional amounts, and conventions) that are summarized in a brief legal document called the *confirmation*.

Example 3.7 A company took out a CZK 200 million 10-year loan with interest defined as 6M Pribor +1.5% payable semiannually. Initially, the float rates were very low, but after 2 years the short-term rates began to rise. The treasurer is afraid of a further increase of the interest rate payments, and (s)he would like to hedge against the risk. The series of uncertain semiannual float payments over the next 8 years can be easily exchanged for a fixed rate using an 8-year IRS. Currently, the company pays the float rate (Pribor +1.5%) on the loan; under the swap, it should pay a fixed rate R_K and receive the float (Pribor) offsetting the float part (Pribor) of the loan payments. The resulting fixed net cash flow paid by the company would be $R_K + 1.5\%$. The swap should be optimally negotiated a few days before the loan interest rate period start date so that the Pribor rates for the loan and for the IRS are fixed on the same day. The quoted ask ("Best sell") 8Y IRS rate is 1.61% (Fig. 3.11). This is an interbank market quote, hence the negotiated rate between the company

Fig. 3.12 The cash flow of the company hedging loan float payments with an IRS

and a profit-seeking bank can be a few basis points higher, e.g., 1.63% on the CZK 200 million notional amount. If the trade date is 13.7.2019 and the start date of the next loan interest rate period is 16.7.2019, then the swap start date can be confirmed on the same day. The confirmed swap maturity is 16.7.2027. Figure 3.12 shows the cash flow between the financing bank A, the company, and the swap bank B. The Pribor rates are set on the same days and the cash flows are matched exactly. What remains is the annual 1.63% p.a. payment to the bank B and the semiannual 1.5% p.a. payment to the bank A. Thus, the annual interest rate cost of the company after the IRS hedge is fixed at (approximately) 3.13% p.a.

3.3.2 IRS Valuation

Like for forward contracts, when a swap is entered into under market conditions, the contract value should be close to zero for both counterparties, as there is no initial premium payment. However, if the market rates change later, then the contract needs to be revalued, and the profit/loss must be accounted for according to valid accounting rules. Outstanding IRS cash flows are not all fixed, and so we cannot directly apply the simple present value principle. Fortunately, there is a simple argument according to which the IRS market value must be equal to a fixed cash flow market value. The first trick is to look at an outstanding IRS (from the perspective of the fixed rate receiver, for example) as at a long position in a fix-coupon bond (with the coupon rate equal to the swap rate, the same notional amount, and maturity) and a short position in a floating rate note (paying the same float-rate on the same notional and maturity). This is correct when we artificially add an exchange of the notional amount A between the counterparties at the swap maturity date (Fig. 3.13). The net cash flow remains unchanged and so, if there is no default risk, the market value remains unchanged as well.

The fixed coupon bond can be valued simply by discounting the known cash flow:

$$Q_{\text{fix}} = \sum_{i=1}^{n} e^{-r(T_i)T_i} C_i. \tag{3.20}$$

The float rate bond (FRN) can also be valued as a discounted fixed cash flow. The standard argument is the following: the payment of $A + R_{\text{fl},n}$ at time T_n with coupon $R_{\text{fl},n}$ set at time T_{n-1} covers the time value of money (amount A) over the period from T_{n-1} to T_n and so the present value discounted to time T_{n-1} is exactly A. In addition at time T_{n-1} there is the payment of $R_{\text{fl},n-1}$ that covers the time value of money over

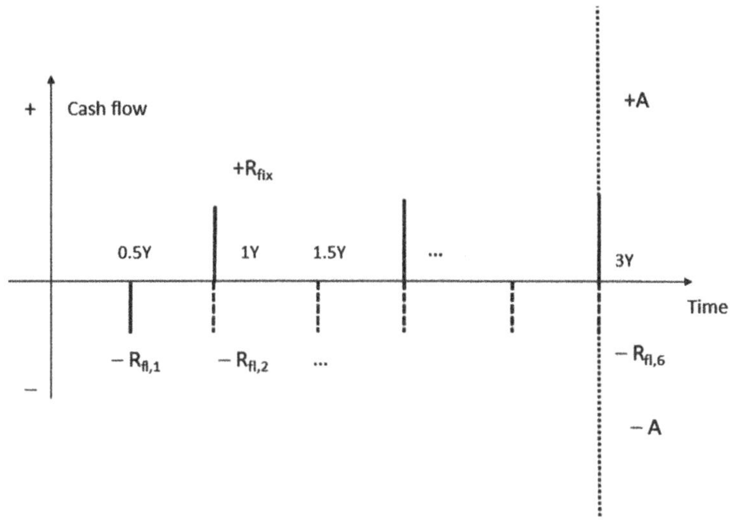

Fig. 3.13 Extended cash flow of a 3-year interest rate swap

the period from T_{n-2} to T_{n-1}, and so we can discount the cash flow to T_{n-2} and get again A, and so on down to T_1. The next float interest rate $R_{\mathrm{fl},1}$ is already set. The swap is valued, generally, from the perspective of the time zero somewhere between two coupon payments, hence $R_{\mathrm{fl},1}$ does not necessarily reflect the time value of money and we simply have to discount $A + R_{\mathrm{fl},1}$ from T_1 to the time 0, i.e.,

$$Q_{\mathrm{float}} = (A + R_{\mathrm{fl},1})e^{-r(T_1)T_1}. \tag{3.21}$$

The swap market value (from the fixed rate receiver's perspective) is then calculated as the difference between the fix rate bond and float rate bond values:

$$V_{\mathrm{swap}} = Q_{\mathrm{fix}} - Q_{\mathrm{float}}. \tag{3.22}$$

Alternatively, the unknown float rates can be replaced with the forward interest rates. The argument is that we could enter a series of FRAs replacing the float rates by the forward rates. Since the FRA contracts are entered into under market conditions, their market value is zero and the value of the original float leg equals to the value of the fixed cash flow given by the forward rates. It can also be algebraically shown that the two approaches are equivalent.

Example 3.8 A 3-year swap with the fixed rate 2.2% and notional €10 million was entered into 3 months ago. The next float rate has been fixed at 1.5%. Value the swap, assuming the zero-coupon rate in continuous compounding is a flat 2.5% for all maturities, that the current float payment period has 183 days and it is just 91 days to the next float rate payment. Then the float leg market value according to Eq. (3.21) is

$$Q_{\text{float}} = €10\left(1 + 0.015\frac{183}{360}\right)e^{-0.025 \cdot 91/365} = €10.014.$$

The fixed leg market value according to Eq. (3.20) is

$$Q_{\text{fix}} = 0.22e^{-0.025 \cdot 0.75} + 0.22e^{-0.025 \cdot 1.75} + 10.22e^{-0.025 \cdot 2.75} = 9.967.$$

Finally, the current market value of the swap is negative for the fixed rate receiver,

$$V_{\text{swap}} = Q_{\text{fix}} - Q_{\text{float}} = -€0.046,$$

and positive for the fixed-rate payer,

$$V_{\text{swap}} = Q_{\text{float}} - Q_{\text{fix}} = €0.046.$$

The valuation formula (3.22) indicates that the market value sensitivity of an IRS position to interest rate movements is similar to the sensitivity of the fixed coupon bond value. In fact, we can use the concept of duration. If D_{fix} is the fix coupon bond (modified) duration and D_{float} the float rate note duration, then the change in swap market value from the fixed-rate receiver perspective can be approximately estimated as

$$\Delta V \cong -(D_{\text{fix}} - D_{\text{float}}) \times A \times \Delta R \tag{3.23}$$

when the rate moves by ΔR. Here, we neglect the difference between the notional amount A and bonds' market value, yet the estimate (3.23) still gives us a good idea of possible losses if there is an adverse movement of the interest rates. Moreover, the float rate bond duration is always less than 0.5 and can be neglected, or estimated by the time to the next float rate payment. Thus, the IRS position sensitivity to interest rate movements is indeed close to the sensitivity of the fixed rate bond position.

Example 3.9 Let us estimate the change in market value of the 3-year swap from Example 3.8 if the rates go up or down by 1%. The fixed coupon bond duration can be estimated as $D_{\text{fix}} = 2.69$ with the yield 2.5% (for example, using the function MDURATION in the Excel spreadsheet) and the float leg duration as $D_{\text{float}} = 0.25$. According to Eq. (3.23), if the rates go up 1%, the fixed rate receiver suffers a loss of around $0.01 \times (2.69\text{–}0.25) \times €10 = €0.244$ million. If the rates go down 1%. then swap market values increase by approximately the same amount.

It follows from Eq. (3.22) that the quoted swap rates should be equal or close to bond yields. When a new swap is contracted, the market value should be zero, and so $Q_{\text{fix}} = Q_{\text{float}}$. Since the market value of a newly issued FRN is, in theory, equal to par (100% of the face value), the same must hold for the fixed coupon bond with the coupon rate equal to the swap rate. This is the case, by definition, if and only if the

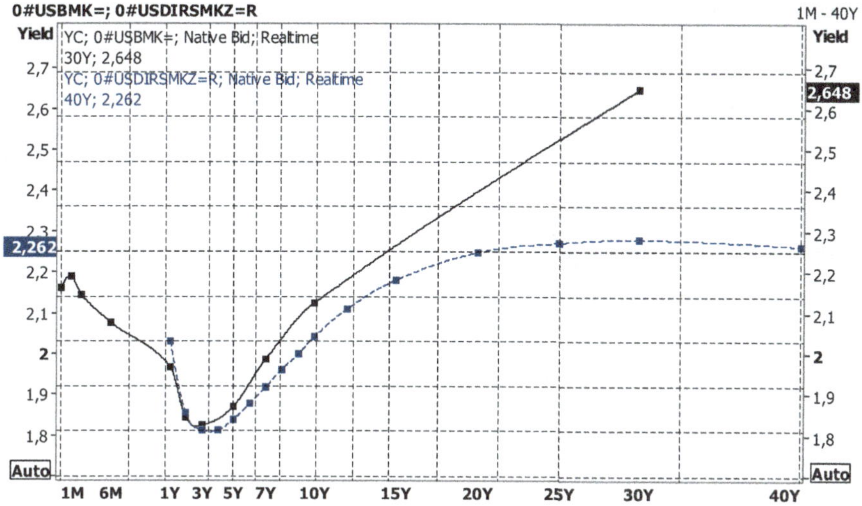

Fig. 3.14 Comparison of the government USD bond yield curve (black solid line) and the IRS rate (blued dashed line) curve (Source: Thomson Reuters, 13.7.2019)

coupon rate equals to the yield to maturity. Consequently, the quoted swap rates should be theoretically equal to the yields of government bonds.

The swap quotations can be used to build a zero-coupon curve similarly to the curve built from government bond prices. If R_{fix} is the quoted IRS swap rate for a maturity, then we use the price $Q_{\text{fix}} = 100\%$ for the bond with the coupon R_{fix} as an input into the bootstrapping procedure. For maturities up to 1 year, the money market rates are typically used. The zero-coupon curve built from the government bonds should be close to the swap zero-coupon curve based on the arguments above. Unfortunately, the reality is slightly different, as already illustrated by Fig. 3.4.

The same picture is obtained by looking at the benchmark US Treasury bonds YTM versus IRS quotations (Fig. 3.14). The swap rates are lower than the bond yields for longer maturities, but higher for shorter maturities. For example, the 10-year Treasury bond yield 2.12% is 8 bps higher than the quoted 10Y swap rate 2.04%. On the other hand, the 1-year Treasury rate 1.97% is 6 bps lower than the 1-year swap rate 2.03% (as indicated by Fig. 3.14) used to build the swap curve.

The discrepancies are caused by the different credit and liquidity risk profiles of the two instruments. Investors in long-term bonds face the credit risk of losing all or a substantial part of the principal amount, and the liquidity risk of not being able to sell the bonds quickly if needed. On the other hand, the credit and liquidity risk of a long (fixed-rate receiving) IRS position is much lower, and it involves only the possibility of losing the positive market value, which is typically just a fraction of the notional amount, if the swap counterparty defaults. Regarding the short-term credit risk, Treasury notes are usually viewed as very liquid with minimum credit risk, but the fixed swap rates replacing the float rates always incorporate the money market rate (e.g., 6M Libor), i.e., also the average credit risk spread of the banking sector

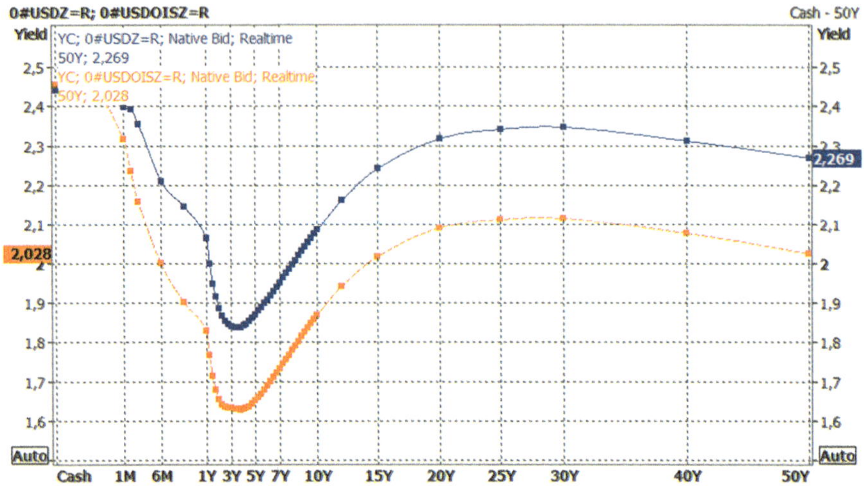

Fig. 3.15 USD bond zero-coupon curve (blue solid line) versus OIS zero-coupon (orange dashed line) curve (Source: Thomson Reuters, 13.7.2019)

that has become more visible since the Global financial crisis. These two opposite factors cause the complex relationship between the two curves that can be observed in Figs. 3.4 and 3.14. In addition, the yield to swap curve spreads might even be much higher for countries where the government's ability to repay its debt has become questionable.

This phenomenon creates a practical problem. Swaps, for example, 10Y contracted today with the currently quoted fixed rate should be valued at zero. But if the zero-coupon curve was obtained from the bond prices as described in Sect. 3.1 then the fixed coupon leg of the swap would be discounted with the 10Y bond zero-coupon rate, which is about 8 bps higher than the swap zero-coupon rate. Thus, the swap market value comes out significantly negative, which would be incorrect. The issue is solved if the zero-coupon curve is constructed from the currently quoted swap rates.

As mentioned above, even the standard swap rates involve some credit risk, and that is why the market has started to use the *Over-Night Index Swap* (OIS) rates to build the zero-curve as a more precise proxy of the theoretical risk-free curve. An OIS, in principle, exchanges a fixed rate against the O/N float rate, and so it is in fact like a plain vanilla interest swap. However, since it is not practical to make the float payments on daily basis, certain reference O/N interest rates (The Federal Fund rate for USD, Eonia for EUR, etc.) are compounded and paid at the swap maturity, or at the same frequency as the fixed rate. Since the credit risk of O/N inter-bank deposits, expressed by an annualized spread, is minimal (see Sect. 5.5 for a more detailed discussion) the OIS rates should be much better proxies of risk-free rates compared to ordinary Libor based IRS rates. The Libor-OIS spread indicates the credit risk of the banking system; it was around 10 bps before the financial crisis, but during the crisis it spiked to more than 350 bps. Figure 3.15 shows that currently the swap USD zero-coupon rates are around 25 bps higher compared to the OIS zero-coupon curve corresponding to the still relatively high levels of the Libor-OIS spread. The OIS zero-coupon curve is nowadays often used as the risk-free rate to discount cash flows, but note that it cannot be used to calculate forward (Libor) interest rates, since it is derived from the OIS rates.

A swap portfolio might have various profiles of sensitivity to interest rates in different maturities. Fixed rate receiving (long) swaps behave like long positions in bonds, and fixed rate paying (short) swaps like short positions in bonds. While it is not straightforward to take a short bond position, a combination of long and short swaps with various maturities is, for an IRS trader, very easy to achieve, and the positions can be changed within minutes. It is important for the trader to monitor continuously the portfolio market value and its risk profile. The total market value equals to the sum of the individual swap values obtained using the swap zero-coupon curve. The zero curve is calculated from the money market and IRS quotes, for example, R_{6M}, R_1, ..., R_{10}, which are continuously updated. Consequently, the portfolio value is a function of the market rates

$$V_{\text{port}} = f(R_{6M}, R_1, \ldots, R_{10})$$

and its change can be estimated by taking the first-order partial derivatives and using the differential

$$\Delta V_{\text{port}} \approx \frac{\partial f}{\partial R_{6M}} \Delta R_{6M} + \frac{\partial f}{\partial R_1} \Delta R_1 + \cdots + \frac{\partial f}{\partial R_{10}} \Delta R_{10}. \tag{3.24}$$

In practice, analysts will calculate rather the basis point value for different maturities. For example,

$$BPV_{10} = f(R_{6M}, R_1, \ldots, R_{10} - 0.01\%) - f(R_{6M}, R_1, \ldots, R_{10})$$

measures the change in portfolio value if the 10Y rate goes 1 bp down and all the other rates remain unchanged. The approximation (3.24) can be the rewritten as

$$\Delta V_{\text{port}} \approx BPV_{6M}\Delta_{6M}^{bps} + BPV_1\Delta_1^{bps} + \cdots + BPV_{10}\Delta_{10}^{bps}, \tag{3.25}$$

where the deltas measure the movements of interest rates in basis points. It is obvious that a T-year swap value is sensitive mostly to the T-year swap rate. However, there are also residual sensitivities with respect to the shorter maturities. To hedge a portfolio of swaps with maturities of up to 10 years the trader should first enter a 10Y swap so that the new portfolio $BPV_{10} = 0$, then a 9Y swap to make $BPV_9 = 0$, and so on until all the coefficients in Eq. (3.25) are close to zero.

A more advanced approach to sensitivity analysis and hedging may also use the Principal Component Analysis (PCA, see, e.g., Jolliffe 2002), which identifies the key term structure movement scenarios, such as parallel shift, twist of the curve, etc.

which explain most of the curve movements using just a few factors (see Sect. 7.2 for a detailed description of the technique).

3.3.3 Cross Currency Swaps

If a company needs to change not only the type of interest rates but also the currency of a loan then *Cross Currency Swaps* (CCS) may be used. This contract, in principle, swaps the cash flows of two loans in different currencies and possibly with different types of interest payments (fix/float). CCS cash flows are outlined in Fig. 3.16.

Counterparty X wants to swap a loan with principal A_X denominated in the currency X, and counterparty Y needs to swap a loan with principal $A_Y = A_X/K$ denominated in the currency Y. The counterparties negotiate the exchange rate K (Y/X) that would normally be close, but not necessarily equal to the spot exchange rate. The loans certainly do not have to exist; in fact, usually at least one of the counterparties is a swap bank trading with CCS on the international financial markets. The key parameters that need to be set are the interest rate R_X on A_X and the rate R_Y on A_Y. The rates can be fixed, but one of them or both might be defined as floats. In fact, the standard quotations are based on float-to-float CCS—see below. Unlike in interest rate swaps, the counterparties exchange the principal amounts A_X and A_Y at the start date and at the maturity date of the swap. That is, the counterparty X will first pass the amount A_X to Y and Y will return it at the maturity date. The same happens with A_Y but in the opposite direction. Between the start and maturity dates, the counterparties exchange interest payments: X will pay the interest R_Y to Y and Y will pay R_X to X. If the counterparty X indeed started with the loan in X, then the new net cash flow will be equivalent to the loan in Y, and similarly for Y.

Cross currency swaps are useful for hedging but as usual they can be used for speculation as well. An open CCS position will bear not only an interest risk but also a significant exchange rate risk. The market value of an outstanding CCS, valued after the initial exchange of principals from the perspective of the counterparty X (Fig. 3.16), can, similarly to IRS, be seen as the difference between the values of two bonds:

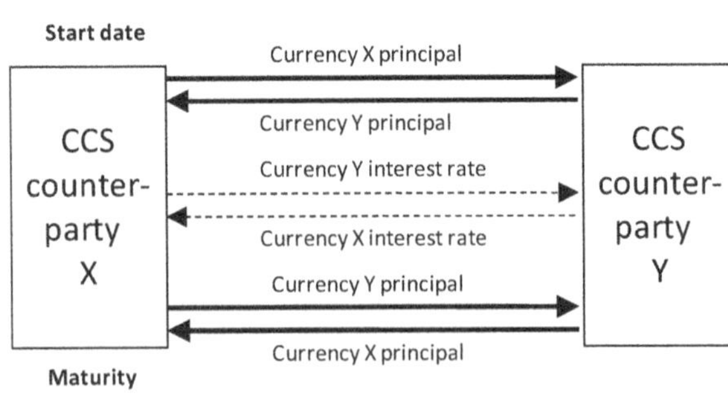

Fig. 3.16 Cross currency swap cash flow

$$V_{\text{CCS}} = Q_X - Q_Y.$$

Q_X is valued as the discounted (domestic) currency X cash flow (3.20), but the cash flow in the currency Y has to be first discounted with the Y zero-coupon rate and then converted to the currency X with the current Y/X exchange rate S:

$$Q_Y = S \times Q_Y^{FC} = S \times \sum_{i=1}^{n} C_i^Y e^{-r^Y(T_1)T_i}.$$

So, the CCS market value $V_{\text{CCS}} = Q_X - S \times Q_Y^{FC}$ expressed in X depends on the exchange rate S and on the interest rates in currencies X and Y.

Example 3.10 A US company would like to swap a 5Y loan of $12 million with the fixed 5% p.a. interest payable semiannually negotiated with a local bank to EUR. This can be easily achieved with EUR/USD cross currency swap. The treasurer negotiates with another US bank a 5Y CCS to swap 5% p.a. on $12 million against 3.5% p.a. on €10 million (i.e., the fixed EUR/USD exchange rate is 1.20). The company will initially pass the $12 million principal and receive €10 million in exchange. The company then gets 5% p.a. on the USD principal and pays 3.5% p.a. on the EUR principal semiannually. After 5 years, the €10 million and $12 million principals are exchanged in the opposite direction. The company's cash flow will be, including CCS, equivalent to €10 million loan with 3.5% p.a. interest. If the loan is used for an investment project and the revenues are in EUR, then there is essentially no exchange rate risk.

Let us, on the other hand, assume that the bank does not hedge the outstanding currency swap position. The market value of the currency swap in USD from the bank's perspective is

$$V_{\text{CCS}} = S_{\text{EUR/USD}} \times Q_{EUR}^{FC} - Q_{USD}.$$

Assume that the bonds' values $Q_{USD} = \$12$ million and $Q_{EUR}^{FC} = €10$ million 1 year later remain unchanged but that EUR depreciates 10% with respect to USD. Then the market value of the swap position will be $V_{\text{CCS}} = 1.08 \times €10 - \$12 = -\$1.2$ million, i.e., the bank suffers a large loss not due to the interest risk but due to the FX risk of the CCS position.

The standard quoted CCS rates are based on float-to-float swaps (also called *Currency Basis Swaps*—CBS) and indicate the spread that is added or subtracted from one of the float rates. The spread may reflect a certain supply and demand imbalance for liquidity in the two currencies and it also gives space for an ordinary bid-ask spread. Therefore, theoretically it should be very small. Nevertheless, since the global financial crisis, CCS spreads have significantly widened. For example, Fig. 3.17 gives an example of recent CCS swap quotes where 3M Euribor is exchanged against 3M USD Libor. The negative quoted spread, deducted from the Euribor side, reflects a shortage of USD liquidity relative to available EUR liquidity.

EURCBS | EUR CBS FOCUS | LINKED | DISPLAYS | MONEY

EURUSD 3MEuribo/3MLibor				DEALING	
1Y	-20	-16	BROKER	GFX	15:52
2Y	-19.875	-15.875	BROKER	GFX	15:52
3Y	-20.5	-16.5	BROKER	GFX	15:52
4Y	-21.75	-17.75	BROKER	GFX	15:52
5Y	-22.75	-18.75	BROKER	GFX	15:52
6Y	-23.875	-19.875	BROKER	GFX	15:52
7Y	-24.875	-20.875	BROKER	GFX	15:52
8Y	-25.375	-21.375	BROKER	GFX	15:52
9Y	-25.5	-21.5	BROKER	GFX	15:52
10Y	-25.5	-21.5	BROKER	GFX	15:52
12Y	-25.125	-21.125	BROKER	GFX	15:52
15Y	-23.25	-19.25	BROKER	GFX	15:52
20Y	-17.875	-13.875	BROKER	GFX	15:52
25Y	-13.375	-9.375	BROKER	GFX	15:52
30Y	-9.875	-5.875	BROKER	GFX	15:52

Fig. 3.17 EUR/USD currency basis swap quotes (Source: Reuters, 13.7.2019)

It may be also related to a higher credit risk perceived on the EUR market relative to the USD market. For example, according to the quotes, a counterparty would provide USD funds for an equivalent EUR amount, receive 3M USD Libor, and pay only 3M Euribor—0.1875% over a 5-year period (based on an ask quote). In other words, USD financing through CCS is around 20 bps above the USD swap rates (since the Libor float rates can be swapped to a fixed rate on the USD IRS market).

The prevailing wide market currency basis spreads create yet another problem in swap valuation. A CCS entered into under market conditions, e.g., based on the quoted mid spread, should have an initial market value equal or at least close to zero. However, if the cash flows are discounted by the IBOR (Libor, Euribor, etc.) zero-coupon curves that do not reflect the CCS spreads, then the market value will significantly deviate from zero. This leads to the concept of currency basis adjusted zero curves that technically solve the issue but create even more confusion with respect to the notion of a theoretically unique risk-free zero-coupon curve (see Sect. 5.5 for a more detailed discussion).

3.3.4 Other Swaps

OTC contracts are generally flexible, and the standard OTC derivatives, including swaps, trade in many different variations. For example, companies need to swap float payments with different periodicity—quarterly, monthly, or even shorter—and so there are interest rate swaps with various float rate periodicities. Alternatively, in a *basis swap*, one periodicity might be exchanged for another float rate periodicity; for example, 1-month Libor can be exchanged for 6-month Libor. As we have seen above, the over-night-index swaps exchanging a compounded O/N rate against a fixed rate have also become an important benchmark market instrument.

Corporations and other institutions hedging their assets and liabilities often need to match a variable principal, for example, due to an amortizing schedule or gradual drawdown of a loan. In an *amortizing swap*, the notional amount is reduced in a regular predetermined way, while in a *step-up swap* the principal increases. *Forward swaps* are swaps with their start date in the future.

All the various swaps mentioned above can be transformed, for the sake of valuation, into fixed cash flows and valued by applying the elementary present value principle. There are, however, many swaps that involve more complex elements, and their precise valuation requires stochastic interest rate modeling. For example, *Libor-in-arrears swaps* are swaps where the Libor rate is not fixed at the beginning, but at the end of each interest rate period. The swaps can be valued similarly to plain vanilla swaps, but a convexity adjustment should be applied (see Sect. 6.4). *Constant-Maturity-Swaps* (CMS) are swaps where the float rate is determined as a constant maturity swap rate and reset for each (e.g., semiannual) period. If the future swap rates are replaced with the forward interest rates, then the valuation is not precise. Again, a convexity adjustment or full interest modeling needs to be applied. The floating swap rates might be capped and/or floored like the corresponding loan interest rates. In this case, valuation of the swaps involves the valuation of interest rate options (see Chap. 6). To conclude, there are many exotic swaps that are traded on the global OTC derivative markets. Exotic swap users should be aware of the dangers in their valuation. The banks offering complex swaps unfortunately often tend to misuse their information and know-how asymmetry. They know better how to hedge and value the contracts, but the derivative user might be much less experienced, unwillingly pay a high cost firstly in terms of the initial contract value, or enter into contracts that do not provide the desired hedging at all (see, for example, Witzany 2009).

Option Markets, Valuation, and Hedging 4

4.1 Options Mechanics and Elementary Properties

Forwards can be used by financial managers to fix the price of an asset in the future. As for other basic assets, derivative traders on the financial markets can be classified as *market-makers* and *market-users*. Market-makers buy and sell the instruments to make a profit on the differences between the buying (bid) and selling (ask/offer) prices. Their existence is important for the liquidity of the markets. Market users, on the other hand, from time to time just use the market to hedge, speculate, or perform arbitrage. By the term hedging, we mean entering into a new contract that will reduce our risk in one or more underlying assets. On the other hand, a speculative transaction will create or increase the risk, while arbitrage would be a combination of two or more transactions that generate a profit without any risk. Let us illustrate these concepts with a few examples.

Example 1.1 If a company enters into a long forward position and if, at maturity, the market price is above the fixed forward price, then everybody is happy. But if the market price falls below the fixed price, then the financial manager might face unpleasant criticism. Thus, it is natural that some financial managers prefer to keep only the upside potential and have an option not to apply the forward rate in the downside scenario. This need is exactly satisfied by the options as illustrated in Example 1.2.

Generally, there is an *option holder* and an *option underwriter*, or seller. The (*call* or *put*) *option* holder has the right, but not an obligation, to buy or sell an underlying asset at a fixed (*exercise* or *strike*) price K. Unlike forwards, there is an initial payment of the option *premium* to the option seller, since the option seller keeps only the downside, while the option holder keeps only the upside. A *European* option can be *exercised* only at maturity (*expiration date*), while an *American* option can be exercised any time up to the expiration date. Options are traded on organized

© The Editor(s) (if applicable) and The Author(s), under exclusive license to Springer Nature Switzerland AG 2020

J. Witzany, *Derivatives*, Springer Texts in Business and Economics,
https://doi.org/10.1007/978-3-030-51751-9_4

exchanges (usually in parallel with futures on the same underlying) and over the counter (OTC)—mostly FX and exotic options.

When options are traded, it is the premium that is negotiated, while the strike price, maturity, and option type are the agreed parameters of the option. Figure 4.1 shows a selection of August 2019 Gold option prices traded on the CME. More options could be shown for other future maturities. There are many options even for the single maturity, although not all possible strike prices are listed and traded on the exchange. The number of options on the OTC market would be potentially unlimited, since option parameters are individually set between any two counterparties. The options in Fig. 4.1 can be generally classified as *in-the-money*, *at-the-money*, and *out-of-the money*. The gold exchange traded options are in fact American futures options. It means that the option holder enters the August 2019 futures position if the option is exercised before maturity. At the same time, there is a cash settlement based on the difference between the option price and the actual futures price. In the case of the gold options quoted in Fig. 4.1, the resulting futures positions are settled physically in August 2019, or maybe closed before maturity. Consequently, exercising a call (put) option is profitable if, and only if, the strike price is less (higher) than the current futures price. In this case the option is called *in-the-money*.

For example, the \$1410 strike price Call (the second option in Fig. 4.1) is in-the-money. Its immediate exercise yields \$1425.4–\$1410 = \$15.4 profit. The profit can be locked in by closing out the futures position right after the exercise of the option. The potential immediate profit is called the *intrinsic value* of the option. The option is quoted at \$23.5, i.e., \$7.9 higher than the intrinsic value. The difference is called the option's *time value*, because there is potentially a higher profit if the option is realized later. On the other hand, the \$1410 strike price Put (the first option in Fig. 4.1) is out-of-the-money—its immediate exercise would yield a loss. Hence, its intrinsic value is zero and the quoted (last) price of \$4 reflects only its time value. Finally, the \$1425 strike price Call and Put options are (nearly) at-the-money, their market prices are approximately equal (\$10.9 and \$10.3), in line with the put-call parity (See Sect. 4.1.2).

Exchange traded options (similarly to futures) use margin accounts, but the mechanism is not the same as in the case of futures. First, in the case of long option positions (buying options) there is no need to make a margin deposit, as there is only an upside potential. Short positions can be *covered* by an offsetting position in the underlying asset or cash. Only *naked* (i.e., uncovered) short positions require a margin deposit. Unlike futures, there is no daily profit/loss settlement, but the required deposit is daily recalculated and additional funds might be required. The calculation adds up the market value of the option and a percentage (around 10–20%) multiplied by the nominal amount. On the other hand, OTC options (likewise forwards) generally do not require any margin deposit, unless mutually agreed between the counterparties.

While exchange traded option premiums are quoted directly, on OTC markets the premiums are quoted indirectly using the so-called *volatility* as shown, for example, in Fig. 4.2 for the EUR/USD European options. The quoted volatility can be, roughly speaking, defined as the annualized standard deviation of future (log)returns

Gold Options Quotes

Globex

View Another Product

Quotes | Settlements | Volume | Time & Sales | Contract Specs | Margins | Calendar

Underlying Future	Charts	Last	Change	Prior Settle	High	Low	Volume	Hi / Low Limit	Updated
Aug 2019		1425.4	-2.7	1428.1	1454.4	1424.0	463,944	No Limit / No Limit	12:54:49 CT 19 Jul 2019

Type: American Options

Expiration: Aug 2019

Strike Range: At The Money

Calls									Puts							
Updated	Hi / Low Limit	Volume	High	Low	Prior Settle	Change	Last	Strike Price	Last	Change	Prior Settle	Low	High	Volume	Hi / Low Limit	Updated
12:45:11 CT 19 Jul 2019	No Limit / 0.10	18	46.90	27.10	25.80	+1.30	27.10	1405.0	2.70	UNCH	2.70	0.80	3.10	1,369	No Limit / 0.10	12:45:12 CT 19 Jul 2019
12:45:11 CT 19 Jul 2019	No Limit / 0.10	114	42.30	22.00	22.10	+1.40	23.50	1410.0	4.00	UNCH	4.00	1.10	4.50	1,001	No Limit / 0.10	12:48:47 CT 19 Jul 2019
12:45:11 CT 19 Jul 2019	No Limit / 0.10	181	35.50	16.30	18.80	-2.50	16.30	1415.0	5.80	+0.10	5.70	1.20	6.10	648	No Limit / 0.10	12:51:22 CT 19 Jul 2019
12:45:11 CT 19 Jul 2019	No Limit / 0.10	184	29.60	13.40	15.80	-2.40	13.40	1420.0	7.70	UNCH	7.70	2.70	8.20	467	No Limit / 0.10	12:46:43 CT 19 Jul 2019
12:45:11 CT 19 Jul 2019	No Limit / 0.10	300	25.50	10.50	13.30	-2.40	10.90	1425.0	10.30	+0.10	10.20	4.00	10.40	549	No Limit / 0.10	12:49:30 CT 19 Jul 2019
Updated	Hi / Low Limit	Volume	High	Low	Prior Settle	Change	Last	Strike Price	Last	Change	Prior Settle	Low	High	Volume	Hi / Low Limit	Updated
12:45:12 CT 19 Jul 2019	No Limit / 0.10	276	27.00	8.70	11.10	-1.90	9.20	1430.0	13.00	UNCH	13.00	4.90	13.40	857	No Limit / 0.10	12:45:11 CT 19 Jul 2019
12:45:12 CT 19 Jul 2019	No Limit / 0.10	498	22.40	6.80	9.20	-2.00	7.20	1435.0	16.40	+0.30	16.10	6.50	16.50	161	No Limit / 0.10	12:45:12 CT 19 Jul 2019
12:53:49 CT 19 Jul 2019	No Limit / 0.10	643	18.30	5.00	7.60	-2.30	5.30	1440.0	20.00	+0.50	19.50	9.90	20.00	139	No Limit / 0.10	12:45:12 CT 19 Jul 2019
12:52:47 CT 19 Jul 2019	No Limit / 0.10	873	13.80	4.10	6.20	-2.10	4.10	1445.0	21.60	-1.50	23.10	12.40	21.60	26	No Limit / 0.10	12:45:12 CT 19 Jul 2019
12:45:12 CT 19 Jul 2019	No Limit / 0.10	2,537	14.00	3.30	5.10	-1.50	3.60	1450.0	25.50	-1.50	27.00	13.00	25.50	65	No Limit / 0.10	12:45:11 CT 19 Jul 2019

Fig. 4.1 Gold options quotes (Source: www.cmegroup.com, Date: 19/7/2019)

EURVOL EUR FX VOL LINKED DISPLAYS MONEY

<EURVIEW> <EURIRS> <EUROIS> <EURFRA> <EURVOL> <0#FEI:> <EUR/1>

	EUR			DEALING	
SW	3.690	4.290	ThomRtrs	TRC	11:57
1M	4.800	5.050	ThomRtrs	TRC	11:47
2M	4.742	4.943	ThomRtrs	TRC	11:57
3M	4.890	5.090	ThomRtrs	TRC	11:57
6M	5.175	5.375	ThomRtrs	TRC	11:50
9M	5.465	5.665	ThomRtrs	TRC	11:47
1Y	5.722	5.923	ThomRtrs	TRC	11:47
2Y	6.250	6.550	ThomRtrs	TRC	11:35
3Y	6.700	7.050	ThomRtrs	TRC	10:48
5Y	7.375	7.775	ThomRtrs	TRC	10:50
7Y	7.825	8.175	ThomRtrs	TRC	09:24
10Y	8.250	8.650	ThomRtrs	TRC	09:24

Fig. 4.2 EUR/USD option volatility OTC quotes (Source: Reuters, 16.7.2019)

in the given maturity horizon. Future returns are unknown today, and so they can be viewed as random variable values, and hence the volatility reflects our uncertainty about the future. In practice, the volatility is derived mainly from past experience, but there is also the anticipation of future market developments. For example, the quoted 1Y bid volatility 5.722% means that the market estimates that the forward-looking 1-year (log) returns of the EUR/USD exchange rate have the standard deviation of 5.722%.

The volatility (usually denoted σ) is used in the Black-Scholes formula (explained in detail later in this chapter) to calculate the premium of a given option. The other inputs of the formula are the option parameters (time to expiration and the strike price) and relevant market factors (underlying asset market price, interest rates, dividend rate or expected income on the asset). As we shall see, the Black-Scholes model is just one of many possibilities. Nevertheless, it has become a market standard used to translate market volatility to the market premium and vice versa.

Option price turns out to be an increasing function of volatility, provided the other parameters and factors are fixed. The argument is that with increasing volatility, the average potential gain goes up, in the case of realization, but the downside (in the case of expiration) does not change. Consequently, there is a one-to-one relationship between the option price and volatility and the calculation can be reversed, at least numerically (see Fig. 4.3). The volatility calculated from a quoted option premium is called the *implied volatility*. OTC market makers quote volatility as an indirect price proxy since there are no standardized options. The traders negotiate option prices in terms of volatility, which is, at the end, translated to the option premium paid by the option buyer. Hence, it is, in a certain sense, the premium that is implied by the *quoted volatility*. Unlike forward prices, which are theoretically determined by

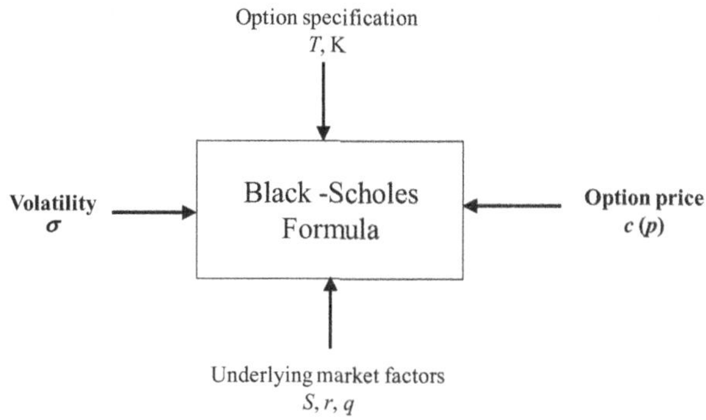

Fig. 4.3 Relationship between volatility, option price, option parameters, and other factors

already existing market factors (spot price, interest rate, income on the asset, etc.), volatility is a completely new market factor. Trading with options is sometimes called volatility trading, since it is about the market view on volatility.

4.1.1 Elementary Properties of Options

As discussed above, an option price (call or put) positively depends on volatility. There is also a certain dependence on the other parameters and market factors that can be qualitatively analyzed without any specific formula. It is obvious that a lower strike price call option is more valuable than a call option with a higher strike price and the same maturity. Similarly, the higher the spot price, the higher the call option value should be. The impact of strike and spot prices on the put option value should apparently be the opposite. Since the domestic currency interest rate positively influences the expected growth of assets, its impact on a call option value should be positive. On the other hand, income on the underlying asset (dividend rate, foreign currency interest rate, etc.) has a negative effect on the expected growth of the asset price, and so the impact on the call option value should be negative. In the case of put options, the situation is the opposite. Finally, there is time to maturity, where the impact is ambiguous. American options with a longer time to maturity are always more valuable. In the case of European options, it depends on the relationship between spot and forward prices (normal or inverted market). If spot prices are expected to grow over time then, ceteris paribus, a call option value will grow with a longer time to maturity. But if spot prices are expected to decline (e.g., due to high dividend payout), then the call option value might decrease with a longer time to maturity. The dependencies are summarized in Table 4.1.

The put and call prices satisfy the well known *put-call parity*. Before we prove the parity, let us look at a few basic inequalities. Table 4.2 gives an overview of the standard notation used throughout this book and in derivatives literature generally.

An American option is always at least as valuable as the corresponding European option, other parameters being equal. On the other hand, an American or European call (put) option can never be worth more than the underlying asset (the strike price amount paid for the underlying asset). Consequently, we have

Table 4.1 Effect of option parameters and market factors on option prices

Variable	Call	Put
Spot price S	+	−
Strike price K	−	+
Time to maturity $T-t$	+ (American)? (European)	+ (American)? (European)
Volatility σ	+	+
Interest rate r	+	−
Asset income q	−	+

Table 4.2 Summary of option valuation notation

Variable	Description
S_0	Current value of the asset
K	Strike price (forward price)
T	Expiration time
t	Current time
r	Domestic currency risk-free rate
q	Income on the underlying asset
S_T	Asset price at time T
c (C)	European (American) call option value
p (P)	European (American) put option value

$$c \leq C \leq S_0, p \leq P \leq K.$$

A European put option, if exercised, pays the strike price K at time T, and so its current value will never be higher than the discounted strike price (assuming $t = 0$)

$$p \leq e^{-rT}K.$$

To obtain the option value lower bounds, more sophisticated arbitrage-based arguments need to be used. Let us assume that the asset does not pay any income. Then the call option lower bound is

$$S_0 - e^{-rT}K \leq c. \tag{4.1}$$

The inequality (4.1) is proved by a classical arbitrage argument. Consider two portfolios corresponding to the left-hand side and to the right-hand side of the modified inequality

$$S_0 \leq c + e^{-rT}K,$$

i.e., a portfolio A with one unit of the asset and a portfolio B with one European call option and $e^{-rT}K$ in cash in a money market account (or invested into zero-coupon bonds) earning the interest rate r in continuous compounding. We need to prove that the current value of A is less than or equal to the current value of portfolio B, i.e.,

$A_0 \le B_0$. It is easy to prove the relationship at time T: If $K < S_T$ then the call option is exercised: the accrued cash amount K is used to get one unit of the asset whose value is S_T. If $K \ge S_T$ then the option expires, and the value of the portfolio B is just K. In both cases, the value of the portfolio A is less than or equal to the value of the portfolio B, i.e., $A_T \le B_T$. The inequality between the values of the two portfolios at time T generally implies the same inequality at time 0. By contradiction, if $A_0 > B_0$ then we could short A (i.e., sell one unit of the asset), and the proceeds would be sufficient to buy B (i.e., to buy one call, and invest $e^{-rT}K$ into a money market account) and to keep a positive profit. At time T the position can be closed by selling off the portfolio B and buying A, i.e., one unit of the asset to close the position. Since $A_T \le B_T$, there is also a nonnegative profit at time T. Altogether we have achieved a positive arbitrage profit, contradicting our assumption about arbitrage-free markets.

Example 4.1 A 6-month European call option on a non-dividend stock is quoted at €3.10. The strike price is €50 and the current stock value is €53. Assume that the interest rate in continuous compounding is 2%. It is easy to verify that (4.1) fails to hold, since $53 - e^{-0.01}50 = 3.5$ is larger than 3.1. It means not only that a theoretical statement is violated, but, moreover, that there is an attractive arbitrage opportunity that can be utilized in practice. A trader of a large investment bank could possibly short one million of stocks, buy one million of calls, and invest the remaining amount in short term bonds. When the position is closed after 6 months, the profit is at least €0.4 million, regardless of whether the calls are realized or not. One might expect that the violation of (4.1) to be only temporary, that the option market value will be corrected, and that the position could be closed with a similar profit much sooner.

The inequality (4.1) can be easily generalized for income paying assets. Let I denote the present value of income paid by the asset until maturity T. Then

$$S_0 - I - e^{-rT}K \le c.$$

To prove the inequality, set portfolio B as above and A as one unit of the asset and $-I$ in cash (i.e., borrow I that will be repaid by the income).

Similar inequalities can be obtained for European put options on non-income (or income) paying assets:

$$e^{-rT}K - S_0 \le p. \tag{4.2}$$

To prove the inequality, set portfolio A to be the cash $e^{-rT}K$ in a money market account with B being one put option and one unit of the asset. It is easy to see that at maturity $A_T \le B_T$, and so $A_0 \le B_0$.

Example 4.2 A 6-month European put option on a non-dividend stock is quoted at €1.20. The strike price is €55, the current stock value is €53, and the interest rate is 2% again. The inequality (4.2) fails, as $e^{-0.01}55 - 53 = 1.45$ is larger than 1.2.

Again, there is an arbitrage opportunity to make a profit around €250,000, if the transactions are performed with one million shares of this stock.

4.1.2 Put-Call Parity

Consider a portfolio A of one long call and one short put European options with the same underlying asset, maturity T, and strike price K. It is easy to see (Fig. 4.4) that the portfolio payoff at time T is

$$\text{payoff}(A) = \max(S_T - K, 0) - \max(K - S_T, 0) = S_T - K.$$

Indeed, if $S_T > K$ then the call option pays $S_T - K$; if $S_T < K$ then the call option is realized and we lose $K - S_T$; finally, if $S_T = K$ then the payoff of both options is zero. But $S_T - K$ is exactly the payoff of a long forward to buy the asset for K with the maturity T. If B is the portfolio consisting of the long forward, then $A_T = B_T$, and so $A_0 = B_0$, repeating the arguments stated above. Today's value of A is $c–p$ and the value of B is given by (2.12), i.e., we have proved the put-call parity in the form

$$c - p = e^{-rT}(F_0 - K), \tag{4.3}$$

where F_0 is the current market forward price of the asset for the maturity T. If the asset pays no income (e.g., non-dividend paying stock), then $F_0 = e^{rT}S_0$ and so (4.3) can be written as:

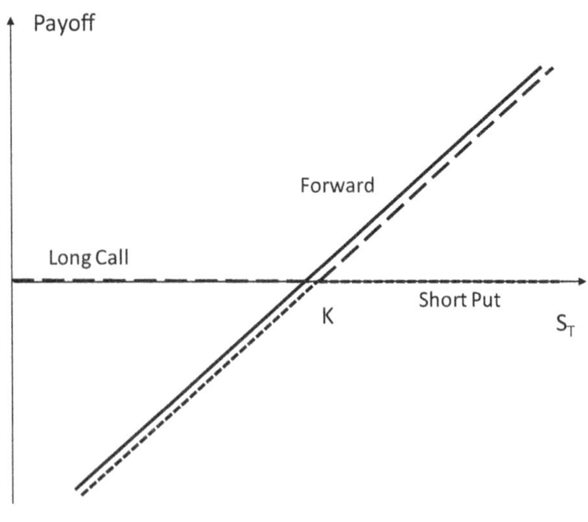

Fig. 4.4 Long call and short put combined payoff

$$c - p = S_0 - e^{-rT}K. \tag{4.4}$$

If the asset pays an income q in continuous compounding, then combining (2.4) and (4.3) we immediately get the generalized put-call parity equation:

$$c - p = e^{-qT}S_0 - e^{-rT}K.$$

The put-call parity allows us to calculate the value of the put option if we know the value of the call option and vice versa. Once we find a formula for c, then (4.3) automatically gives a formula for p. The put-call parity is alternatively often written and proved in the form

$$c + e^{-rT}K = p + S_0. \tag{4.5}$$

If market option quotes do not respect (4.5) then the portfolios A and B corresponding to the left-hand side and right-hand side can be formed. Again, it is easy to verify that $A_T = B_T$. If the left-hand side value is lower than the right-hand side, then B should be shorted, and A should be bought to get an arbitrage profit. If the left-hand side value is higher than the right-hand side then we short A, and invest into B.

Example 4.3 A 3-month European call option on a non-dividend stock is quoted at €1.20 while the corresponding put option is €2.50. The strike price for both options is €55, the current stock value is €53, and the interest rate is 2%. The left-hand side of the put-call parity (4.5) is $12 + e^{-0.005}55 = 55.93$, while the right-hand side is $2.5 + 53 = 55.5$. Thus, the put-call parity is violated, indicating an arbitrage opportunity. The arbitrage can be achieved by shorting the "left-hand side" portfolio (or its multiple) for €55.93 and investing €55.5 into the "right-hand side" portfolio. The position should be closed in 3 months with zero payoff and we can keep the €0.43 (per one stock) risk-free arbitrage profit. Specifically, we can sell 100,000 call options for €120,000 and borrow €$5,472,569 = e^{-0.005}5,500,000$. At the same time, we buy 100,000 put options for €250,000 and 100,000 stocks for €5,300,000. The remaining amount of €42,569 can be set aside as the arbitrage profit. After 3 months, either the call option or the put option is realized (or neither of them, if the spot price equals to the strike price). If the call option is realized, then we sell 100,000 stocks for €5,500,000 and repay the loan, the put option expires, and the position is closed with zero payoff. If the put option is realized, then the stocks are again sold for €5,500,000 and the loan is repaid. Finally, if the spot price equals exactly to €55, then neither the call nor the put has to be realized. The stocks can be sold on the spot market for €5,500,000 and the position is closed again.

4.1.3 American Versus European Options

The put-call parity has been shown for European options, but does it hold for American options? It turns out that the valuation and analysis of European options are simpler than in the case of American options. Put-call parity equation does not hold for American options, but it can be shown that for non-income paying assets a similar inequality holds,

$$S_0 - K \leq C - P \leq S_0 - e^{-rT}K. \tag{4.6}$$

We may follow arbitrage arguments analogous to the European put-call parity case, but the complication is that both options might be realized at different times before maturity. To prove the first inequality, consider the portfolio A of one stock and one American put and portfolio B consisting of one call and initial cash K. The outcome at time T must take into account the possible realization of the options before maturity. To prove the second inequality the portfolio A has one call and e^{-rT} of the initial cash, while the portfolio B consists of one put and one stock.

Example 4.4 The 1-year American call option on a non-dividend stock is priced at €4.50 while the American put is sold for €3. The strike price for both options is €55, the current stock value is €57, and the interest rate is 2%. The first inequality in (4.6) fails to hold while the second is satisfied since $S_0 - K = 2$, $C - P = 1.50$, and $S_0 - e^{-rT}K = 3.09$. The arbitrageur will sell the put for €3 and short one stock for €57. The proceeds will be used to buy one call for €4.50 and deposit €55. The remaining amount of €0.50 should be the guaranteed arbitrage profit. If the put is exercised at a time $t \leq T$ then we have to pay out €55 for the stock; the stock can be used to close the short stock position and we, in addition, have the remaining nonnegative accrued interest and a potential profit on the call option. If the put option is not exercised until maturity T then we can exercise the call option, buy the stock for €55, and close the short position. Again, we have closed the position and ended up with a positive cash balance.

The analysis of (4.6) can be simplified if we realize that it is never optimal to exercise early a call on a non-income paying asset. If the call is realized at a time $t < T$ then we end up with one unit of the asset at time T. The price paid for the stock from the time T perspective is $e^{r(T-t)}K > K$ (provided positive interest rates), and so it is always better to wait until the maturity of the call. We have shown that for non-income paying assets $c = C$ and so

$$C - P = c - P \leq c - p = S_0 - e^{-rT}K.$$

However, it might be rational to exercise an American call option early, if there is an income, e.g., large dividend paid out before maturity.

The same conclusion, generally, does not hold for put options—it may be optimal to exercise an American put option early even on a non-income paying asset, hence

$p < P$. Assume that the asset value S_t is very low at time $t < T$ and so the put option is deeply in-the-money. Then it might be better to realize the option immediately and get the payoff $K - S_t$ that accrues to $(K - S_t)e^{r(T-t)}$ at time T. If S_t is very small, then the accrued payoff will be larger than the maximal payoff K, if the put was realized at maturity.

4.1.4 Option Strategies

Options are used for hedging, speculation, or arbitrage. A typical hedging application of options has already been shown in Example 1.2. An arbitrage with options has been illustrated in Examples 4.1, 4.2, 4.3, 4.4. There are other possible arbitrage strategies, related, for example, to replication-based option pricing theory. Regarding speculation, options can be combined in many ways to create new trading strategies. The strategies may speculate on a range of future asset values, or even, in a sense, on future market volatility. For example, a long position in a call and a put with identical strike forming a so-called *straddle* (Fig. 4.5) could be based on the

Fig. 4.5 Profit from a long straddle (long call and long put with the same strike price)

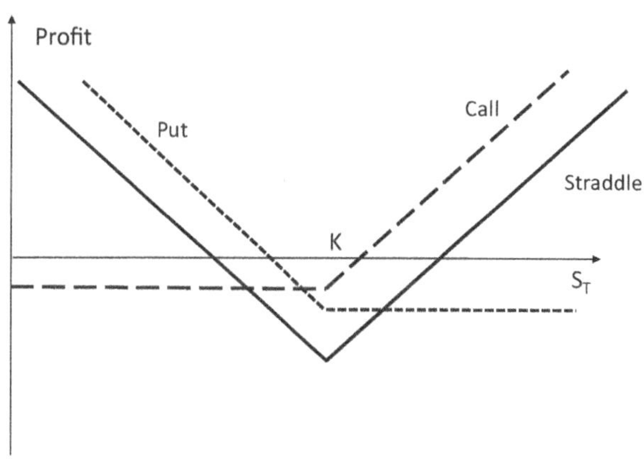

Fig. 4.6 Profit from a long strangle (the put strike is less than the call strike price)

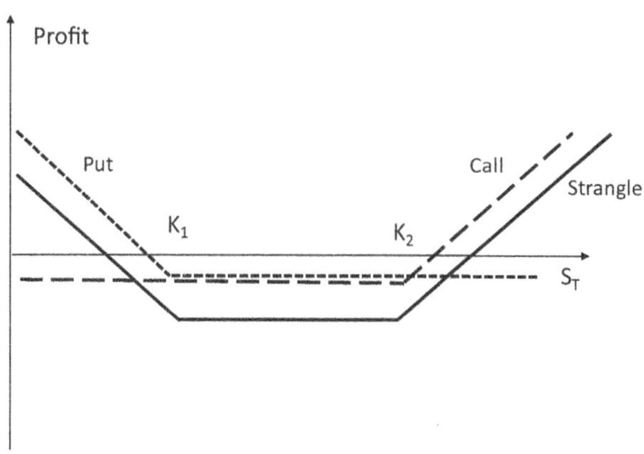

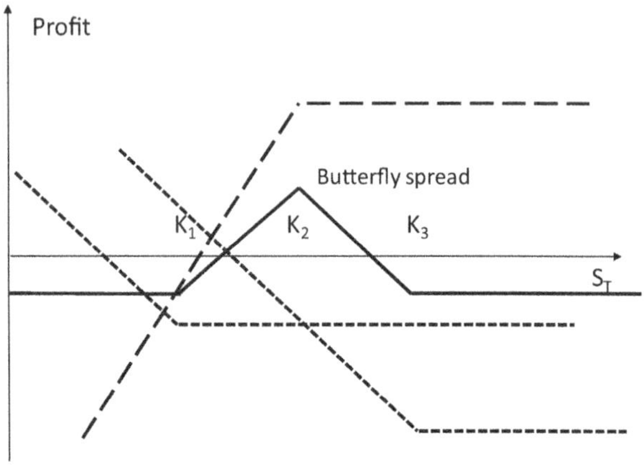

Fig. 4.7 Butterfly spread obtained as a combination of one long put (K_1), two short puts (K_2), and one long put (K_3)

expectation of a certain important event that will cause the price to go significantly up or down. We do not speculate on the direction but, in a sense, on the volatility.

The straddle position can be made cheaper by enlarging the call strike price and reducing the put strike price. The *strangle* profit/loss profile (Fig. 4.6) is similar but not identical to the straddle. The strategy is cheaper; on the other hand, there must be a larger increase or decrease in the asset price to make the strategy profitable.

If a trader believes that the price is going to remain in a limited range, then the strangle or straddle strategy could be shorted. In that case, there is a limited upside and unlimited downside, so traders might prefer to use a strategy like a *butterfly spread* (Fig. 4.7) where the downside is limited.

The strategies are so popular that there are special Thomson Reuters' pages quoting the prices of selected option strategies. The quotes in Fig. 4.8 show not only ordinary EUR/CZK volatilities, but also the butterfly spread, and *risk reversal* prices. The quotes are given via certain specific conventions. For example, "6 M Fly 25% Delta" quote of 0.15%/0.65% indirectly indicates the price of a 6-month maturity butterfly based on $K_1 < K_2 < K_3$ strike prices corresponding to put option deltas 25%, 50%, and 75% (see Sect. 4.3). Specifically, the quote equals to the difference between the average of 25% and 75% delta put volatilities, and the 50% delta put volatility. The fact that the quoted asset volatility may depend on the strike price and option maturity is related to the concept of *volatility smile* and, more generally, *volatility surface* (see Sect. 8.2).

The quotes in Fig. 4.8 also indicate the prices of *risk reversals*. A risk reversal, whose profit/loss profile is shown in Fig. 4.9, consists of a long out-of-the-money call and short out-of-the money put, both with the same maturity. The strategy can be used if a trader does not want to speculate on a moderate growth but only on a large growth in the asset price. The long call is presumably financed by the short put, so there is almost no cost in entering the risk reversal. The reversal yields a zero payoff if the price change is only moderate. But if the price increases significantly, there is an unlimited upside potential. On the other hand, if the price falls, there is also unlimited downside potential. The "RR" quotes in Fig. 4.8 indicate the difference

EURCZKVOL=

FX OPTION VOLATILITY SURFACES

	EURCZK FX Ops ATM		EURCZK FX Ops RR 25% Delta		EURCZK FX Ops B Fly 25% Delta	
	BID ASK	TIME	BID ASK	Time	BID ASK	TIME
ON	0.001 / 11.175	08:16	-4.400 / 5.200	23:21	-2.081 / 2.819	23:21
SW	0.001 / 5.363	08:36	-0.387 / 1.112	07:07	-1.013 / 1.288	07:07
1M	2.050 / 3.450	11:18	0.325 / 1.075	23:40	-0.050 / 0.550	23:40
2M	2.150 / 3.500	11:11	0.375 / 1.125	23:59	0.063 / 0.563	23:59
3M	2.325 / 3.675	11:18	0.450 / 1.200	06:03	0.100 / 0.600	06:02
6M	2.675 / 3.875	11:18	0.500 / 1.250	06:03	0.150 / 0.650	23:21
9M	2.925 / 4.025	11:11	0.538 / 1.288	07:25	0.194 / 0.694	07:25
1Y	3.100 / 4.100	11:11	0.600 / 1.350	06:03	0.250 / 0.750	06:02
2Y	3.125 / 4.975	23:40	-1.313 / 3.287	17:25	-0.619 / 1.831	00:01

EURCZK=
25.573/25.582
16JUL19 13:26:50

EURCZK FX Ops RR 10% DELTA		EURCZK FX Ops B Fly 10% Delta		
BID ASK	TIME	BID ASK	TIME	
-8.475 / 9.275	23:59	-1.244 / 7.606	23:59	ONB
-0.325 / 1.675	23:59	-0.738 / 3.963	07:03	SWB
0.637 / 1.638	23:21	0.516 / 1.266	01:42	1MB
-0.387 / 2.862	23:59	0.762 / 1.513	23:59	2MB
-0.313 / 3.088	23:21	0.784 / 1.534	01:42	3MB
-0.325 / 3.525	06:03	0.350 / 2.050	01:42	6MB
-0.512 / 3.787	23:40	0.506 / 2.456	23:40	9MB
-0.575 / 4.175	06:04	0.494 / 2.744	23:22	1YB
-4.900 / 7.350	23:22	-0.500 / 5.800	23:22	2YB

Fig. 4.8 Volatility quotes for EUR/CZK options, butterfly spreads, and risk reversals (Source: Thomson Reuters, Date: 16/7/19)

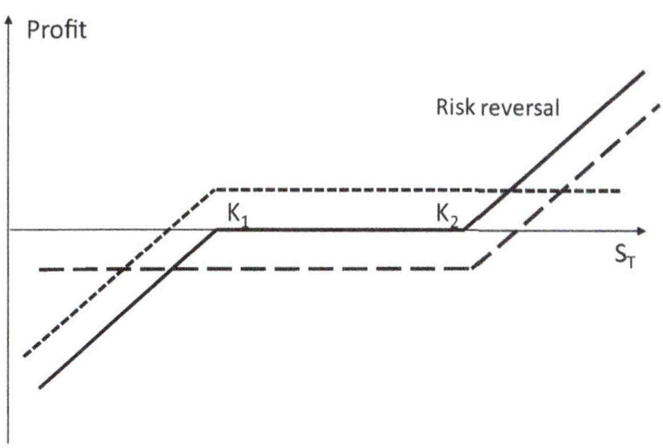

Fig. 4.9 Risk reversal profit loss

between out-of-the-money call and out-of-the-money put volatilities. For example, the 1Y RR 25% Delta quote of 0.60%/1.35% says that the 25% delta call volatility (with the strike price greater than ATM strike) is higher, by this percentage, than the

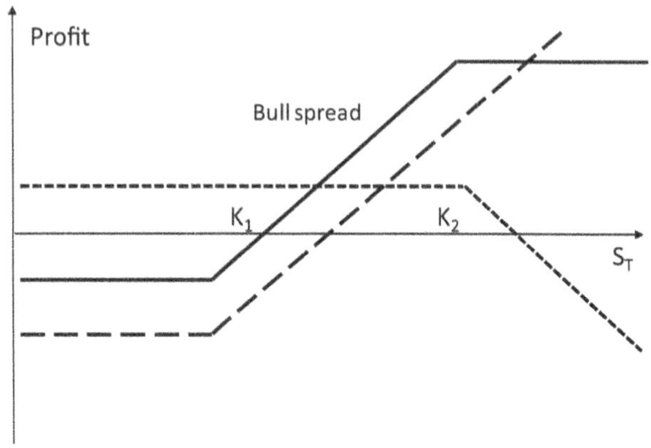

Fig. 4.10 Bull spread profit/loss profile set up from two call options

25% delta put volatility (with the strike price lower than the ATM strike). In practice, it means that the call is more expensive than the put (a large depreciation of CZK with respect to EUR is viewed as more probable than a large appreciation) and the risk reversal cost is positive. A short (opposite) risk reversal strategy could be used if a trader expects a possible extreme fall in the market prices rather than an increase.

Simple complementary strategies called *bull spread* and *bear spread,* where the upside and downside are limited, should also be mentioned. A bull spread can be set by entering into a low strike price (K_1) long call and a higher strike price (K_2) short call (see Fig. 4.10). The lower exercise price will often be in-the-money. The same bull spread can be obtained from a short in-the-money put (with the strike K_2) and a long out-of-the-money put (K_1). As an exercise, use the put-call parity to verify the equivalence of the two definitions. An investor who expects the market to grow, i.e., expects a "bull" market, may speculate on the growth by buying a bull spread. The advantage is that the cost of the long call is reduced by the short call, and, moreover, the downside potential is limited. On the other hand, the upside potential profit is limited as well.

An investor who is pessimistic regarding the future growth of the market, i.e., (s)he expects a "bear" market, might invest in *a bear spread.* A bear spread is defined as precisely the opposite of the bull spread, i.e., a low strike price short call and a higher strike price long call. In this case, the short call will *usually* be in-the-money and the long call out-of-the-money. Hence, there would be an initial cash inflow.

There are certainly many other variations of the option strategies that can be used to speculate on the volatility and/or direction of the market. It should be pointed out that if the market were efficient, then none of the strategies could lead to a systematic profit. It is questionable whether popularity of the option strategies proves the inefficiency of the markets, or whether we can say rather that speculators never learn from their losses.

4.2 Valuation of Options

As discussed in the previous section, the valuation of options is not, unfortunately, as simple as the valuation of forwards. The value of an option depends on the distribution of the underlying asset prices in the future. Option contracts are, in a sense, like insurance products—there is an insurance seller and an insurance buyer, and the value could be, at least approximately, estimated using classical actuarial methods. Based on historical data, an option value might be intuitively estimated as the discounted expected payoff (average insurance loss).

The main goal of this chapter is to show that the famous Black-Scholes formula (representing, in fact, a properly discounted expected payoff) provides a precise and consistent tool to value options. However, it should be pointed out that options used to be traded and valued using intuition, market opinion, or actuarial, or other, methods long before the model was published by Black and Scholes (1973) and, independently, by Merton (1973) in the early 1970s. The formula is based on a stochastic (geometric Brownian motion) model of the underlying asset behavior. The principles of the model can be relatively easily explained in the discrete set up of binomial trees (Cox et al. 1979). We shall then use the concept of infinitesimals to extend the discrete model to a continuous time set-up. The Black-Scholes formula can be arrived at from two directions: firstly as a properly discounted expected payoff using the risk-neutral valuation principle, and secondly as a solution of the Black-Scholes partial differential equation. Both approaches are important from the theoretical and practical points of view, but we will focus on the former and outline the latter only briefly.

4.2.1 One-Step Binomial Trees

A *binomial tree* is a diagram representing the paths of an asset price in time assuming that it follows a *random walk*. In each step the price goes, with certain probability, up or down. Let us first look at the simplest *one-step binomial tree*. The initial price at time 0 is S_0, and at time T it moves up to S_0u with probability p, or down to S_0d with probability $1-p$ (Fig. 4.11). The tree is determined by the three coefficients: $u > 1$, $d < 1$, and the probability $p \in (0, 1)$.

Obviously, it is very simplifying to model future price scenarios over a longer time horizon with only a one-step binomial tree. However, we will argue that a binomial tree with many steps corresponding to very short time intervals approximates the price development well, provided, of course, that the tree parameters are properly calibrated.

Let us assume, for the time being, that there are only two scenarios given by the one-step binomial tree and let f denote the current (unknown) price of a European option (call or put) with maturity T. We can easily calculate the option values f_u and f_d in the "up" and "down" scenarios at time T as the option's payoff given S_T.

An ingenious arbitrage argument can be used to determine the option value f at time 0. The idea is to set-up a riskless portfolio Π combining a short position in the

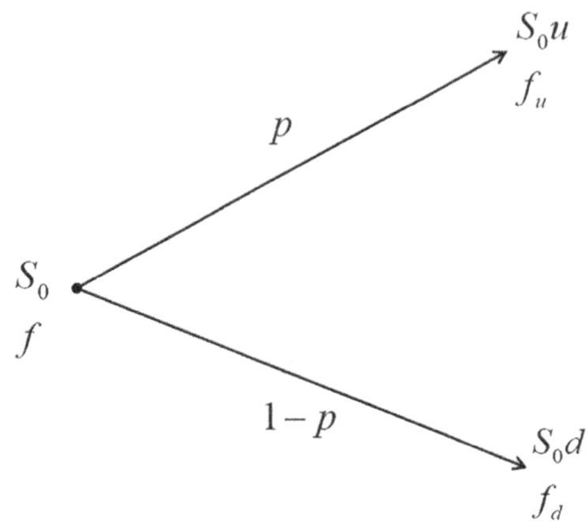

Fig. 4.11 One-step binomial tree

option and a position in "Delta" units of the asset. By a riskless portfolio (or asset) we mean a portfolio whose value is the same in all future scenarios—it means that there is no uncertainty regarding its future value. The basic riskless asset is a bank deposit or government bond investment that yields the risk-free rate r (in continuous compounding) in all scenarios. We also assume the ability to borrow funds at the same risk-free rate r. Any riskless portfolio Π must, therefore, yield exactly r, not less or more. If the yield μ of Π were higher than r (and the value of Π was positive), then we could invest into Π and finance it by borrowing the cash for r. The arbitrage return would be $\mu - r$ of the invested value. If $\mu < r$ then we could short Π and deposit the proceeds for r. The arbitrage return would be $r - \mu > 0$. Assuming there are no arbitrage opportunities we conclude that $\mu = r$.

Let us see whether we can really set up a riskless portfolio of one short option and Δ units of the asset. We assume that the asset is arbitrarily divisible, and so Δ can be fractional. The initial (time 0) portfolio value is

$$\Pi_0 = -f + \Delta \times S_0.$$

The value at time T depends on the two scenarios, and we just need to find an appropriate value Δ so that $\Pi_u = \Pi_d$, i.e., assuming that the asset pays no income:

$$-f_u + \Delta \times S_0 u = -f_d + \Delta \times S_0 d.$$

The single equation with one unknown Δ is easily solved by

$$\Delta = \frac{f_u - f_d}{S_0 u - S_0 d}. \tag{4.7}$$

Note that we are able to obtain this perfectly risk-free portfolio only because there are just two future scenarios and only one parameter Δ that we need to take account at time T. The expression (4.7) for the delta has an interpretation that will be useful later: The numerator is the option value variation corresponding to the underlying asset price variation in the denominator. Consequently, the fraction approximates the partial derivative of the option price with respect to the underlying price:

$$\Delta \approx \frac{\partial f}{\partial S}. \tag{4.8}$$

According to the general argument above, the yield of the portfolio must be equal to the risk-free rate r, $\Pi_u = \Pi_d = e^{rT}\Pi_0$, i.e.

$$-f_u + \Delta \times S_0 u = e^{rT}(-f + \Delta \times S_0). \tag{4.9}$$

Since Δ is given by (4.7), the Eq. (4.9) can finally be solved for the unknown initial price of the option

$$f = \Delta \times S_0\left(1 - e^{-rT}u\right) + e^{-rT}f_u. \tag{4.10}$$

Example 4.5 Let us have a 6-month call option on a non-dividend stock with €50 strike price. Assume that the current stock value is €50, the interest rate is 2%, and the future stock price behavior is approximately modeled by the one-step binomial tree with $u = 1.1$, $d = 0.9$, and 60% up-move probability. The up and down payoffs are:

$$f_u = \max(55 - 50, 0) = 5 \quad \text{and} \quad f_u = \max(45 - 50, 0) = 0.$$

The delta according to (4.7) is

$$\Delta = \frac{5 - 0}{55 - 45} = 0.5,$$

therefore, one short call can be hedged with 0.5 stocks. In order to eliminate the fractional number, let us consider rather, for example, 10 short calls and 5 long stocks whose initial value is $-10f$ + €250. Indeed, the portfolio value in the "up" scenario $-$€50 + €275 equals to the portfolio value €225 in the "down" scenario. The risk-free yield Eq. (4.9) can, in this case, be written and solved as

$$225 = e^{0.01}(-10f + 250), \text{ i.e.}$$
$$f = \frac{250 - e^{-0.01}225}{10} = 2.72.$$

Hence, if the binomial tree perfectly models the future scenarios, then €2.72 would be the only correct price that we should pay for the call option.

4.2.2 Risk-Neutral Valuation

According to the traditional actuarial principle, the option value could be estimated as the probability-weighted average of the time T values discounted to time 0, i.e.,

$$f = e^{-\widetilde{r}T} E_p[f(T)] = e^{-\widetilde{r}T}(pf_u + (1-p)f_d). \tag{4.11}$$

The discount rate $\widetilde{r}$ is not necessarily the risk-free rate, since the payoff is not riskless and investors generally require a higher return for a riskier investment. Hence $\widetilde{r} > r$ and the question is, what the appropriate price of risk is that should be incorporated into the discount rate $\widetilde{r}$.

When we analyze the valuation formula (4.10), there are two surprising conclusions: firstly, the formulas (4.10) and (4.7) do not depend on the probability p, and secondly, the only discount rate used is the risk-free rate r. The option price given by (4.10) can, in fact, after a few algebraic manipulations, be expressed in the form (4.11) as

$$f = e^{-rT} E_q[f(T)] = e^{-rT}(qf_u + (1-q)f_d), \text{where } q = \frac{e^{rT} - d}{u - d}. \tag{4.12}$$

The probability p used in (4.11) is called *real world* (*objective* or *physical*), since we assume that it captures the real future development. According to (4.12), it can be replaced with an artificially defined probability q, called the *risk-neutral probability*, which allows the discounting of the expected payoff only by with the risk-free interest rate r. To explain the notion of risk-neutrality, let us look at the expected stock return over the period T. In the real world, the return μ given by the equation:

$$S_0 = e^{-\mu T} E_p[S(T)] = e^{-\mu T}(pS_0u + (1-p)S_0d), \tag{4.13}$$

or alternatively the probability p is given by the equation:

$$p = \frac{e^{\mu T} - d}{u - d},$$

should be higher than r, since the underlying asset is a risky investment. On the other hand, the return on the stock, when the probability p is replaced by q from (4.12), turns out to be simply the risk-free rate r. Therefore, we call the world where the probability p is replaced by q the *risk-neutral world*. Note that $q < p$ according to (4.12) and (4.13), since $r < \mu$.

We have just seen that investors in the risk-neutral world require only a risk-free return on the risky stock investment, and (4.12) means that they require a risk-free return on any other derivatives with the same source of risk (depending on the same set of scenarios). In this artificial world, investors are risk neutral, they do not require any compensation for risk, and the price of risk is zero. Here we can calculate the expected payoff, discount it by the risk-free rate, and conclude that the resulting price is the correct derivative price applicable in the real world.

We have proved (in the context of a one-step binomial tree) the risk *neutral valuation principle,* which is of paramount importance for the valuation of derivatives in general: *to value an option (or another derivative) we may assume that investors are risk neutral—their price of risk is zero. The derivative can be valued as the expected payoff discounted by the risk-free interest rate. The resulting valuation then also applies in the real world.*

Example 4.6 The real-world probability $p = 60\%$ given in Example 4.5 has not been indeed used in the call option valuation at all. Let us calculate the risk-neutral probability

$$q = \frac{e^{rT} - d}{u - d} = \frac{e^{0.01} - 0.9}{1.1 - 0.9} \approx 55\%.$$

The value obtained in Example 4.5 is, as expected, the same as the result based on the risk-neutral valuation principle (4.12):

$$f = e^{-rT} E_q[f(T)] = e^{-0.01}(q \times 5 + (1 - q) \times 0) = 2.72$$

4.2.3 Multistep Binomial Trees

It is obvious that a one-step binomial tree cannot capture properly the random behavior of asset prices that range over many possible values and change at very short time intervals. The one-step binomial model can, however, be easily extended to a general n-step binomial tree (Fig. 4.12). There are n time-steps of length $\delta t = T/n$, and at each step the asset price goes up or down by the multiplicative factors u and d, and with the same probabilities. The assumption of constant multiplication factors and constant branching probabilities can be relaxed, but in this basic setup the advantage is that the tree is recombining: if we go up and down, then the price change is the same as if we go down and up since $ud = du$. The tree looks like a lattice, and after n steps it has only $n + 1$ end-nodes corresponding to different modeled future asset prices.

An option can be valued repeating the one-step binomial tree valuation formula going backward from the end-nodes to $(n-1)$-step nodes down to the root of the tree. The option values at maturity are known, e.g., in the case of a European call option

$$f_{u^{n-k}d^k} = \max\left(S_0 u^{n-k} d^k - K, 0\right).$$

Given f_{u^n} and $f_{u^{n-1}d}$ get $f_{u^{n-1}}$ by applying (4.10) with T replaced by δt, etc. Finally, get f from f_u and f_d. In the case of European options, it is not necessary to

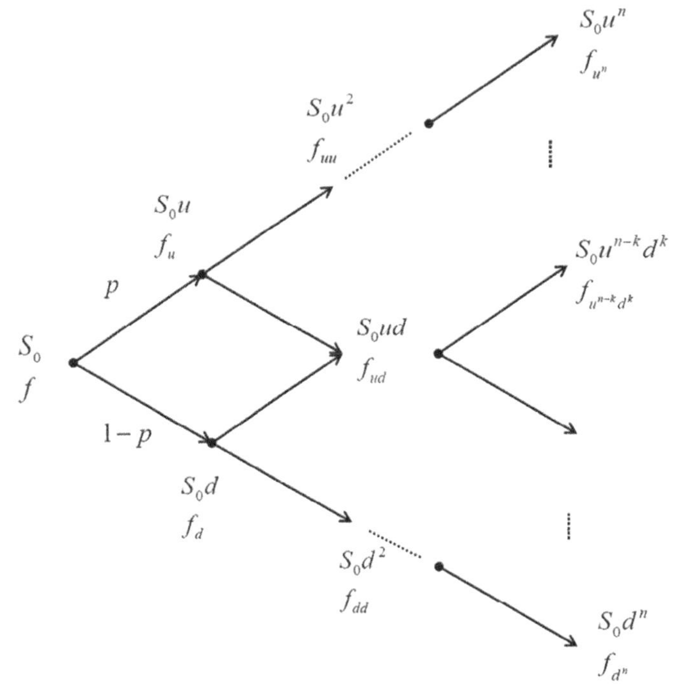

Fig. 4.12 A general n-step binomial tree

repeat the one-step binomial argument again and again. The calculation is significantly simplified by introducing the risk-neutral probability

$$q = \frac{e^{r\delta t} - d}{u - d} \tag{4.14}$$

which does not change over the tree. According to the risk-neutral valuation principle $f = e^{-rT}E_q[f(T)]$, where $f(T)$ denotes the option payoff at time T modeled as a random variable with values given by the end-nodes on the tree and $E_q[\cdot]$ is the expected value with respect to the risk-neutral probabilities. The risk-neutral probability of the node corresponding to $n-k$ ups and k downs is $\binom{n}{k}q^{n-k}(1-q)^k$ with $\binom{n}{k} = \frac{n!}{k!(n-k)!}$ denoting the binomial number n over k (number of unordered k-tuples from a set with n elements), and so we have an explicit formula for the option value:

$$f = e^{-rT}\sum_{k=0}^{n}\binom{n}{k}q^{n-k}(1-q)^k f_{u^{n-k}d^k}. \tag{4.15}$$

In practice, the parameters u and d must be chosen to match the volatility of the asset prices observed (or expected) on the market. In fact, the goal is to match the first two moments of the return distribution over the elementary time step δt. The first moment, i.e., the mean return, would be matched by choosing an appropriate physical probability p, while the second, i.e., the standard deviation of returns, is matched by choosing appropriate u and d. Since we apply the risk-neutral valuation principle, the result does not depend on p; only u and d matching the standard deviation of returns need to be set.

Volatility σ is defined so that $\sigma^2\delta t$ is the variance of the asset returns over a period of length δt. The definition implicitly assumes that the price process satisfies the Markov property—that is, the future development depends only on the present value of the asset, not on the past development. It means, in particular, that the price changes over nonoverlapping time intervals are independent, and the return variances can be added up. Thus, the variance over a 1-year period would be σ^2. The binomial model has the Markov property: the probabilities of going up or down do not depend on the past.

According to Cox et al. (1979), if σ is the volatility over the period T, then the (CRR) parameters can be set as follows[1]:

$$u = e^{\sigma\sqrt{\delta t}}, d = e^{-\sigma\sqrt{\delta t}}. \tag{4.16}$$

This parametrization is in fact quite easy to explain. As discussed below, asset log-returns over short time periods (e.g., daily) are known to have approximately the normal distribution, i.e.,

$$\ln\frac{S(t+\delta t)}{S(t)} = \eta \sim N\left(\mu\delta t, \sigma^2\delta t\right),$$

in other words, $S(t+\delta t) = S(t)e^{\eta}$. The variable η can be roughly approximated by the Bernoulli variable, which takes just two values that are equal to $\pm$ standard deviation of η, i.e., to $\pm\sigma\sqrt{\delta}$. However, according to the Central Limit Theorem (see, e.g., Kopp et al. 2013) when the number of steps increases, the probability distribution of $\ln S(T)/S(0)$ approaches the normal distribution, which is what we need. Regarding the first moment of the distribution, it is easy to show that if $p = \frac{e^{\mu\delta t}-d}{u-d}$ then the variance of the asset returns over the one-step period is $\sigma^2\delta t$ plus a term of order δt^2 that becomes negligible when the number of steps n is large (i.e., δt small). The same conclusion holds for the risk-neutral probability (4.14), and thus the change of probability implies a modified return, but the volatility σ remains essentially

[1]Note that $e^{\sigma\sqrt{\delta t}} \doteq 1+\sigma\sqrt{\delta t}$, for a small δt. In fact, the factors $u = 1+\sigma\sqrt{\delta t}$ and $d = 1-\sigma\sqrt{\delta t}$ could be alternatively used with the same asymptotic results.

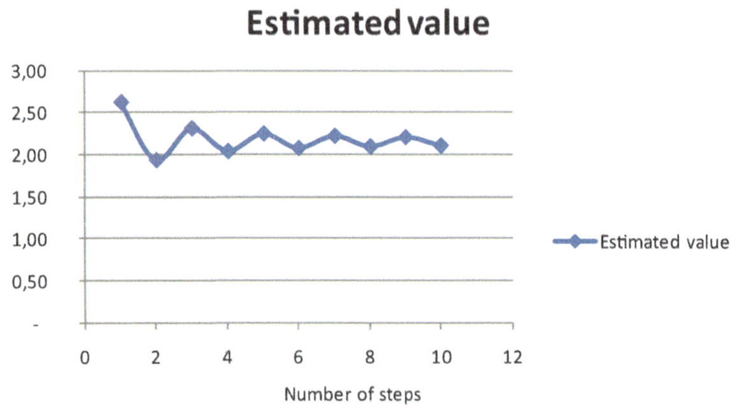

Fig. 4.13 Option value estimated on binomial trees with different numbers of steps

unchanged (for a large number of steps n). Given a volatility estimate σ we may, hopefully, refine our pricing given by (4.15) and (4.16) with $\delta t = T/n$ and a large n.

Example 4.7 Let us consider the same option as in Example 4.5, i.e., a 6-month call option on a non-dividend stock with €50 strike price. The current stock value is €50, the interest rate is 2%, and the estimated volatility is 13.5%. For the one-step binomial tree, the up and down parameters according to (4.16) are: $u = e^{0.135\sqrt{0.5}} = 1.1$ and $d = 1/u = 0.91$. The option value given by the one-step tree formula (4.12) is $f = 2.62$. For the two-step tree $\delta t = 0.25$, $u = e^{0.135\sqrt{0.25}} = 1.07$, and $f = 1.94$. There is a significant difference between the one- and two-step binomial tree valuation, so it is important to enlarge the number of steps further. Fig. 4.13 shows the calculated values when the number of steps goes up from 1 to 10.

The estimated values apparently approach a level between 2.1 and 2.2. If (4.15) is evaluated for 499 and 500 steps, then we get 2.154 and 2.152. Consequently, if the multistep binomial tree model is correct, then the option should be valued around €2.153.

It has been proved in general by Cox et al. (1979) that the binomial tree valuation indeed converges to the Black-Scholes formula result.

4.2.4 Valuation of American Options

So far, we have considered European options. Their numerical valuation using binomial trees is an interesting numerical exercise, but a precise value can be calculated directly by the Black-Scholes formula. However, there is no explicit formula for American options that can be valued only numerically using the binomial trees or other methods (e.g., solving a partial differential equation). Therefore, the trees are very useful for these types of options.

To value an American option, let us first start with the one-step tree in Fig. 4.11. Nothing happens between the time 0 and T, and so we can assume that the option is exercised either at maturity T or at time 0. If the option is exercised at time T then we easily calculate the payoff values f_u and f_d. At time 0 we decide either to exercise the option, and collect the payoff from the early exercise, or decide to wait until maturity T. In the latter case, the option becomes, in fact, European and can, at time 0, be valued as a European option. Consequently, a rational investor will compare the European option value f^E with the early payoff, and the initial American option value will be

$$f = \max\left(f^E, \text{early exercise payoff}\right). \tag{4.17}$$

This one-step valuation principle has to be repeated going backward from maturity T in the n-step binomial tree shown in Fig. 4.12. In this case, there is no single formula like (4.15). The valuation algorithm must go through all the nodes and check (4.17) to decide whether early exercise is optimal or not. The numerical procedure is still feasible, even if done manually, since the number of nodes in an n-step recombining binomial tree is "only" $n(n + 1)/2$.

Example 4.8 The strike price of a 6-month American put option is €52, the current stock value is €50, the interest rate is 2%, and the volatility 13.5%. The two-step binomial tree with CRR parameters (4.16) is shown in Fig. 4.14. The put option is initially in-the-money, and to decide whether it is optimal to exercise it early or not, we have to work from the end. The two-step node values on the right-hand side are simply the put option payoffs conditional on the simulated values. The one-step up node simulated stock value is €53.5, and so the option is out of the money, and its value €0.95 is simply the risk-neutral discounted weighted average of 0 and €2. The situation is more interesting in the one-step down node, where the early exercise payoff is €5.26. If the option were not exercised at this node, then its value would be the risk-neutral discounted weighted average of €2 and €8.31, i.e., only €5. This value is replaced in the tree by €5.26 according to (4.17), and the node can be marked as "early exercise." Finally, at time 0 the immediate exercise payoff would be €2, but without the early exercise the option's value is €3.01, and so we do not exercise. The American put option value estimated by the two-step model is €3.01 and its European counterpart's value is slightly lower: €2.88.

The two-step tree valuation restricts early exercise only to the times 0 or 3 months. Therefore, it is important to run the numerical procedure for a larger number of steps and verify its convergence. It turns out that after 100 steps or more the American option price converges to a value around €2.85, while the European option price converges just €2.76.

So far, we have considered binomial trees only for non-income paying assets. The model can be easily extended for income paying assets. Expected dividend payments

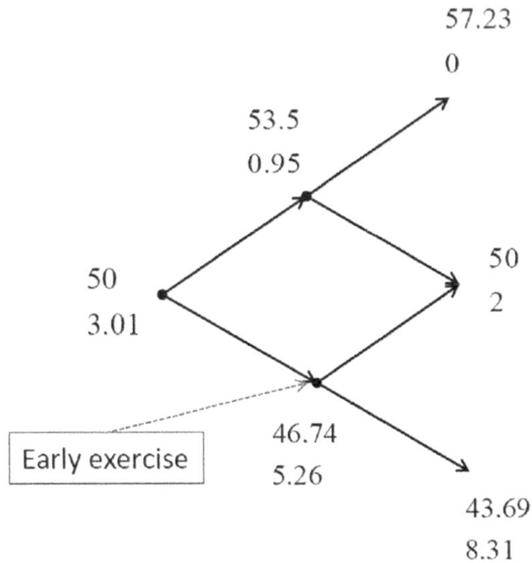

Fig. 4.14 American put option binomial tree

must be incorporated in nodes corresponding to the payment dates.[2] This is important for American call options that can be valued using the Black-Scholes formula only if there is no income paid by the asset (see Sect. 4.1) and a numerical procedure like the binomial tree model is needed in the general case.

4.2.5 The Binomial Tree as a Finite Probability Space

Before we go on to continuous time price modeling, it will be useful to formulate the binomial tree model in the context of elementary probability theory (see also Shreve 2005).

By a *finite probability space*, we mean a nonempty finite set Ω and a function $P:\Omega \to [0,1]$ assigning a probability $P(\omega)$ to each element ω of Ω so that the sum of all probabilities equals to one, i.e., $\sum_{\omega\in\Omega} P(\omega) = 1$. The set Ω typically represents a collection of possible outcomes of an experiment.

Most key probability concepts can be developed in the context of finite probability spaces. An *event* is defined as a set of possible outcomes $A \subseteq \Omega$ and its probability is defined by $P(A) = \sum_{\omega\in A} P(\omega)$. A *random variable* is a real valued

[2]Note that the recombining property will unfortunately be violated in the step following the dividend payment even if the multipliers u and v are constant. The recombining property will hold in the subsequent steps but the number of nodes will double.

function $X : \Omega \rightarrow R$, typically representing a measured value $X(\omega)$ in the case of an outcome $\omega \in \Omega$. The *expected value* (*the mean value or the first moment*) of the random variable X is defined as the probability-weighted average $E[X] = \sum_{\omega \in \Omega} P(\omega)X(\omega)$. Sometimes we are interested in *the expected value of X conditional on an event A* defined as

$$E[X|A] = \frac{1}{P(A)} \sum_{\omega \in A} P(\omega)X(\omega).$$

It is easy to see that the expectation operator is linear, i.e., if X_1 and X_2 are two random variables and c_1 and c_2 are two constants then

$$E[c_1X_1 + c_2X_2] = c_1E[X_1] + c_2E[X_2].$$

Another key concept is the notion of the *variance* of X defined as the mean squared difference of the random variable X and its expected value:

$$\mathrm{var}[X] = E\left[(X - E[X])^2\right] = \sum_{\omega \in \Omega} P(\omega)(X(\omega) - \mu)^2, \text{ where } \mu = E[X].$$

The standard deviation of X is defined as the square root of the variance $\sigma(X) = \sqrt{\mathrm{var}[X]}$. It can be interpreted as an average deviation of X from its expected value. Generally, the n-th moment of X is defined as $E[X^n]$. In the case of variance and the second moment, it is easy to see that

$$\mathrm{var}[X] = E[X^2] - (E[X])^2.$$

Having reviewed the key finite probability concepts, let us focus on binomial trees. An outcome of a one-step binomial tree can be viewed as the result of coin tossing where the head side and the tail side do not necessarily have equal probabilities. Let us consider the one-step binomial tree and encode the head (up) by the letter U and the tail (down) by D. The probability space $\Omega = \{U, D\}$ with objective probabilities $P(U) = p$ and $P(D) = 1 - p$ represents the set of two possible outcomes where we model the underlying asset value as the random variable S with values $S(U) = S_0u$ and $S(D) = S_0d$. We can also introduce an option payoff function as a random variable, e.g.,

$$f_{\text{payoff}}(\omega) = \max(S(\omega) - K, 0), \text{ i.e.}$$

$$f_{\text{payoff}}(U) = \max(S_0u - K, 0) \quad \text{and} \quad f_{\text{payoff}}(D) = \max(S_0d - K, 0).$$

The initial option value can be expressed according to (4.12) if we change the probability measure to Q, defining $Q(U) = q$ and $Q(D) = 1 - q$ where q is given by (4.12).

A general n-step binomial tree can be represented as the set of sequences of heads and tails of length n, i.e., formally by $\Omega_n = \{U, D\}^n$. Each element $\omega \in \Omega_n$ represents a scenario or a path, i.e., a sequence of ups and downs until the time $T = n\delta t$. Since the up and down moves are independent with probabilities p and $1 - p$, respectively, the probability will depend only on the number of ups and downs, i.e.,

$$P(\omega) = p^{\mathrm{ups}(\omega)}(1-p)^{n-\mathrm{ups}(\omega)}$$

where $ups(\omega)$ denotes the number of occurrences of U in ω. The random variable S_n modeling the asset price at time $T = n\delta t$ is defined similarly by

$$S_n(\omega) = S_0 u^{\mathrm{ups}(\omega)} d^{n-\mathrm{ups}(\omega)}.$$

The scenarios have a time structure. If $\omega \in \Omega_n$ is restricted to the first k moves, denoted as $\omega \restriction k$, then we get the partial information known at time $k\delta t$. The simulated asset value can be calculated along the path ω:

$$S_k(\omega) = S_k(\omega \restriction k) = S_0 u^{\mathrm{ups}(\omega \restriction k)} d^{k-\mathrm{ups}(\omega \restriction k)}. \tag{4.18}$$

The sequence of random variables S_0, S_1, . . ., S_n is an example of an adapted stochastic process. Generally, a sequence of random variables $X_0, X_1, \ldots, X_n$ on Ω_n is called an *adapted stochastic process*, if for every $\omega \in \Omega_n$ and $k \le n$ the value $X_k(\omega)$ depends only on $\omega \restriction k$, i.e., only on the information known at time $t = k\delta t$.

The binomial tree Ω_n itself is not recombining (Fig. 4.12), but the asset value random variable (4.18) is (it depends only on the number of ups and downs). By applying the risk-neutral principle, a European option with known payoff function f_n at maturity $T = n\delta t$ can be valued as the discounted expected value, i.e.,

$$f_0 = e^{-rT} E_q[f_n] \tag{4.19}$$

changing the measure P to Q setting $Q(\omega) = q^{\mathrm{ups}(\omega)}(1 - q)^{n - \mathrm{ups}(\omega)}$ with q given by (4.14). Note that the *change of measure* does not change the scenarios—the set of sequences Ω_n, the variables S_n, and f_n remain unchanged—we change only the probabilities of individual scenarios and the corresponding probability distributions of the random variables. It can be easily shown that (4.19) is equivalent to (4.15) in the case that u and d are constant. If the two parameters change across the tree, then the valuation (4.19) remains applicable.

Finite binomial trees can be used to explain easily the notions of conditional expectation and martingale. Let X be a random variable on Ω_n and $k < n$. If $\omega \in \Omega_k$ is a sequence of length k representing a partial information known at time $t = k\delta t < n\delta t = T$ then the expected value $E[X|\omega]$ conditional on ω is the probability-weighted average of X over all scenarios $\omega' \in \Omega_n$ that start with ω:

$$E[X|\omega] = E[X|\{\omega' \in \Omega_n, \omega' \upharpoonright k = \omega\}].$$

The function assigning the conditional expectation $E[X| \omega]$ to $\omega \in \Omega_k$ is called the *conditional expectation* operator and denoted $E[X| \Omega_k]$ or $E_k[X]$.

An adapted stochastic process $M_0, M_1, \ldots, M_n$ is called *a martingale* if $M_k = E[M_m| \Omega_k]$ whenever $k < m \leq n$. It turns out that discounted asset values are martingales with respect to the risk-neutral probability measure. The Eq. (4.19) following from the one-step binomial tree argument can be transposed to any $k < m \leq n$:

$$f_k = e^{-r(m-k)\delta t} E_Q[f_m|\Omega_k].$$

Let us define $M_k = e^{-rk\delta t} f_k$, then $M_0, M_1, \ldots, M_n$ is a martingale since

$$M_k = e^{-rk\delta t} f_k = e^{-rk\delta t} e^{-r(m-k)\delta t} E_Q[f_m|\Omega_k] = E_Q[e^{-rm\delta t} f_m|\Omega_k] = E_Q[M_m \Omega_k].$$

The discounted underlying asset value $e^{-rk\delta t} S_k$ is a martingale as well, since $S_k = e^{-r(m-k)\delta t} E_Q[S_m| \Omega_k]$ with respect to the risk-neutral probability measure.

4.2.6 The Wiener Process and the Geometric Brownian Motion

Modern financial theory models dynamic asset prices using continuous time stochastic processes. We have already introduced discrete time stochastic processes where the variable can change only at certain fixed points in time, while *continuous time stochastic processes* can change at any point in time. To avoid technically the difficult mathematical theory of stochastic processes, many authors often characterize the continuous time process intuitively as a limiting case of the discrete time stochastic processes (see, e.g., Hull 2018 or Wilmott 2006). This is in principle correct, but it is difficult to explain what is exactly meant by the limit of the discrete stochastic processes. Instead, we will use the concept of *infinitesimal numbers* that allows us to extend easily the finite binomial trees to infinite (hyperfinite) binomial trees with infinitesimal time steps representing well the continuous time processes.

The notion of infinitesimals was used as early as in the seventeenth century by Leibniz and Newton, who discovered differential and integral calculus. Many key concepts and theorems have been formulated and proved using infinitesimals. Mathematicians later abandoned the notion of infinitesimals, which, in some cases, had caused a number of errors when used without caution. More recently, mathematicians laid a proper foundation for infinitesimals (Robinson 1966). According to these results, the set of "standard" real numbers R can be extended to a larger set R^* including standard real numbers, but also "non-standard" numbers, in particular infinitesimal numbers that are smaller, in absolute value, than any standard positive number, as well as infinite numbers that are larger, in absolute value, than any standard real number. The extended set of (hyper) real numbers R^*, with the extended operations of addition and multiplication, has the same properties as the set of (standard) real numbers R. Similarly, any function or more complex mathematical structure can be extended to its nonstandard counterpart while preserving the original properties. The extended objects can be used to work easily with many mathematical concepts; for example, the extended real numbers and functions can be used to perform integration simply by summing up hyperfinite series and rounding the result to the nearest standard real number to obtain the classical integral. In fact, this corresponds well to the integral sign $\int$, which originally represented an elongated S from the Latin word *Summa.*

There are many research papers and textbooks on differential and integral calculus based on the concept of infinitesimals (see, e.g., Keisler 1976, 2013, or Vopěnka 2010, 2011) as well on the theory of probability and stochastic processes (Nelson 1977; Albeverio et al. 1986; Cutland et al. 1991; Witzany 2008; Herzberg 2013, etc.). A more exact introduction to elementary stochastic calculus is be given in the Appendix A. In this chapter, we are going to use the concept of infinitesimals rather intuitively.

The continuous time processes will be built on a hyperfinite N-step binomial tree $\Omega_T = \Omega_N$ where N is an infinite integer and the elementary time-step $\delta t = T/N$ is infinitesimal. The changes of the modeled variables can take place at any moment on the infinitesimal time step scale $\mathcal{T} = \{0, \delta t, 2\delta t, \ldots, N\delta t\}$, and so, in fact, at any time from the standard point of view. We will use the letter t, possibly with an index, exclusively for elements of the time scale $\mathcal{T}$.

The key building block of financial stochastic processes is the *Wiener process* (also referred to as *Brownian motion*) in which, starting from the zero initial value, movement at each step up and down is independent of the past, so that the mean value of the changes equals zero, and the variance equals the length of the time interval. For the time being, let us assume that the up and down probabilities are equal both to 0.5, and that the Wiener process can then be generated by the elementary equation

$$\delta z = \pm\sqrt{\delta t}, \tag{4.20}$$

where "+" applies if the path goes up, and "–" if the path goes down. The Eq. (4.20) can be viewed as a script for a virtual machine that is able to generate randomly hyperfinite sample paths and calculate iteratively $z(t + \delta t) = z(t) + \delta z$ starting from $z(0) = 0$. Formally, z is not a single function on $\mathcal{T}$, but a family of functions indexed by the paths $\omega \in \Omega_T$, i.e., $z : \Omega_T \times \mathcal{T} \to R^*$.

Note that the mean of the one-step increment δz is 0 and the variance is δt. Summing up a series of independent elementary increments over a time interval from t_1 to t_2 we get the increment $z(t_2) - z(t_1)$ with mean 0 and variance $t_2 - t_1$. Moreover, if the number of elementary steps between t_1 and t_2 is infinite then the random variable $z(t_2) - z(t_1)$ (from the perspective of time t_1) will, according to the Central Limit Theorem, be normally distributed. The statement, in fact, holds up to an infinitesimal error. Since we are interested, at the end, in standard real values,

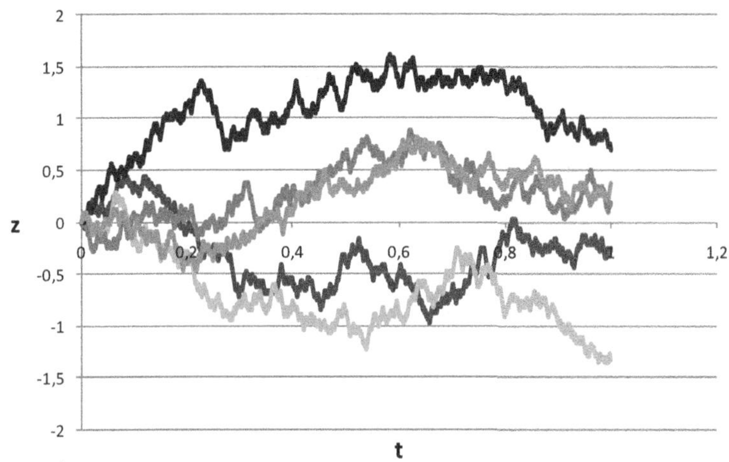

Fig. 4.15 Five sample paths of the Wiener process ($T = 1, N = 1000$)

these infinitesimal errors can be neglected. If dt is an infinite multiple[3] of δt, then the corresponding dz is again normally distributed, up to an infinitesimal error of a higher order with respect to dt (i.e., the error is infinitely smaller compared to dt). We will use the notation dt and dz either for the elementary time step and the Wiener process increment generated by (4.20) or for a general infinitesimal time step, which is, at the same time, an infinite multiple of δt, and dz for the corresponding normally distributed change of z.

Figure 4.15 shows a few sample paths of the Wiener process for the time $T = 1$. The process z is a family of functions, but since we cannot show all the paths, the figure gives just a few samples. The sample paths have been generated according to (4.20) using Excel generated random numbers with $N = 1000$ and $\delta t = 1/1000$. Thus, in fact, the time step is not infinitesimal, but very small, and the number of steps is very large—what we get is a discrete Wiener process approximation.

A generalized Wiener process x can be built by multiplying the Wiener process by a constant and adding a deterministic drift term. It can be described by a simple *stochastic differential equation (SDE)* that describes how to calculate the increment dx given the time increment dt and the Wiener process increment dz:

$$dx = adt + bdz \tag{4.21}$$

starting from an initial value $x(0) = x_0$. If $dt = \delta t$ is the elementary time step, then the process can be generated by the equation

[3]If ϵ is an infinitesimal, then $\sqrt{\varepsilon}$ is infinitely larger, yet still infinitesimal. Consequently, there are infinite multiples of δt that are still infinitesimal.

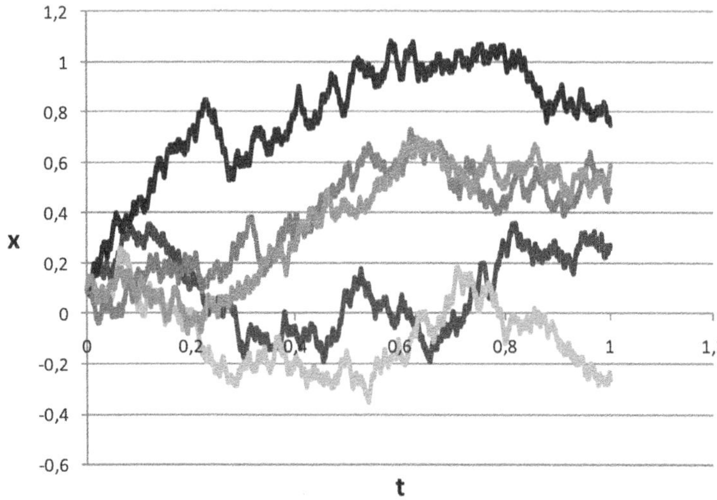

Fig. 4.16 Five sample paths of the generalized Wiener process ($T = 1$, $N = 1000$, $x_0 = 0.1$, $a = 0.3$, $b = 0.5$)

$$\delta x = a\delta t \pm b\sqrt{\delta t}$$

where we use "+" if the path goes up and "–" if the path goes down. Again, the process x is not a single function, but a family of functions indexed by $\omega \in \Omega_T$. It is easy to see that for a given path $\omega \in \Omega_T$ and $t \in \mathcal{T}$, we have $x(\omega, t) = x_0 + at + bz(\omega, t)$ or briefly $x = x_0 + at + bz$.

Consequently, the generalized Wiener process increment

$$x(t_2) - x(t_1) = a(t_2 - t_1) + b(z(t_2) - z(t_1))$$

over the time interval from t_1 to t_2 is normally distributed with mean $a(t_2 - t_1)$ and variance $b^2(t_2 - t_1)$. Hence, the term a is known as the *drift rate* representing the mean change over a unit of time, while b^2 is called the *variance rate* corresponding to the variance over a unit of time. Figure 4.16 shows a few generalized Wiener process paths corresponding exactly to those in Fig. 4.15 but with the initial value $x_0 = 0.1$, the drift rate $a = 0.3$, and the variance rate $b^2 = 0.5^2 = 0.25$.

The generalized Wiener process is still not general enough to capture satisfactorily the behavior of asset prices. One obvious problem is that it can attain negative values (see Fig. 4.16), but asset prices are never negative. Moreover, observing daily, weekly (or another regular time interval) price changes, there is an evidence of an approximately normal distribution of returns, i.e., of relative price changes, not increments (absolute price changes). If S_{i-1} and S_i denote observed prices at the end of the periods $i - 1$ and i then $\Delta S_i = S_i - S_{i-1}$ denotes the absolute increment while $u_i = \frac{\Delta S_i}{S_{i-1}}$ denotes the relative return. Figure 4.17 shows the histogram of the Czech

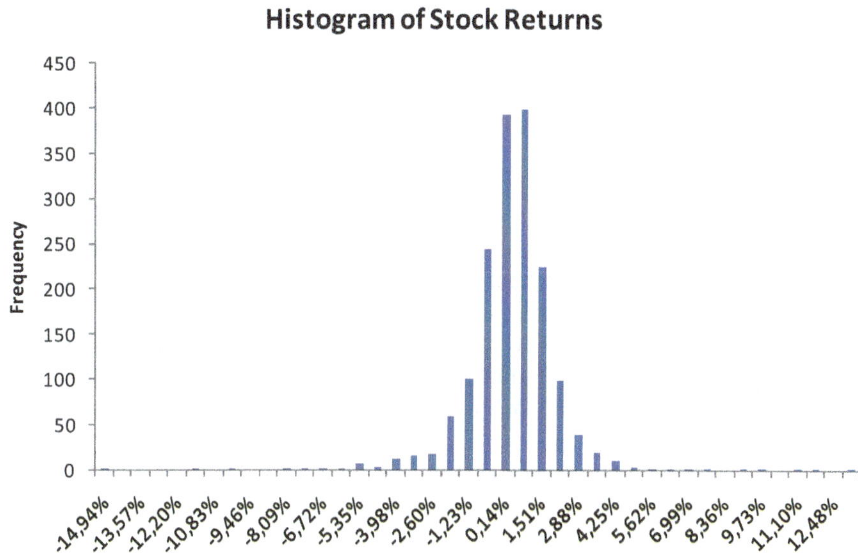

Fig. 4.17 Histogram of the PX stock index daily returns (3.1.2002–11.2.2011)

stock index PX daily returns over a period of more than 9 years. The returns appear visually, and can be statistically tested, to be (more or less) normally distributed. However, the same conclusion could not be drawn for absolute returns, since the level of the index changes over time.

The returns over nonoverlapping periods also turn out to be statistically (almost) independent. Consequently, the appropriate stock or other asset price process model could be realistically described by the stochastic differential equation

$$\frac{dS}{S} = \mu dt + \sigma dz,$$

or equivalently

$$dS = \mu S dt + \sigma S dz. \tag{4.22}$$

The *drift* parameter μ is the expected annualized return of the asset price and σ corresponding to the standard deviation of annualized return is referred to as the *volatility* of the asset price. The stochastic process is known as the *geometric Brownian motion*. It can, on the infinitesimal time scale, be generated by the equation

$$\delta S = \mu S \delta t \pm \sigma S \sqrt{\delta t}. \tag{4.23}$$

Note that in order to calculate $S(t + \delta t) = S(t) + \delta S$, the already known value $S(t)$ is used on the right-hand side of (4.23), i.e.,

$$S(t+\delta t)=S(t)\left(1+\mu\delta t\pm\sigma\sqrt{\delta t}\right).$$

In practice, we use rather a discrete time model with a very small (but not infinitesimal) time step Δt generating the sample paths according to the equation

$$\Delta S=\mu S\Delta t+\sigma S\varepsilon\sqrt{\Delta t},$$

where $\varepsilon\sim N(0,1)$ is normally distributed with mean 0 and variance 1. The starting value $S(0)=S_0$ should be the actual asset price. If the parameters, i.e., the drift and volatility, are properly calibrated, based on historical data and/or on our future market behavior prediction, then the distribution of future prices $S(t)$ for a fixed time t should have a realistic probability distribution. In order to characterize the distribution, we need to use the *Ito's lemma*, which is of key importance in elementary stochastic calculus.

4.2.7 Ito's Lemma and the Lognormal Property

Ito's process is a stochastic process x following the stochastic differential equation:

$$dx=a(x,t)dt+b(x,t)dz. \tag{4.24}$$

The coefficients a and b can depend on the last known value of x and on t. On the level of an elementary time step the already known values $x(t)$ and t are used to calculate $a(x(t),t)$ and $b(x(t),t)$, so

$$x(t+\delta t)=x(t)+a(x(t),t)\delta t\pm b(x(t),t)\sqrt{\delta t}. \tag{4.25}$$

If the functions a and b are reasonable (continuous plus some other properties), then the process is well and uniquely defined (i.e., satisfies the general SDE (4.24) if sampled according to (4.25)).

The Ito's lemma tells us what happens when an Ito's process x is transformed by a function of two variables, $G=G(x,t)$. The transformed process (also denoted as G) assigns the value $G(x(\omega,t),t)$ to a given scenario ω and time t. In other words, the function $G(x,t)$ is used to transform each individual path of x to a path of the new process G. For example, we may ask what sort of process e^x is, if x is the generalized Wiener process, or alternatively what type of process $\ln(S)$ is, if S is the geometric Brownian motion, etc.

The *Ito's lemma* claims that if x satisfies (4.24) and if $G=G(x,t)$, as a function of two variables, is sufficiently differentiable, then the transformed stochastic process G is again an Ito's process satisfying the stochastic differential equation called the *Ito's formula*:

$$dG = \left(\frac{\partial G}{\partial x}a + \frac{\partial G}{\partial t} + \frac{1}{2}\frac{\partial^2 G}{\partial x^2}b^2\right)dt + \frac{\partial G}{\partial x}bdz. \tag{4.26}$$

Ito's lemma is not so difficult to prove using Taylor's theorem and the concept of infinitesimals. Before we outline the proof, let us apply the lemma, in order to characterize the geometric Brownian motion.

Let S be the geometric Brownian motion satisfying the Eq. (4.22) and let $G(S, t) = \ln S$. Since S is an Ito's process, the transformed process $G = \ln S$ must also be an Ito's process satisfying (4.26). Our guess is that the function is $\ln S$ since the returns over a short time interval can be approximated by the log returns

$$u_i = \frac{\Delta S_i}{S_{i-1}} \approx \ln\frac{S_i}{S_{i-1}} = \ln S_i - \ln S_{i-1}.$$

Consequently, absolute increments of $\ln S$ should be normally distributed and the process $\ln S$ should follow the generalized Wiener process (i.e., with SDE constant coefficients no longer depending on S). Indeed, according to the Ito's formula (4.26) the coefficients on the right-hand side of the equation

$$d\ln S = \left(\frac{1}{S}\mu S + 0 + \frac{1}{2}\times\frac{-1}{S^2}\sigma^2 S^2\right)dt + \frac{1}{S}\sigma S dz = \left(\mu - \frac{1}{2}\sigma^2\right)dt + \sigma dz$$

are constant. Consequently, $\ln S$ is the generalized Wiener process with the drift rate $\mu - \frac{1}{2}\sigma^2$ and the variance rate σ^2. Alternatively, let x be a generalized Wiener process satisfying (4.21) and $G(x, t) = e^x$, then according to the Ito's lemma

$$dG = \left(e^x a + 0 + \frac{1}{2}e^x b^2\right)dt + e^x bdz = \left(a + \frac{1}{2}b^2\right)Gdt + bGdz$$

and so the exponential process e^x is a geometric Brownian motion with the mean rate of return $a + \frac{1}{2}b^2$ and variance rate b^2. In particular, if $a = \mu - \frac{1}{2}\sigma^2$ and $b = \sigma$, then the geometric Brownian motion parameters are μ and σ. This shows that the geometric Brownian motion values are always (with 100% probability) positive, which was implicitly assumed when we used the transformation $G = \ln S$.

We have shown that $\ln S$ increments are normally distributed, and, in particular, that

$$\ln S(T) - \ln S(0) \sim N\left(\left(\mu - \frac{1}{2}\sigma^2\right)T, \sigma^2 T\right), \text{ or equivalently}$$

$$\ln S_T \sim N\left(\ln S_0 + \left(\mu - \frac{1}{2}\sigma^2\right)T, \sigma^2 T\right). \tag{4.27}$$

The distribution of $S_T = S(T)$ characterized by (4.27) is a known parametric distribution called the *lognormal distribution*. Figure 4.18 shows an example of the lognormal distribution (4.27) density function.

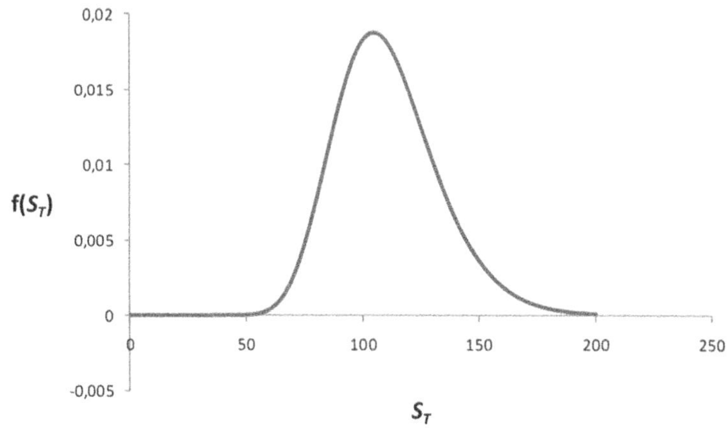

Fig. 4.18 Lognormal distribution ($S_0 = 100, \mu = 0.1, \sigma = 0.2, T = 1$)

The future asset price S_T characterized by (4.27) as a lognormal distribution can be handled quite well analytically. It can be shown that

$$E[S_T] = S_0 e^{\mu T}. \tag{4.28}$$

It would be tempting to say that the mean of S_T equals to

$$\exp(E[\ln S_T]) = S_0 e^{\mu T - \sigma^2 T/2},$$

but this is not correct, since the functions exp and ln are nonlinear (convex and concave).[4] The variance S_T can be shown[5] to be:

$$\text{var}[S_T] = S_0^2 e^{2\mu T}\left(e^{\sigma^2 T} - 1\right). \tag{4.29}$$

Example 4.9 The lognormal distribution with the density shown in Fig. 4.18 is characterized by the equation

$$\ln S_1 \sim N\left(\ln 100 + \left(0.1 - \frac{1}{2}0.2^2\right), 0.2^2\right) = N(4.685, 0.04),$$

[4]According to Jensen's inequality, if X is a random variable and $\varphi(X)$ a convex function, then $E[\varphi(X)] \geq \varphi(E[X])$.

[5]The mean $E[S_T]$ and variance $\text{var}[S_T] = E\left[S_T^2\right] - E[S_T]^2$ are obtained simply by integrating the lognormal density function multiplied by S_T and S_T^2. We will perform a similar integration when proving the Black-Scholes formula.

i.e., the log of the lognormally distributed variable S_1 has the normal distribution shown above. The relation can be used to determine certain critical values of $\ln S_1$, and, therefore, of S_1. For example, we may need to know the critical value S_c, below which the future asset price S_1 will not fall with 99% probability. Such a critical value is easily calculated for a normal distribution $N(m, s^2)$ as

$$m + sN^{-1}(0.01) = m - sN^{-1}(0.99),$$

where $N^{-1}(\alpha)$ is the inverse cumulative probability distribution function[6] for a standardized normal variable, i.e., $N(0,1)$. Therefore the $\ln S_1$ critical value on the 99% probability level is

$$S_c = 4.685 - 0.2 \times N^{-1}(0.99) = 4.685 - 0.2 \times 2.326 = 4.22.$$

Hence, we have $\ln S_1 \geq 4.22$ with 99% probability. Since the exponential is an increasing function, we can conclude that $S_1 \geq e^{4.22} = 68.03$ with 99% probability.

The formulas (4.28) and (4.29) can be used to obtain the expected value $E[S_1] = 100e^{0.01} = 110.52$ and the variance

$$\mathrm{var}[S_1] = 100^2 e^{2\cdot 0.1}\left(e^{0.2^2} - 1\right) = 498.46.$$

Therefore, the standard deviation of the asset price in 1 year is $\sqrt{\mathrm{var}[S_1]} = \sqrt{498.46} = 22.33$. We have to keep in mind that it is not correct to multiply this standard deviation by standardized inverse normal distribution values (quantiles) in order to obtain critical values of S_1.

4.2.8 Proof of Ito's Lemma

We are going to outline a proof of Ito's lemma using the Taylor's series expansion and infinitesimals. If $G = G(x, t)$ is a sufficiently differentiable function, then its increment $\Delta G = G(x + \Delta x, t + \Delta t) - G(x, t)$ at a point (x, t) can be expressed by a series involving partial derivatives of G and powers of Δx and Δt called the *Taylor's expansion*:

$$\Delta G = \frac{\partial G}{\partial x}\Delta x + \frac{\partial G}{\partial t}\Delta t + \frac{1}{2}\frac{\partial^2 G}{\partial x^2}\Delta x^2 + \frac{1}{2}\frac{\partial^2 G}{\partial t^2}\Delta t^2 + \frac{\partial^2 G}{\partial x \partial t}\Delta x \Delta t + \cdots \tag{4.30}$$

If $\Delta x = dx$ and $\Delta t = dt$ are infinitesimal (and of the same order), then the higher-order powers of dx and dt can be neglected and the expansion can be written as:

[6]The cumulative distribution function $N(x) = \Pr[X \leq x]$ where $X \sim N(0, 1)$ can be evaluated in Excel as NORMSDIST while the inverse function $N^{-1}(\alpha)$ is evaluated as NORMSINV. The cumulative distribution function and its inverse are also often denoted as Φ and Φ^{-1}.

$$dG = \frac{\partial G}{\partial x}dx + \frac{\partial G}{\partial t}dt. \tag{4.31}$$

Let dx be the Ito's process increment given by the equation

$$dx = a(x,t)dt + b(x,t)dz.$$

It is tempting to apply (4.31), but we have to take into account, that dx is not of the same order as dt. If dt is the elementary time step, then $dz = \pm\sqrt{dt}$ and so dx is of the order of $\sqrt{dt}$ (infinitely larger than dt). To get the correct expansion we need to use (4.30) where all the terms with higher powers of dt can be neglected, but we have to keep the term

$$dx^2 = a^2dt^2 + 2abdtdz + b^2dz^2 = a^2dt^2 + 2abdtdz + b^2dt \approx b^2dt. \tag{4.32}$$

The key point is that $dz^2 = \left(\pm\sqrt{dt}\right)^2 = dt$ and so, when (4.32) is plugged into (4.30), the first two terms on the right-hand side of (4.32) can be neglected, but the last one must be kept:

$$\begin{aligned} dG &= \frac{\partial G}{\partial x}(adt + bdz) + \frac{\partial G}{\partial t}dt + \frac{1}{2}\frac{\partial^2 G}{\partial x^2}(adt + bdz)^2 = \\ &= \frac{\partial G}{\partial x}(adt + bdz) + \frac{\partial G}{\partial t}dt + \frac{1}{2}\frac{\partial^2 G}{\partial x^2}b^2dt = \\ &= \left(a\frac{\partial G}{\partial x} + \frac{\partial G}{\partial t} + \frac{1}{2}b^2\frac{\partial^2 G}{\partial x^2}\right)dt + b\frac{\partial G}{\partial x}dz. \end{aligned}$$

This completes the proof. The lemma and the proof can be easily generalized to transformations of a multidimensional Ito's process with several sources of uncertainty (independent or correlated Wiener processes).

4.2.9 The Black-Scholes Formula

Let S be the price of a non-income paying asset modeled by the geometric Brownian motion

$$dS = \mu Sdt + \sigma Sdz$$

on a hyperfinite binomial tree Ω_T. Our initial up and down probabilities are set to $p = 0.5$, $1 - p = 0.5$ and the probability of any particular path ω is infinitesimal, $P(\omega) = 0.5^N$, as N is infinite. Nevertheless, calculating the expected values like $E_P[\max(S_T - K, 0)]$ we do not have to deal with the point probabilities, but we can use the fact that S_T has the lognormal distribution given by (4.27). In order to discount the expected values properly we still need to prove the risk-neutral principle. But there is no work to do; we have already proved the principle for the finite binomial trees, and the same conclusion holds for the hyperfinite binomial trees. In this case, the up and down parameters are $u = 1 + \mu\delta t + \sigma\sqrt{\delta t}$ and $u = 1 + \mu\delta t - \sigma\sqrt{\delta t}$, hence, according to (4.14), the changed up-move probability is

$$q = \frac{e^{r\delta t} - d}{u - d} = \frac{e^{r\delta t} - 1 - \mu\delta t + \sigma\sqrt{\delta t}}{2\sigma\sqrt{\delta t}} = 0.5 - \frac{\mu - r}{\sigma}\frac{\sqrt{\delta t}}{2}.$$

Note that q is infinitesimally close the original up probability $p = 0.5$, in addition $q < 0.5 < 1 - q$ provided the price of risk $\frac{\mu-r}{\sigma}$ is positive (see also Sect. 6.1), which is usually the case. The new probabilities of individual paths now depend on the number of ups and downs:

$$Q(\omega) = q^{\mathrm{ups}(\omega)}(1 - q)^{N - \mathrm{ups}(\omega)}.$$

We have not changed the values of S on the binomial tree Ω_T, but again we have just changed the probability measure P to a new probability measure Q. The key conclusion is that the drift rate with respect to Q is now the risk-free interest rate r, while the volatility remains unchanged, thus symbolically

$$dS = rSdt + \sigma Sdz \quad \text{and} \tag{4.33}$$

$$df = rfdt + \sigma fdz \tag{4.34}$$

where f is the price of any derivative depending only on S. In particular, if we know the payoff f_T at time T, then the derivative value at time 0 is

$$f_0 = e^{-rT}E_Q[f_T]. \tag{4.35}$$

According to (4.33), S_T has, with respect to the probability measure Q, the lognormal distribution given by

$$\ln S_T \sim N\left(\ln S_0 + \left(r - \frac{1}{2}\sigma^2\right)T, \sigma^2 T\right), \tag{4.36}$$

and so there is a good chance to evaluate (4.35) analytically, if f_T is a simple payoff function. This is the case of European call and put options. For a European call option with the strike price K and maturity T, the payoff function is $c_T = \max(S_T - K, 0)$. By integrating the expected value and rearranging the results (see Sect. 4.2.10) we obtain the famous *Black-Scholes pricing formula*

$$c_0 = S_0 N(d_1) - Ke^{-rT}N(d_2), \quad \text{where} \tag{4.37}$$

$$d_1 = \frac{\ln(S_0/K) + (r + \sigma^2/2)T}{\sigma\sqrt{T}}, \tag{4.38}$$

$$d_2 = \frac{\ln(S_0/K) + (r - \sigma^2/2)T}{\sigma\sqrt{T}} = d_1 - \sigma\sqrt{T}, \tag{4.39}$$

and $N(x) = \Phi(x)$ is the standard normal distribution cumulative probability function. Similarly, for a European put option with the payoff $p_T = \max(K - S_T, 0)$ we get the formula

$$p_0 = Ke^{-rT}N(-d_2) - S_0N(-d_1) \tag{4.40}$$

where d_1 and d_2 are given by (4.38) and (4.39).

Example 4.10 The current value of a non-dividend paying stock is €100, the interest rate is 2% (in continuous compounding and for all maturities), and an in-the-money 6-month European call on the stock with strike price €95 is traded for €8 while an out-of-the-money call with the strike €105 is offered for €5. Are the prices acceptable or could we even make an arbitrage profit by buying an undervalued option or selling an overvalued option? The key question can be answered by applying the Black-Scholes formula. First, we must find out what the volatility is. Let us assume that we believe that the (constant) volatility, in the context of the model, will be 20%. Then all we need to do is to plug in the parameters and market factors into the formulas (4.37)–(4.39):

$$d_1 = \frac{\ln(100/95) + (0.02 + 0.2^2/2)0.5}{0.2\sqrt{0.5}} = 0.469,$$

$$d_2 = 0.469 - 0.2\sqrt{0.5} = 0.327,$$

$$c(K = 95) = 100N(0.469) - 95e^{-0.1}N(0.327) = 8.944.$$

The cumulative distribution function has been evaluated with the Excel function NORMSDIST. Similarly, for the out-of-the-money option we obtain $c(K = 105) = 3.98$. Hence, according to our model, the in-the-money option quoted at €8 is underpriced, while the out-of-the-money option quoted at €5 is overpriced. If our goal is to hedge a stock position, then we can simply buy the in-the-money option and be happy with the price. On the other hand, we may decide to use the opportunity and just go short in the out-of-the money call options with the strike price €95, and sell for €5, since we believe that the fundamental value is less than €4. If we sell the options and do nothing until maturity, then we may have good luck and suffer no loss, or we may have a bad luck and suffer a significant loss at maturity of the contract. If we want to fix the profit believed to be over €1 then we have to perform the so-called dynamic *delta-hedging* that lies at the heart of the binomial tree argument. At each instant of time until maturity we need to be long in $\Delta = \frac{\partial c}{\partial S}$ stocks to cover the risk of the short call—see (4.8). Since the partial derivative changes with time and with the stock price, we have to rebalance our hedging position continuously (in infinitesimal time intervals). In practice, the rebalancing can certainly not be done continuously, but only at relatively short time intervals. Thus, the delta-hedging will be only approximate. According to the general theoretical argument, we should end up with €1.02 (plus accrued interest and a hedging error) irrespective of the stock price development.

The delta-hedging described above is an example of a trading strategy that can be formalized easily in the context of binomial trees. A trading strategy with the risk-free asset (zero-coupon bonds) and a risky asset starts with an initial wealth V_0 in the risk-free asset. It tells us what to do at each time $t < T$ and on each path $\omega \in \Omega_t$, i.e., how much of the risky asset should be bought or sold. The proceeds from the sale of the risky asset are kept in the risk-free asset, and any purchase of the risky asset is financed by the sale of the risk-free asset (possibly going short). The binomial tree argument means that if we start with cash $V_0 = f_0$ equal to an option value and delta hedge the short option until maturity T, and then the value of the hedging portfolio at time T will exactly offset the short option payoff, i.e., $V_T - f_T = 0$, or equivalently $V_T = f_T$. The last equation shows that the delta hedging strategy precisely *replicates* the option payoff—we call this a *replication strategy*.

It is important to keep in mind that the Black-Scholes formula and the replication argument are based on a set of rather idealistic assumptions:

1. The asset price follows the geometric Brownian motion process with constant drift and volatility (lognormal returns).
2. There is no income paid by the asset.
3. The risk-free interest rate r is constant.
4. We can lend and borrow at the same rate and without any restrictions.
5. There are no transaction costs or taxes.
6. Assets are arbitrarily divisible.
7. The short selling of securities is possible without restrictions.
8. There are no arbitrage opportunities.
9. Security trading is continuous; we can trade at infinitesimal time intervals.

It is obvious that the real market is less perfect in almost all of these categories. We will see that some of the assumptions can be relaxed (for example, 2 and 3). Others are difficult to deal with. In spite of the differences between reality and the theoretical assumptions the Black-Scholes model has become a market standard, but the market makes its own corrections that will be discussed later.

With the option pricing formula, we can significantly improve our analysis (Table 4.1) of option value dependence on the various input parameters.

Example 4.11 Let us consider European call and put options with the strike $K = 100$ and time to maturity $T = 0.5$. Assume that the interest rate $r = 2\%$ and the volatility $\sigma = 20\%$ The formulas (4.37) and (4.40) can be used to plot the

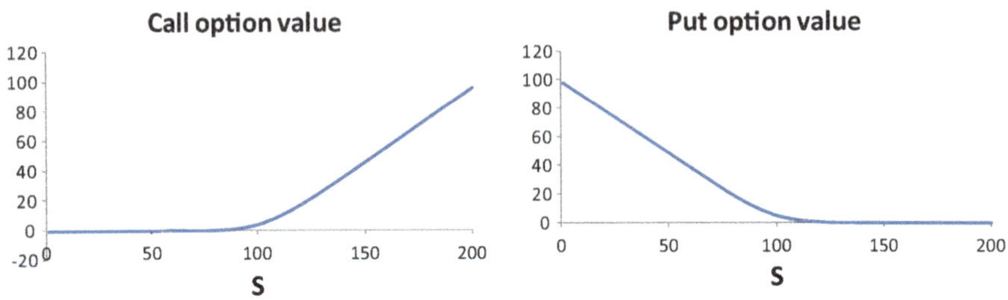

Fig. 4.19 Dependence of the European call and put options' value on the underlying asset price S

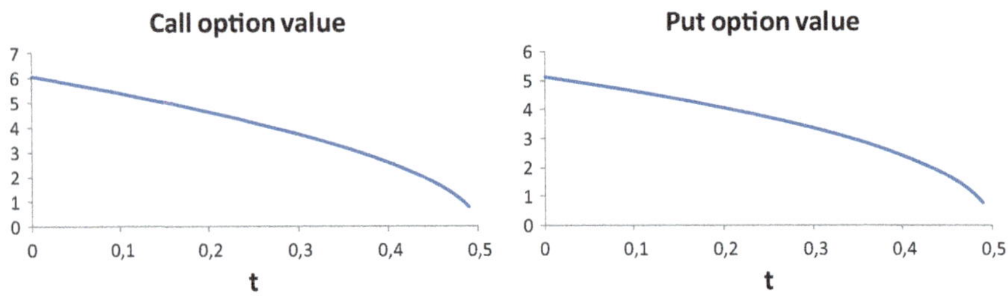

Fig. 4.20 Dependence of European call and put option value on the time t ($S_0 = 100$, $K = 100$, $r = 2\%$, $\sigma = 20\%$)

dependence of the values on the underlying asset price S keeping all the other inputs fixed (Fig. 4.19).

We could continue analyzing dependence on the other parameters, namely interest rate r, volatility σ, and strike price K. Let us look at the time to maturity dependence, where Table 4.1 contains a question mark. In this case, we have to replace the time to maturity T in (4.37)–(4.40) by the difference $T-t$ so that T can be fixed, and t goes from 0 to T, i.e., $c = c(t, S, r, \sigma; T, K)$ and $p = p(t, S, r, \sigma, T, K)$ where T and K are the only fixed option parameters, while t, S, r change over time, and σ is the model parameter that should be theoretically fixed, but in practice changes over time as well. Figure 4.20 shows that for the given input values and $S_0 = 100$, the call and put option values decrease with time approaching maturity.

However, the put option value can theoretically increase with time approaching maturity if the interest rate is high and the volatility is low, as shown in Fig. 4.21. The high drift rate tends to "beat" the impact of low volatility, but this advantage disappears when t approaches T.

4.2.10 Derivation of the Black-Scholes Formula

Let $g(S)$ be the risk-neutral probability density of the lognormally distributed variable $S = S_T$ with parameters given by (4.36), i.e.,

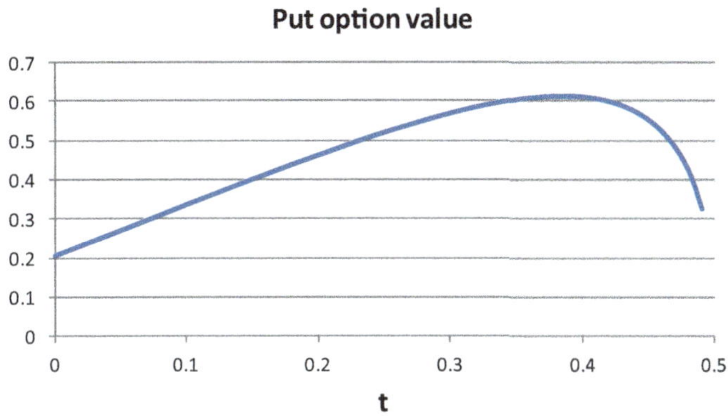

Fig. 4.21 Dependence of European put option value on the time t ($S_0 = 100$, $K = 100$, $r = 15\%$, $\sigma = 10\%$)

$$\ln S \sim N\left(m, w^2\right), \text{where } m = \ln S_0 + \left(r - \frac{1}{2}\sigma^2\right)T \text{ and } w^2 = \sigma^2 T. \tag{4.41}$$

To verify the call option pricing formula, we need to evaluate

$$E[\max(S - K, 0)] = \int_K^{\infty} (S - K) g(S) dS.$$

Let us transform S to the standardized normal, $N(0, 1)$, variable $X = \frac{\ln S - m}{w}$ and use the fact that the density function of X is $\varphi(X) = \frac{1}{\sqrt{2\pi}} e^{-X^2/2}$. The probability $g(S)dS$ must be equal to $\varphi(X)dX$, and so

$$\begin{aligned} E[\max(S - K, 0)] &= \int_{(\ln K - m)/w}^{\infty} \left(e^{Xw+m} - K\right)\varphi(X)dX = \\ &= \int_{(\ln K - m)/w}^{\infty} \frac{1}{\sqrt{2\pi}} e^{(-X^2+2Xw+2m)/2} dX - K \int_{(\ln K - m)/w}^{\infty} \frac{1}{\sqrt{2\pi}} e^{-X^2/2} dX. \end{aligned} \tag{4.42}$$

Alternatively, we could use the relation $g(S) = \varphi(X)\left|\frac{dX}{dS}\right|$ to express the lognormal distribution density function $g(S) = \varphi(X)\frac{1}{Sw} = \frac{1}{S\sqrt{2\pi w^2}} e^{-\frac{(\ln S - m)^2}{2w^2}}$. The integrals on the right-hand side of (4.42) can be expressed analytically, at least in terms of the cumulative standard normal distribution function

$$N(x) = \Phi(x) = \Pr[X \le x] = \int_{-\infty}^{X} \varphi(X)dX.$$

Since $\varphi(-X) = \varphi(X)$ and

$$\int_{x}^{\infty} \varphi(X)dX = \int_{-\infty}^{-x} \varphi(X)dX = N(-x)$$

the second integral on the right-hand side of (4.42) simply equals to $N(-(\ln K - m)/w)$ and in order to evaluate the first integral, we just need to complete the square in the exponent

$$\frac{-X^2 + 2Xw + 2m}{2} = \frac{-(X-w)^2 + 2m + w^2}{2}.$$

Therefore,

$$\int_{(\ln K - m)/w}^{\infty} \frac{1}{\sqrt{2\pi}} e^{\left(-X^2 + 2Xw + 2m\right)/2} dX = e^{m + w^2/2} \int_{(\ln K - m)/w}^{\infty} \frac{1}{\sqrt{2\pi}} e^{-(X-w)^2/2} dX =$$
$$= e^{m + w^2/2} N(w - (\ln K - m)/w) \ .$$

It is easy to check, using the definition (4.41) of the variables m and w, that

$$w - (\ln K - m)/w = \frac{-(\ln K - m) + w^2}{w} = \frac{-\ln K + \ln S_0 + rT - \sigma^2 T/2 + \sigma^2 T}{\sigma\sqrt{T}} =$$
$$= \frac{\ln S_0/K + (r + \sigma^2 T)/2}{\sigma\sqrt{T}} = d_1,$$

$$-(\ln K - m)/w = \frac{\ln S_0/K + (r - \sigma^2 T)/2}{\sigma\sqrt{T}} = d_2, \text{ and}$$

$$e^{m + w^2/2} = e^{\ln S_0 + rT} = S_0 e^{rT}.$$

Finally, according to (4.35) we get the call option Black-Scholes valuation formula

$$c = e^{-rT}\left(S_0 e^{rT} N(d_1) - KN(d_2)\right) = S_0 N(d_1) - e^{-rT} KN(d_2) \ .$$

The put option formula can be verified similarly, or simply by using the put-call parity relationship.

4.2.11 The Black-Scholes Partial Differential Equation

Black and Scholes (1973) derived the formula in their original paper by setting up and solving a partial differential equation (PDE). The argument leading to the differential equation is almost the same as the one for risk-neutral pricing. But to solve the PDE one needs to have some experience with these equations. Interestingly, the Black-Scholes differential equation turns out to be, after a few substitutions, the *heat-transfer equation* well-known in physics. Although the derivation of the Black-Scholes formula through the PDE is technically more difficult, there are some advantages. The Black-Scholes PDE holds for many other general derivatives and the only difference consists in the boundary conditions. If the PDE does not have an analytical solution, there are relatively efficient numerical methods to solve these equations. The methods are usually much faster than the Monte Carlo simulations typically used in the risk-neutral numerical valuation approach.

To derive the Black-Scholes PDE let us consider an asset price process S driven by the geometric Brownian motion equation

$$dS = \mu S dt + \sigma S dz \tag{4.43}$$

and a derivative depending only on S, i.e., the derivative value $f = f(S, t)$ depends on S and time t. According to the Ito's formula

$$df = \left(\frac{\partial f}{\partial S}\mu S + \frac{\partial f}{\partial t} + \frac{1}{2}\frac{\partial^2 f}{\partial S^2}\sigma^2 S^2\right)dt + \frac{\partial f}{\partial S}\sigma S dz. \tag{4.44}$$

In order to set up a riskless portfolio, we need to eliminate the source of uncertainty dz by combining appropriately the Eqs. (4.43) and (4.44). This is done when we combine one short derivative contract with $\Delta = \frac{\partial f}{\partial S}$ units of the underlying asset. The portfolio value is

$$\Pi = -f + \frac{\partial f}{\partial S}S$$

and

$$\begin{aligned} d\Pi &= -df + \frac{\partial f}{\partial S}dS = \\ &= -\left(\frac{\partial f}{\partial S}\mu S + \frac{\partial f}{\partial t} + \frac{1}{2}\frac{\partial^2 f}{\partial S^2}\sigma^2 S^2\right)dt - \frac{\partial f}{\partial S}\sigma S dz + \frac{\partial f}{\partial S}\mu S dt + \frac{\partial f}{\partial S}\sigma S dz = \\ &= -\left(\frac{\partial f}{\partial t} + \frac{1}{2}\frac{\partial^2 f}{\partial S^2}\sigma^2 S^2\right)dt \end{aligned} \tag{4.45}$$

Since the delta-hedged portfolio is riskless (over the very short or infinitesimal time period of length dt) and there are no arbitrage opportunities, we must have

$$d\Pi = r\Pi dt \tag{4.46}$$

Putting the two Eqs. (4.45) and (4.46) together we obtain the Black-Scholes partial differential equation:

$$-\left(\frac{\partial f}{\partial t} + \frac{1}{2}\frac{\partial^2 f}{\partial S^2}\sigma^2 S^2\right)dt = r\Pi dt = r\left(-f + \frac{\partial f}{\partial S}S\right)dt, \quad \text{i.e.}$$

$$-\frac{\partial f}{\partial t} - \frac{1}{2}\frac{\partial^2 f}{\partial S^2}\sigma^2 S^2 = -rf - \frac{\partial f}{\partial S}rS, \quad \text{or}$$

$$\frac{\partial f}{\partial t} + \frac{\partial f}{\partial S}rS + \frac{1}{2}\frac{\partial^2 f}{\partial S^2}\sigma^2 S^2 = rf. \tag{4.47}$$

The Black-Scholes PDE (4.47) is a *linear parabolic partial differential equation*. The meaning of "linear" is that a linear combination of any two solutions is again a solution of the PDE. There are, in fact, infinitely many solutions, and to specify the one that values a given option we must set up certain boundary conditions. In the case of a European call option, the key condition[7] is given by the payoff function:

$$f(S, T) = \max(S - T, 0). \tag{4.48}$$

For example, the underlying asset price $f(S, t) = S$, money market account value $f(S, t) = e^{rt}$, or a forward contract value $f(S, t) = S - Ke^{-r(T-t)}$ all solve (4.47), but do not satisfy the boundary condition (4.48). Note that the Eq. (4.47) and the boundary condition (4.48) does not contain μ, and so the solution does not depend on the drift μ as expected.

The Eq. (4.47) can be, after an appropriate substitution, transformed to the *heat diffusion equation* in the form

$$\frac{\partial u}{\partial t} = c\frac{\partial^2 u}{\partial x^2}, \tag{4.49}$$

where the function $u(x, t)$ represents the temperature in a bar at a spatial coordinate x and time t. The partial derivative $\frac{\partial u}{\partial t}$ measures the change of temperature and so $\frac{\partial u}{\partial t}dxdt$ is proportional to the change of heat in the piece of length dx over the time dt. On the other hand, the first-order derivative $\frac{\partial u}{\partial x}$ measures the spatial gradient of temperature and it is proportional to the flow of heat, and so the second-order derivative multiplied by dx and dt, i.e., $\frac{\partial^2 u}{\partial x^2}dxdt$, is proportional to the heat retained by the piece dx over dt proving (4.49).

There is a variety of analytical and numerical methods for solving the famous heat equation, e.g., using the Green's function, Fourier transformation, similarity

[7]The other technical boundary conditions are $f(S, t) \to 0$ as $S \to 0$ and $f(S, t)/S \to 1$ as $S \to \infty$ for $t \in [0, T]$.

reduction, or numerically with a finite-difference method. In the case of boundary conditions of the type (4.48), there is a general analytical solution that leads to the Black-Scholes formula.

Alternatively, we can just verify that $c(S, t)$ given by (4.37) with T replaced by $T - t$, i.e.,

$$c(S,t) = SN(d_1) - Ke^{-r(T-t)}N(d_2), \quad \text{where} \tag{4.50}$$

$$d_1 = d_1(S,t) = \frac{\ln(S/K) + (r + \sigma^2/2)(T-t)}{\sigma\sqrt{T-t}},$$

$$d_2 = d_2(S,t) = \frac{\ln(S/K) + (r - \sigma^2/2)(T-t)}{\sigma\sqrt{T-t}} = d_1 - \sigma\sqrt{T-t},$$

solves the differential Eq. (4.47) and satisfies (4.48). We have to do some algebraic work in order to find the partial derivatives of $c(S, t)$, but this investment will pay back in Sect. 4.3 on Greeks. By applying the chain rule (i.e., differentiating d_1, d_2, and c) and simplifying the formulas, we obtain the following results:

$$\frac{\partial c}{\partial S} = N(d_1), \tag{4.51}$$

$$\frac{\partial^2 c}{\partial S^2} = \frac{N'(d_1)}{S\sigma\sqrt{T-t}}, \quad \text{and} \tag{4.52}$$

$$\frac{\partial c}{\partial t} = -rKe^{-r(T-t)}N(d_2) - SN'(d_1)\frac{\sigma}{2\sqrt{T-t}}, \tag{4.53}$$

where $N'(x) = \varphi(x) = \frac{1}{\sqrt{2\pi}}e^{-x^2/2}$ is the standard normal probability density function (pdf). It is now easy to verify that the Black-Scholes Eq. (4.47) holds. Regarding the boundary condition (4.48) we have to define $c(S, T)$ as the limit when t approaches T since d_1 and d_2 are undefined for $t = T$. Using the concept of infinitesimals, let $T - t$ be infinitesimally small. Then, obviously, d_1 and d_2 are positive infinite if $S > K$, negative infinite if $S < K$, and infinitesimally close to zero if $S = K$. Since $N(-\infty) = 0$, $N(\infty) = 1$, and $N(0) = 0$, the limits are $c(S, T) = 0$ for $S \leq K$ and $c(S, T) = S - K$ for $S > K$. Consequently, the boundary condition (4.48) holds.

The argument used to get the Black-Scholes PDE is based on the delta hedging idea that we also used to prove the risk-neutral pricing principle defining the risk-neutral probability measure. In fact, the PDE can be alternatively derived using the risk-neutral measure Q and the concept of martingales (Shreve 2004). The Eq. (4.35) can be, through the construction of the risk-neutral measure on the hyperfinite binomial tree Ω_T, put into the more general form

$$f(S(t),t) = e^{r(t'-t)}E_Q[f(S(t'),t')|\Omega_t], \quad \text{i.e.,}$$

$$e^{-rt}f(S(t),t) = E_Q\left[e^{-rt'}f(S(t'),t')|\Omega_t\right] \quad \text{for any} \quad t < t'.$$

Hence, $M(S,t) = e^{-rt}f(S,t)$ is a martingale. According to Ito's lemma, the process satisfies the stochastic differential equation

$$dM = \left(\frac{\partial M}{\partial S}\mu S + \frac{\partial M}{\partial t} + \frac{1}{2}\frac{\partial^2 M}{\partial S^2}\sigma^2 S^2\right)dt + \frac{\partial M}{\partial S}\sigma S dz.$$

Since M is a martingale, its drift rate, i.e., the coefficient of dt, must be zero. Consequently,

$$\frac{\partial M}{\partial S}\mu S + \frac{\partial M}{\partial t} + \frac{1}{2}\frac{\partial^2 M}{\partial S^2}\sigma^2 S^2 = 0.$$

When the partial derivatives of M are expressed in terms of f and the discount factor e^{-rt} we get

$$e^{-rt}\frac{\partial f}{\partial S}\mu S + e^{-rt}\frac{\partial f}{\partial t} - re^{-rt}f + \frac{1}{2}e^{-rt}\frac{\partial^2 f}{\partial S^2}\sigma^2 S^2 = 0. \tag{4.54}$$

Divided by e^{-rt} the Eq. (4.54) becomes the Black-Scholes PDE.

4.2.12 Options on Futures and Income Paying Assets

So far, we have considered European options on non-income paying assets. However, most assets pay some income—foreign currencies pay interest, stocks pay dividends, bonds pay coupons, and commodities bear storage costs, and possibly provide a lease or a convenience yield. Moreover, exchange traded futures are often based on futures prices that do not follow the same process as the spot prices.

We will first consider an asset paying a continuous yield q, i.e., paying $qSdt$ over a time interval of length dt where S is the current asset value. This is, for example, the case of a foreign currency paying foreign interest r_F. Broad equity indexes with many stocks paying dividends at different times over a year are usually assumed to pay a *continuous dividend yield q*. Of course, this is only an approximation that is closer to reality in the case of US stock indices, where dividends are paid quarterly, than in the case of European indices where stocks usually pay dividends annually.

In a risk-neutral world, an asset paying a continuous yield q must still have a total return (including the income q) equal to r. Consequently, the drift rate must be $r - q$ and the price should follow the process:

$$dS = (r - q)Sdt + \sigma Sdz.$$

Let us introduce a new transformed process $U(t) = e^{-q(T-t)}S(t)$. It follows from the Ito's lemma that

$$dU = rUdt + \sigma Udz.$$

The process $U(t)$ can be interpreted as a reinvestment strategy portfolio value where we start with the portfolio having e^{-qT} units of the asset and continuously reinvest the income $qUdt$ back into the asset, i.e., multiply the holding by $1 + qdt$ so that at time t we hold $e^{-q(T-t)}$ units of the asset S. Moreover, at maturity $U(T) = S(T)$ and so at the maturity T the payoff of a European put or call on the asset U will be exactly the same as the payoff on S. Since U pays no income and follows the drift r geometric Brownian motion (in the risk-neutral world), the call options are valued by the Black-Scholes formulas (4.37) and (4.40) with S_0 replaced by $U(0) = e^{-qt}S_0$. After rearranging the formulas appropriately, we get

$$c_0 = S_0 e^{-qT} N(d_1) - K e^{-rT} N(d_2), \tag{4.55}$$

$$p_0 = K e^{-rT} N(-d_2) - S_0 e^{-qT} N(-d_1), \tag{4.56}$$

where

$$d_1 = \frac{\ln(S_0/K) + (r - q + \sigma^2/2)T}{\sigma\sqrt{T}}, \tag{4.57}$$

$$d_2 = \frac{\ln(S_0/K) + (r - q - \sigma^2/2)T}{\sigma\sqrt{T}} = d_1 - \sigma\sqrt{T}. \tag{4.58}$$

These results for dividend paying stocks were first obtained by Merton (1973). In the case of FX options to buy or sell foreign currency in terms of domestic currency, the rate q is replaced by the foreign currency interest rate r_F. The formula for FX options is usually credited to Garman and Kohlhagen (1983).

Example 4.12 Let us consider a European 1-year call option on a price return stock index with the strike $K = \$100$. The current index value is $I_0 = \$100$, the interest rate $r = 1\ \%$, the index dividend yield is $q = 1\%$, and the index market volatility $\sigma = 15\%$. Likewise, index futures, index options are settled financially based on the difference $\max(I_T - K, 0)$. If we neglect the effect of the dividends, one index call option value would be \$6.46, while with the effect of dividend yield according to (4.55) it is \$5.92, i.e., relatively almost 10% less. It is important that the index is price-return, i.e., it copies the value of the index portfolio not including the dividends paid-out. Some indexes are calculated as a total return (corresponding to the variable U defined above), and in that case we can use the option formula for non-income paying assets. Figure 4.22 shows that the dependence of a call option value on the

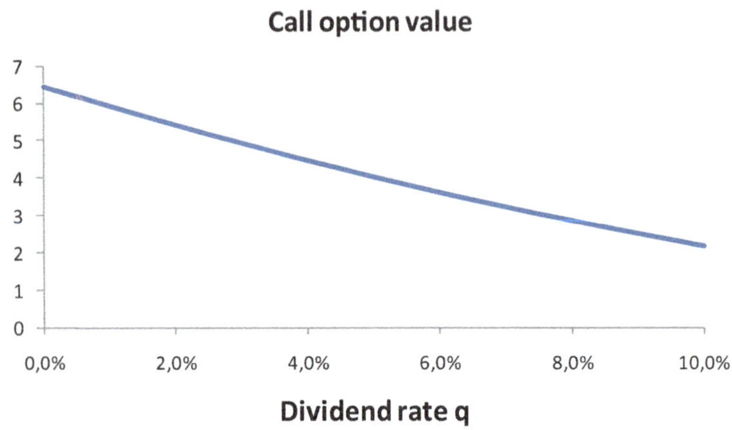

Fig. 4.22 Dependence of a European call option on the dividend rate q

dividend rate is quite significant, and so it is important to set up and estimate q properly.

Most exchange traded options are settled as futures options with payoff depending on the futures price F rather than on the spot price S. Technically, a call option, if exercised, is settled by entering into a long futures contract with the immediate cash settlement of $F_T - K$. The futures contract maturity can be longer than the option contract maturity; it is up to the new position holder whether it is closed immediately or later, after the option exercise date. The option payoff is, in any case, $F_T - K$.

To value the options on futures one needs to model the process for futures prices. The key argument is that the drift of a futures price in the risk-neutral world is zero. By entering into a futures position we do not invest any amount, we only take a risk. Hence, the return on the initial futures price (when we, in fact, have not invested anything in cash) must be zero in the risk-neutral world. Therefore, the futures stochastic price can be modeled in the risk-neutral world by the stochastic differential equation without a drift

$$dF = \sigma F dz = (r - r)F dt + \sigma F dz.$$

It means that for the purpose of derivatives valuation, the futures price can be treated as if there were a continuous financial yield $q = r$. Thus, in the formulas (4.55)–(4.58) the rate q is replaced by r, and S_0 is replaced by F_0 yielding the Black's formula (Black 1976):

$$c_0 = e^{-rT}[F_0 N(d_1) - KN(d_2)], \tag{4.59}$$

$$p_0 = e^{-rT}[KN(-d_2) - F_0N(-d_1)], \tag{4.60}$$

where

$$d_1 = \frac{\ln(F_0/K) + \sigma^2 T/2}{\sigma\sqrt{T}},$$

$$d_2 = \frac{\ln(F_0/K) - \sigma^2 T/2}{\sigma\sqrt{T}} = d_1 - \sigma\sqrt{T}.$$

Example 4.13 Let us value a European Light Sweet Crude Oil options traded on the CME. According to the contract specification "On expiration of a call option, the value will be the difference between the settlement price of the underlying Light Sweet Crude Oil Futures and the strike price multiplied by 1000 barrels, or zero, whichever is greater..." Consider an option on futures maturing in 6 months with the strike \$90 and assume that $r = 1\%$, $\sigma = 15\%$, and $T = 1/3$. Assume that the actual sweet crude oil spot price is \$88.2 while the quoted futures price is \$89.20. It would be a mistake to value the option with the basic Black-Scholes formula (4.37) giving \$2.38. The correct pricing formula (4.59) with $F_0 = 89.2$ gives \$2.70, i.e., a price that is significantly higher in terms of practical trading.

Finally, let us consider options on assets that pay a known income at certain known times in the future, for example, a stock paying known dividends. The asset can be decomposed into two parts: a riskless component that pays the known income and the remaining risky component. The riskless component is valued as the present value of the known cash flow, while the risky component evolves according to the geometric Brownian motion model. Therefore, the Black-Scholes formula (4.37) is correct if S_0 equals the risky component of the stock (i.e., the spot price minus discounted income paid until the maturity of the option), and σ is the volatility of the risky component.

Example 4.14 Let us consider a 1-year European put option on a stock currently quoted at €100. The strike price is €90, market volatility 15%, and interest rate 1%. It is known that the stock will pay a dividend of €6 in 3 months. If the put were priced according to (4.40) with non-adjusted $S_0 = 100$ then the put option value would be €1.79. Nevertheless, the correct valuation should be based on the dividend adjusted value $\widetilde{S}_0 = S_0 - D = 100 - 6e^{-0.01/4} = 94.01$ and the resulting value €3.36 is almost twice as high. Theoretically, the volatility of the dividend adjusted price process and the dividend non-adjusted price process are not the same. If $\sigma = 15\%$ were the volatility of the non-adjusted price process (e.g., obtained from historical data) then the risk component volatility would be approximately equal to

$$\sigma\frac{S_0}{S_0-D}=15\%\frac{100}{94.01}=15.96\%.$$

The volatility adjustment would increase the put option value to €3.69. The volatility adjustment is not needed if the volatility is already based on dividend adjusted prices (being calculated, for example, as the implied volatility).

4.2.13 Estimating the Volatility

To apply the Black-Scholes formula, we need to enter one completely new market factor—the asset return volatility. If there is no existing option market, then the first natural proposal would be a historical volatility estimate. For example, if we want to value a 1-year option, then we can use the series of daily returns over the last year, calculate their standard deviation, and annualize it to get a volatility estimate. If we believe that the volatility of the market will remain the same during the next year, then the result will be a reasonable volatility estimate that can be entered into the Black-Scholes formula.

Specifically, let S_i, $i = 0, \ldots, n$ be a series of observed prices at the ends of regular time periods (e.g., days, weeks, or months) of equal length Δt. The market prices change only during business days, and so holidays should not be taken into account. For example, we can set $\Delta t = 1/252$ in the case of daily returns assuming that there are 252 business days in a year. The returns should be calculated in line with the model as log-returns, i.e.,

$$u_i = \ln\frac{S_i}{S_{i-1}},\ \text{for}\ i = 1, \ldots, n.$$

However, for short time intervals, it is not a big mistake to calculate ordinary linear returns, $u_i = (S_i - S_{i-1})/S_{i-1}$. The sample estimate of the returns standard deviation is then

$$\widehat{s}=\sqrt{\frac{1}{n-1}\sum_{i=1}^{n}(u_i-\overline{u})^2}, \tag{4.61}$$

where

$$\overline{u}=\frac{1}{n}\sum_{i=1}^{n}u_i.$$

For short time intervals, e.g., daily observations, the mean return $\overline{u}$ is almost negligible and in practice often set at zero. If the process volatility were σ then the theoretical log-return volatility over an interval of length Δt would be $\sigma\sqrt{\Delta t}$. Consequently, the corresponding annualized volatility estimate is

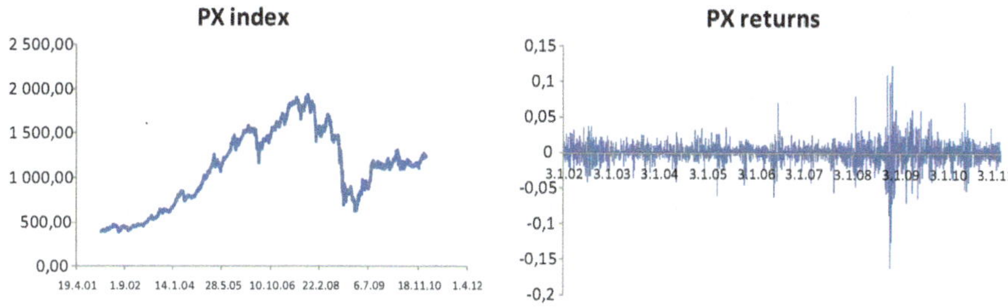

Fig. 4.23 The series of PX index values and daily returns

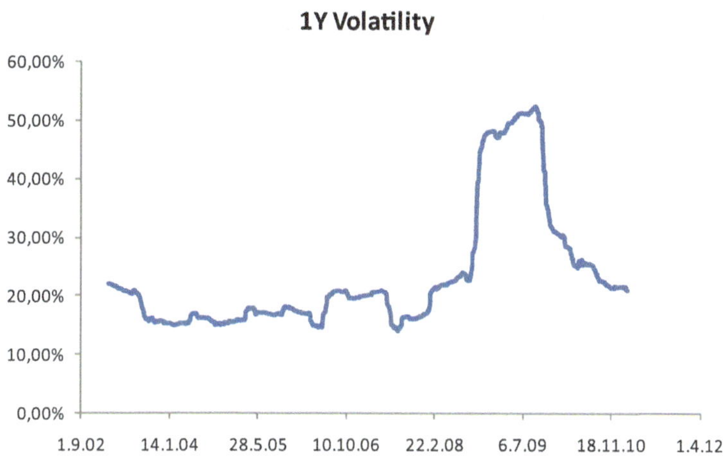

Fig. 4.24 Historical 1-year volatility of the PX index returns

$$\widehat{\sigma} = \frac{\widehat{s}}{\sqrt{\Delta t}}.$$

Example 4.15 Figure 4.23 shows the series of the Czech stock index PX values and index returns. Let us use the historical data to estimate the volatility over the next year.

The standard deviation calculated from the last 252 available returns according to (4.61) is $s = 1.32\%$ and the annualized volatility estimate is

$$\widehat{\sigma} = 1.32\% \times \sqrt{252} = 21.01\%.$$

If the volatility were constant, then this would be a good forward-looking volatility estimate, but, inspecting the series of returns, it seems obvious that the volatility was not constant in the past. Figure 4.24 shows 1-year historical volatility calculated retrospectively on a 252-day moving window. Although the volatility remained around 20% in the years 2002 through 2007, the historical data-based estimate would be completely wrong at the beginning of the financial crisis in 2008. In fact, the market volatilities at that time went up much faster than the historical volatilities based to a large extent on returns from the normal period.

The example above shows that the historical volatility is a useful estimate but cannot be the only input into our estimation of future volatility.

One popular way to make the historical volatility estimate (4.61) more sensitive to the recent behavior of the market is to give more weight to the most recent data. The popular exponentially moving average (EWMA) model is a particular case of this idea where the weights are $1, \lambda, \lambda^2, \ldots, \lambda^{n-1}$ for $0 < \lambda < 1$ starting from the most recent observations and going back to the oldest ones. A typical value for λ would be around 0.95. The weights must certainly be normalized (divided) by their sum

$$1 + \lambda + \cdots + \lambda^{n-1} = \frac{1 - \lambda^n}{1 - \lambda} \approx \frac{1}{1 - \lambda}$$

if n is large.

Thus, the EWMA volatility estimation is given by the formulas

$$s = \sqrt{(1 - \lambda) \sum_{i=1}^{n} \lambda^{n-i} (u_i - \overline{u})^2}, \tag{4.62}$$

where

$$\overline{u} = (1 - \lambda) \sum_{i=1}^{n} \lambda^{n-i} u_i.$$

Example 4.16 Figure 4.25 compares 1-year equal weighted and EWMA ($\lambda = 0.97$) historical volatility. The EWMA volatility went up much faster than the equal-weighted volatility at the beginning of the crisis. On the other hand, the last EWMA volatility estimation of 16.5% looks rather optimistic and should be treated carefully. The chart shows that the EWMA volatility estimates are themselves rather volatile, reacting to the most recent developments, which often do not extend into the future.

More advanced volatility estimation methods are discussed in Sects. 5.3 and 8.4. The reader is also referred to specialized econometrics textbooks like Cipra (2008) or Arlt and Arltová (2007).

If there is an existing option market, analysts should certainly compare the historical volatility estimates with the quoted or *implied market volatility*. Given a quoted call option price c_{market} with parameters K, T, underlying asset price s, market interest rate r, and continuous asset income q, we need to solve for σ the equation

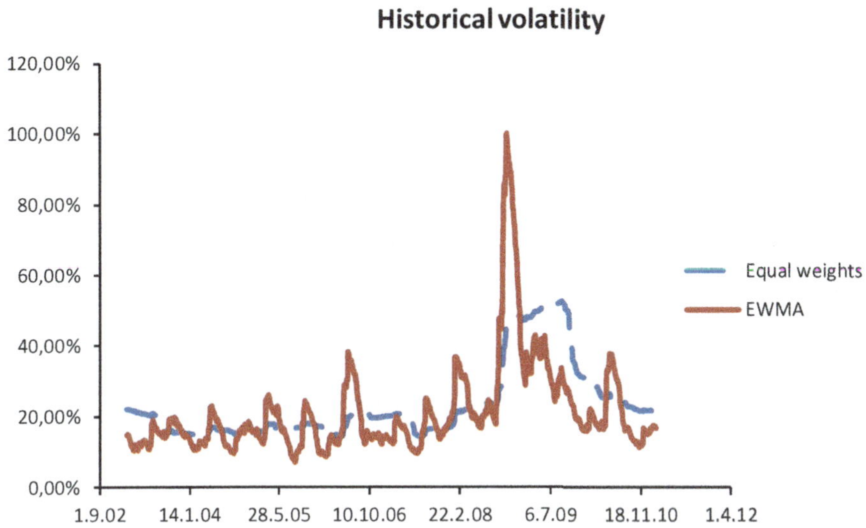

Fig. 4.25 Comparison of equal weighted and EWMA historical 1-year PX index volatility

$$c_{\text{market}} = c(S, r, q, \sigma, K, T)$$

with $c(S, r, q, \sigma, K, T)$ given by (4.55). The solution must be found numerically, for example in Excel using the tool "Solver."

Example 4.17 In Example 4.13, we have valued a European Light Sweet Crude Oil Futures call option with the parameters $K = 90$, $r = 1\%$, $\sigma = 15\%$, $T = 1/3$, and the actual futures price $F_0 = 89.2$. The Black-Scholes formula (4.59) gave us the price \$2.70. However, we observe that the market quotation is \$4.37. The problem is in our volatility estimate. It might be obtained from the historical data, but the market opinion regarding future volatility is apparently different. The implied volatility is extracted from the quoted premium by solving the equation.

$$4.37 = c(\sigma, 89.2, 0.01, 90, 1/3).$$

with one unknown variable σ and the function c given by (4.59). The numerical solution $\sigma_{\text{implied}} = 23.15\%$ is not too far from our initial 15% volatility estimate, but the impact on the option price is quite dramatic. Figure 4.26 showing the strong (and almost linear) dependence of the call value on volatility input parameter σ illustrates the importance of good volatility estimates.

4.3 Greek Letters and Hedging of Options

Options are used by hedgers to hedge their underlying positions (Example 1.2). On the other hand, option traders who sell and buy many options with different strike prices and maturities end up with complex and often risky option portfolios that need

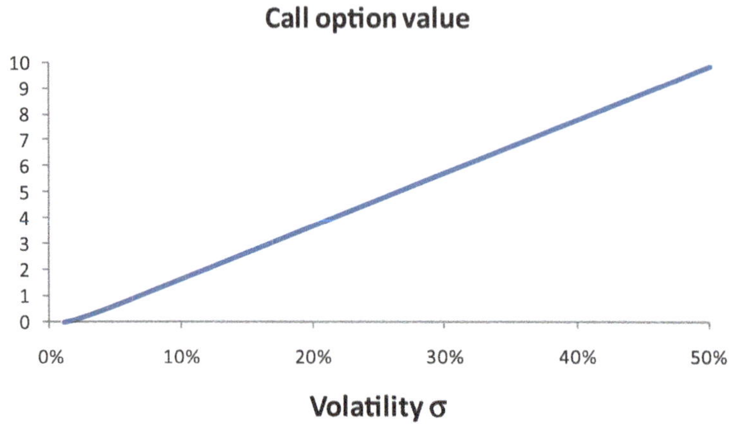

Fig. 4.26 Dependence of call option value on volatility ($K = 90$, $r = 1\%$, $T = 1/3$, $F_0 = 89.2$)

to be properly managed. The key strategy is based on the *delta-hedging* principle. Let us consider a portfolio consisting of European options, forwards or futures, cash, and possibly asset positions, with a single underlying non-income paying asset. The portfolio value depends on the asset spot price S and other parameters (volatility σ and risk-free rate r) that are supposed to be fixed according to the model

$$\Pi(S) = \sum_{i=1}^{n} f_i(S),$$

where $f_i(S)$ is the value of the i-the instrument in the portfolio. The sensitivity of the portfolio to the underlying price changes can be measured by the derivative of $\Pi(S)$ with respect to S which can be decomposed into derivatives, i.e., deltas, of the individual positions:

$$\frac{\partial \Pi(S)}{\partial S} = \sum_{i=1}^{n} \frac{\partial f_i(S)}{\partial S} = \sum_{i=1}^{n} \Delta_i.$$

Thus, the total *delta* of the portfolio $\Delta_\Pi = \frac{\partial \Pi(S)}{\partial S}$ is just the sum of deltas of the individual options, forwards, futures, cash, or spot positions in the portfolio. Since the change of portfolio value $\Delta\Pi$ caused by a change in the underlying price ΔS can be approximated according to the equation

$$\Delta\Pi \approx \Delta_\Pi \times \Delta S, \tag{4.63}$$

the goal of the delta-hedging is simply to keep the delta of the portfolio close to zero, i.e., $\Delta_\Pi = 0$. Recall that according to (4.51) a call option delta is given by

$$\Delta_{\text{call}} = \frac{\partial c}{\partial S} = N(d_1).$$

Similarly, we can obtain a formula for the put option delta

$$\Delta_{\text{put}} = \frac{\partial p}{\partial S} = 1 - N(d_1).$$

The long forward or futures delta on one asset unit is simply

$$\Delta_{\text{forward}} = \frac{\partial}{\partial S}\left(S - Ke^{-r(T-t)}\right) = 1,$$

and analogously the delta of one unit of the asset held is

$$\Delta_{\text{spot}} = \frac{\partial}{\partial S}(S) = 1.$$

The delta of a cash position held is obviously zero $\Delta_{\text{cash}} = 0$ since the value does not depend on the asset price. Hence, the forwards, futures, or spot contracts can be used to adjust the portfolio's delta. If the initial Δ_Π is positive then we just sell Δ_Π units of the asset on the spot or forward market, while if it is negative then we buy $-\Delta_\Pi$ units of the asset on the spot or forward market. While the deltas of linear contracts (spot, forward, and futures) remain constant, they change in the case of options where the value depends non-linearly on the spot price, i.e., $\Delta_\Pi = \Delta_\Pi(S)$. Therefore, it is not sufficient to perform delta hedging only once; the rebalancing of the portfolio has to be reiterated any time the delta moves too far from zero. We speak, therefore, about a *dynamic delta-hedging strategy*.

Example 4.18 Let us consider a trader who has just sold an at-the-money straddle on 1000 units and bought an out-of-the money call on 1500 units of a non-dividend paying stock. The actual stock price is $S = 50$, we assume a constant interest rate $r = 1\%$ and volatility $\sigma = 15\%$. All three European options have 6 months to maturity, i.e., $T = 0.5$, the strike price of the straddle call and put options is $K = 50$, and the strike of the out-of-the money call is $K = 60$. The trader has received a net initial premium of €5000 and currently (s)he is in profit by around €940. However, as shown in Fig. 4.27, a relatively small movement of the stock price will cause a loss, namely if S goes down €4 or up €5, the total portfolio value becomes negative. On the other hand, if S increased to €70 or more, the portfolio value would be positive again due to the long out-of-the money call option. The actual delta of the portfolio is almost zero relative to the nominal position in 1000 or 1500 stocks, $\Delta_\Pi(50) = 29.5$ and so initially we more or less do not have to delta-hedge. However, the chart on the right-hand side of Fig. 4.27 shows that the portfolio's delta might change quickly if the stock price moved up or down.

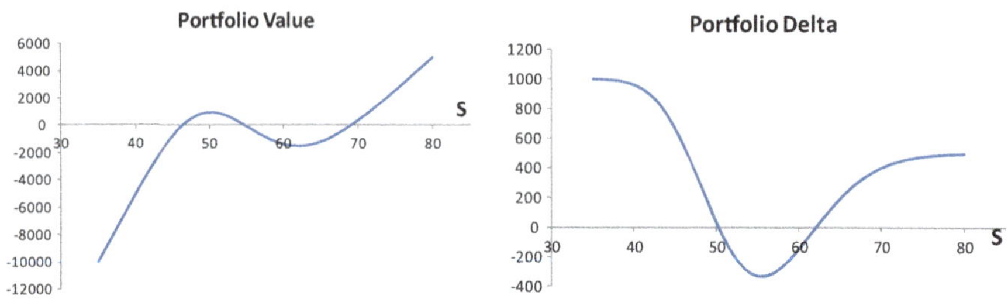

Fig. 4.27 Development of an option portfolio value and delta depending on the underlying stock price

Hence, it is necessary to monitor the delta closely and rebalance the portfolio if necessary. A possible strategy would be to delta-hedge on a daily basis, buying or selling (shorting) the stocks on the spot market. Table 4.3 gives a simulation example of the process when the stock price gradually goes up over 10 days. Without dynamic delta-hedging the portfolio value would fall below negative €1100. With the delta hedging the portfolio value is preserved at around €900. The columns "Port. delta" and "Or. port. Value" show the delta and the value of the original portfolio without hedging. The column "Delta pos." is the required hedging position in the stock, i.e., minus delta rounded to units. The trader, at the beginning of each day, recalculates the required delta position and buys or sells an appropriate amount of stocks. The first day he sells 14 stocks, the next day morning he buys $112 = 98 - (-14)$ stocks, etc. The cumulative cost of the delta position is shown in the second column from the left, and the total portfolio value, including the delta position market value and the cumulative cost, is shown in the last column. The calculation, for simplicity, does not include the accrued interest, but it should be considered, if the simulation is done over a longer time horizon.

The delta-hedging simulation shows that the total portfolio value does not remain constant, but slightly fluctuates. This is caused by the fact that we do not perform perfect continuous hedging. The daily rebalancing is only an approximate delta-hedging that should, in theory, be performed continuously (at infinitesimal time intervals). If the rebalancing were performed every hour, minute, or even second, then the resulting hedged portfolio value should be very close to the initial portfolio value (plus accrued interest). In practice, this is not possible due to the existence of transaction costs (bid/ask spread and commissions).

The initial position shown in Fig. 4.27 is particularly risky because the delta is almost zero, but it may change very fast when the stock price goes up and down. Sometimes it might be virtually impossible to rebalance the portfolio in time, and it could suffer a significant loss. This is one reason why traders monitor not only the

Table 4.3 Simulation of dynamic portfolio delta-hedging

Day	S	Port. delta	Or. port. value	Delta pos.	Buy/sell	Cost	Cum. cost	Total port.
1	50.00	14.33	947.40	−14	−14	700.00	700.00	947.40
2	51.00	−98.56	889.67	98	112	−5712.00	−5012.00	875.67
3	51.50	−148.12	821.45	148	50	−2575.00	−7587.00	856.45
4	52.00	−192.30	730.40	192	44	−2288.00	−9875.00	839.40
5	53.00	−262.86	491.53	262	70	−3710.00	−13,585.00	792.53
6	53.50	−288.71	349.93	288	26	−1391.00	−14,976.00	781.93
7	54.00	−308.10	197.86	308	20	−1080.00	−16,056.00	773.86
8	55.00	−327.77	−124.85	327	19	−1045.00	−17,101.00	759.15
9	57.00	−297.45	−760.01	297	−30	1710.00	−15,391.00	777.99
10	59.00	−197.32	−1249.13	197	−100	5900.00	−9491.00	882.87

delta (the first-order derivative) but also the second-order derivative of the portfolio value with respect to the underlying asset price,[8] called the *gamma* of the portfolio

$$\Gamma_\Pi = \frac{\partial^2 \Pi(S)}{\partial S^2}.$$

The Taylor's approximation (4.63) can then be improved with the second-order term

$$\Delta\Pi \approx \Delta_\Pi \times \Delta S + \Gamma_\Pi \times \Delta S^2/2. \tag{4.64}$$

Therefore, if gamma is negative, then the delta-gamma approximation (4.64) is always lower than the delta approximation (4.63). In particular, if delta is zero, gamma negative and ΔS large, positive or negative, then the loss according to (4.64) can be significant, although (4.63) indicates that there is no risk.

The portfolio gamma is again calculated as the sum of the individual instruments' gammas:

$$\Gamma_\Pi = \sum_{i=1}^{n} \frac{\partial^2 f_i(S)}{\partial S^2} = \sum_{i=1}^{n} \Gamma_i.$$

According to (4.52) and the put-call parity a European call and put options' gamma on a non-income paying asset is given by the formula

$$\Gamma_{call} = \Gamma_{put} = \frac{N'(d_1)}{S\sigma\sqrt{T-1}}.$$

Note that the gamma of a long call or put option is always positive. Equivalently, the market value of a long call or put is a convex function of S (Fig. 4.19). On the other hand, short call and put option positions correspond to concave market value functions, creating the risk of losses even in the case of zero delta.

The gamma of a forward, futures, or spot position is zero, since the delta is constant, and consequently the gamma of an option portfolio can be hedged only with options. In practice, some option maturities and strikes prices are less liquid than others. Certain (OTC) options in the portfolio could be tailored and sold to clients based on their specific needs. In this case, the trader can use the most liquid options to adjust the gamma of the portfolio of options that are not normally traded.

Example 4.19 The gamma of the portfolio from Example 4.18 at $S = 50$ can be calculated as the sum of the gammas of the three options multiplied by the number of stocks:

[8] Some dealers managing large and complex portfolios monitor even the third-order derivative of the portfolio value with respect to the underlying price, which is called the *speed*.

$$\Gamma_{\Pi} = -1000 \times 0.075 - 1000 \times 0.075 + 1500 \times 0.0195 = -120.72.$$

The negative gamma means that the portfolio can easily suffer a loss even if the delta is hedged exactly. According to (4.64), if the stock price moves just €1 up or down, the portfolio will lose more than €60; if the price changes €2 or more, then the loss exceeds €240. Such a price movement easily happens during a day when the delta is not rebalanced. This explains the value deterioration that occurs during the daily delta hedging that can be observed in Table 4.3. With negative gamma, when the price moves, there is always a loss, and after rebalancing the delta, we do not get the loss back, unless we come to a region with positive gamma. Let us try to gamma-hedge the portfolio with liquid 1-month at-the-money put and call options. The gamma of the call or put option ($K = 50$, $T = 1/12$, $S = 50$, $\sigma = 15\%$, $r = 1\%$) is $\Gamma = 0.184$, and so we need approximately $656 = 120.72/0.184$ options to offset the negative gamma of the portfolio. We can use either calls or put, since the gamma is the same. We note that our goal is to stabilize the portfolio value in the region around the current stock price $S = 50$. In order to achieve that we will try rather to offset the 6-month short straddle, i.e., the reversed U-shape in Fig. 4.27, by a long 1 month straddle. Therefore, we buy 328 one-month at the money calls and 328 one-month at the money puts. The value of the options according to the Black-Scholes formula is €563, but we pay €580, a little bit more, being on the "ask" side. The delta of the gamma hedged portfolio turns out to be $\widetilde{\Delta}_{\Pi}(50) = 22.46$ and so, in addition, we short 22 stocks to delta-hedge the portfolio. Figure 4.28 shows that we have succeeded quite well in stabilizing the portfolio value. It remains in the region €800–€1000 if the stock price stays between €47.5 and €55. The delta stays relatively low as well; however, if the stock price moved outside the region, another delta hedging and possibly gamma hedging would be needed.

Another interesting Greek letter monitored by traders and related to gamma is *theta*. It is the rate of change of the portfolio value with respect to the passage of time with all the other factors remaining constant. It can be also interpreted as the *time decay* of the portfolio. Consider, for example, an out-of-the money call. To profit on the option, we need the underlying price to go up. If time passes and the price stays constant, we are losing the option's time value. The theta of a long option position is always negative, and the theta of a short option position is positive. Although the

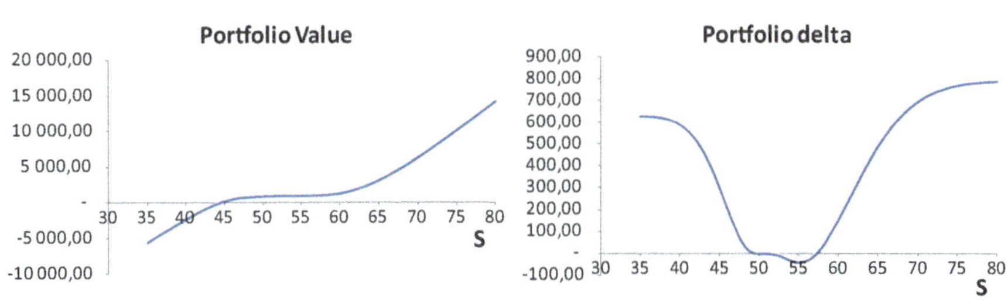

Fig. 4.28 Dependence of the gamma hedged option portfolio value and delta on the underlying price

passage of time is fully predictable, no movement on the market might present a risk that is measured by the theta. By differentiating the Black-Scholes formula (4.50) with respect to t it can be shown that

$$\Theta_{\text{call}} = -\frac{SN'(d_1)\sigma}{2\sqrt{T-t}} - rKe^{-r(T-t)}N(d_2) \quad \text{and}$$

$$\Theta_{\text{put}} = -\frac{SN'(d_1)\sigma}{2\sqrt{T-t}} + rKe^{-r(T-t)}N(-d_2) = \Theta_{\text{call}} + rKe^{-r(T-t)}.$$

The close relationship between theta and gamma follows from the Black-Scholes partial differential Eq. (4.47) rewritten using the Greek letters:

$$\Theta + rS\Delta + \frac{1}{2}\sigma^2 S^2\Gamma = r\Pi. \tag{4.65}$$

If the portfolio is delta-hedged, i.e., $\Delta = 0$, then (4.65) becomes

$$\Theta + \frac{1}{2}\sigma^2 S^2\Gamma = r\Pi. \tag{4.66}$$

Moreover, if $r\Pi$ is relatively small (which is usually the case), then $\Theta = -\frac{1}{2}\sigma^2 S^2\Gamma$, in other words, if gamma is large and negative, then theta tends to be large and positive, and vice versa. It also means that a portfolio that is delta-hedged and gamma-hedged will have a relatively small theta, and so theta does not need to be hedged independently of delta and gamma. To calculate the theta of a portfolio,

$$\Theta_\Pi = \frac{\partial \Pi}{\partial t} = \sum_{i=1}^{n} \frac{\partial f_i}{\partial t} = \sum_{i=1}^{n} \Theta_i,$$

we also need to take into account the theta of forwards and futures

$$\Theta_{\text{forward}} = \frac{\partial}{\partial t}\left(S - Ke^{-r(T-t)}\right) = -rKe^{-r(T-t)}.$$

Similarly, the theta of a cash position C earning the risk-free rate r is simply

$$\Theta_{\text{cash}} = rC.$$

It is customary and more intuitive to express theta in terms of time measured in business or calendar days, i.e., as $\Theta_\Pi/252$ or $\Theta_\Pi/365$, so that it measures the change in portfolio value over one day when everything else remains unchanged.

Example 4.20 One business day theta of the portfolio from Example 4.18 without gamma hedging is 13.28. This means that the portfolio will gain €13.28, if all the market factors remain unchanged. This is pleasant news, but we know that it is outweighed by a significant gamma risk. When the portfolio is gamma and delta-hedged as proposed in Example 4.19, then the theta of the portfolio is reduced to 0.11 in line with the Eq. (4.66). Thus, by reducing the negative gamma we have lost the (relatively small) advantage of a positive theta.

So far, we have stayed within the Black-Scholes model assuming constant volatility and a constant interest rate. However, the option pricing formulas can also be differentiated with respect to the volatility parameter σ, defining the Greek[9] called *vega*, and with respect to the risk-free rate r defining the Greek called *rho*. The two measures of risk are sometimes called *out-of-model Greeks*. In practice, the two parameters (σ and r) change over time, and the sensitivity of the option's value with respect to them presents a risk that must be monitored as well. Interest rates are indeed market factors that change randomly over time. Later, we will generalize the Black-Scholes model allowing for stochastic interest rates (Chap. 6). Regarding volatility, we may still believe in the model with constant volatility, and, at the same time, accept that the implied or quoted volatility changes from one day to another. Volatility is an uncertain parameter characterizing the future price process estimated by the market given a limited amount of information. So, one can say that the market estimates change over time, even though the unknown objective volatility of the process remains constant. Another way to reconcile changing market volatility with our valuation model is to introduce the concept of stochastic volatility (see Chap. 8). The stochastic volatility models are maybe more realistic, but technically much more difficult to handle. In any case, as indicated in Fig. 4.26, vega presents an important risk factor that needs to be monitored closely.

Differentiating the Black-Scholes formulas we obtain a formula for call and put vega

$$\mathcal{V}_{\text{call}} = \mathcal{V}_{\text{put}} = S\sqrt{T-t}N'(d_1). \tag{4.67}$$

Obviously the vega of a long position is positive, and for a short position it is negative. The term $N'(d_1) = e^{-d_1^2/2}/\sqrt{2\pi}$ in (4.67) is the standard normal density function value that takes maximum values when d_1 is zero and becomes negligible if d_1 is large, positive or negative. Hence vega is relatively large for options that are at-the-money and small for options that are deeply in-the-money or out-of-the money. An option portfolio vega is again obtained as the sum of individual options' vega

$$\mathcal{V}_{\Pi} = \frac{\partial \Pi}{\partial \sigma} = \sum_{i=1}^{n} \frac{\partial f_i}{\partial \sigma} = \sum_{i=1}^{n} \mathcal{V}_i.$$

It is useful to quote vega as the estimated change $\mathcal{V}_{\Pi}/100$ of the portfolio value when the volatility goes up one percentage point. Since the vega of forwards, futures, and spot positions is zero, the portfolio's vega can be hedged only with other

[9]Note that there is no real Greek letter called "vega." The name might have been introduced erroneously since the Greek letter ν looks like a Latin "Vee."

options. Unfortunately, by hedging gamma automatically we do not hedge vega. Similarly, the buying or selling of options to hedge vega will change gamma. These conflicting goals might be resolved by using two options with different proportions between gamma and vega, and solving two equations with two unknowns as illustrated in Example 4.21. The unknowns are the numbers (weights) of the hedging options, and the equations set the target gamma and vega to zero.

The rho of a put or call option measuring sensitivity with respect to the interest rate r can be calculated according to the following formulas:

$$\rho_{\text{call}} = K(T-t)e^{-r(T-t)}N(d_2),$$

$$\rho_{\text{put}} = -K(T-t)e^{-r(T-t)}N(-d_2) = \rho_{\text{call}} - K(T-t)e^{-r(T-t)}.$$

Calculating a portfolio rho,

$$\rho_\Pi = \frac{\partial \Pi}{\partial r} = \sum_{i=1}^{n} \frac{\partial f_i}{\partial r} = \sum_{i=1}^{n} \rho_i, \tag{4.68}$$

we must also take into account the interest rate sensitivity of forward and futures contracts,

$$\rho_{\text{forward}} = \frac{\partial}{\partial r}\left(S - Ke^{-r(T-t)}\right) = (T-t)Ke^{-r(T-t)}.$$

The interest rate sensitivity of a cash position accruing the instantaneous interest rate is zero, by definition. Like vega, we often quote rho as the change of value per one percentage point, $\rho_\Pi/100$. Rho is usually the least important factor, in particular, if the options' maturities are short or medium-term. Rho can be adjusted, if needed, using delta-hedging forwards with appropriate maturity and/or with plain vanilla money market instruments.

So far, we have assumed that there is only one interest rate, but the generalized Black-Scholes formula (Chap. 6) uses the maturity specific interest rates. The sensitivity of a portfolio calculated according to (4.68) is then interpreted as the sensitivity with respect to parallel shifts of the risk-free rates across all maturities.

All the Greeks introduced can be used to estimate the change in portfolio value when the factors move up or down:

$$\Delta\Pi \approx \Delta_\Pi \times \Delta S + \Gamma_\Pi \times \Delta S^2/2 + V_\Pi \Delta\sigma + \rho_\Pi \Delta r + \Theta_\Pi \Delta t. \tag{4.69}$$

Example 4.21 The vega (per one percentage point) of the gamma-hedged portfolio from Example 4.19 turns out to be -188.82. This means that the portfolio would lose approximately €188.82 if the market volatility increased just by 1%. This is a real risk, if we plan to liquidate our position. But even if the portfolio is intended to be kept until maturity, it must be revalued based on the market prices, and the

Table 4.4 Gamma and vega of the portfolios and the options to be used for hedging

	Gamma	Vega
Portfolio	−120.72	−226.35
One-month put/call	0.184	0.058
One-year put/call	0.053	0.198

Table 4.5 Greeks of the portfolio after gamma, vega, rho, and delta hedging

	Value	Delta	Gamma	Theta	Vega	Rho
Hedged Portfolio	948.07	0.57	0.11	−0.02	0.10	0.04

changes in the market value must be accounted for. Consequently, the vega risk appears to be very serious since the market volatility could easily go up by 5–10% during a market turmoil. To hedge the gamma and vega at the same time, we can combine liquid options with short and long maturities. Let us assume that the 1-month and 1-year options with strike $K = 50$ are available at favorable prices on the market. Table 4.4 shows the gamma and vega of the hedging options and of the original portfolio before gamma-hedging (Example 4.18).

Our goal is to buy (or sell) a certain number w_1 of the 1-month and w_2 1-year options in order to diminish the gamma and, at the same time, vega of the portfolio. The proportions between gamma and vega for the 1-month and 1-year options are essentially opposite and so it is sufficient to solve the system of two equations with two unknowns:

$$0.184w_1 + 0.053w_2 = 120.72$$
$$0.058w_1 + 0.198w_2 = 226.35.$$

The solutions after rounding to the nearest integers are $w_1 = 358$ and $w_2 = 1038$. Again, we buy, rather, 179 one-month put and call options, and 519 one-year put and call options in order to match the reversed "U" shaped profile of the short straddle in the original portfolio. The Greeks, delta, gamma, theta, and vega are now close to zero. Rho (per one percentage point) remains relatively low 11.5. This means that we would lose €11.5 if interest rates went up by 1%. The positive rho can be easily eliminated by a €1150 one-year money market deposit. The resulting Greeks of the portfolio after gamma, vega, delta, and rho hedging are shown in Table 4.5. The value of the portfolio is rather optimistic, being based on the assumption that the hedging options can be bought exactly at their market value. In practice, there would be certain transaction costs.

According to (4.69), the sensitivity of the portfolio to reasonably small changes of any of the pricing parameters should be under control. Figure 4.29 shows that the gamma-vega hedged portfolio is relatively stable with respect to moves in the underlying asset price and volatility. However, the nice-looking volatility dependence assumes fixed $S = 50$ and will change when S moves. The portfolio value is a

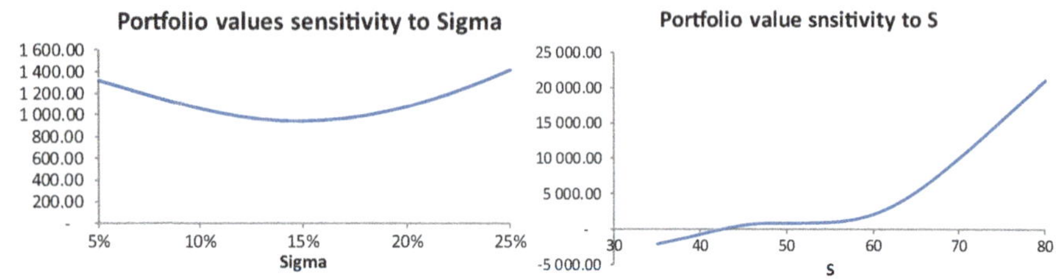

Fig. 4.29 Dependence of the gamma-vega hedged option portfolio value on the market volatility and the underlying price

Table 4.6 General Greek letter formulas for European options on assets paying continuous income q

Greek letter	Call	Put
Delta	$e^{-q(T-1)}N(d_1)$	$e^{-q(T-t)}(1-N(d_1))$
Gamma	$\frac{e^{-q(T-t)}N'(d_1)}{S\sigma\sqrt{T-1}}$	$\frac{e^{-q(T-t)}N'(d_1)}{S\sigma\sqrt{T-t}}$
Theta	$-\frac{e^{-q(T-t)}SN'(d_1)\sigma}{2\sqrt{T-1}}$ $-rKe^{-r(T-t)}N(d_2)+qe^{-q(T-t)}SN(d_1)$	$-\frac{e^{-q(T-t)}SN'(d_1)\sigma}{2\sqrt{T-1}}$ $+rKe^{-r(T-t)}N(d_2)-qe^{-q(T-t)}SN(d_1)$
Vega	$e^{-q(T-1)}S\sqrt{T-t}N'(d_1)$	$e^{-q(T-t)}S\sqrt{T-t}N'(d_1)$
Rho	$K(T-t)e^{-r(T-t)}N(d_2)$	$-K(T-t)e^{-r(T-t)}N(-d_2)$
Rho(q)	$-S(T-t)e^{-q(T-t)}N(d_1)$	$S(T-t)e^{-q(T-t)}N(-d_1)$

function of several arguments, S, σ, and the other variables. Therefore, it is not a "hedge and forget" solution. The portfolio will have to be rebalanced, if there is a larger move in any of the factors.

The Greek letter formulas applied above assumed that the underlying asset does not pay any income. If there is a continuously paid income at the rate q, then the formulas must be slightly modified to differentiate the Black-Scholes formulas (4.55)–(4.58) including q. Table 4.6 summarizes the general Greek letters formulas. The last row shows the derivative of the option value with respect to the parameter q. It can be called *rho with respect to q* or *Rho*(q). The expected continuous income can change over time as well. The second rho is regularly monitored in the case of foreign currency options where $q = r_{\text{foreign}}$, i.e., for FX options rho measures sensitivity with respect to the foreign interest rate movements.

Market Risk Measurement and Management

5

Qualified market risk management of asset and liability portfolios or trading activities is of key importance for banks and financial institutions. It is not only about quantitative measurement of the risks, but also about organizational and regulatory principles. Besides relatively simple market risk measures, we will define and explain various approaches to Value at Risk (VaR) and Conditional VaR estimation and backtesting. The two risk measures are used for day-to-day management as well as key regulatory capital concepts, or as tools of strategic risk management based on economic capital allocation. Market risk management, in particular in case of OTC derivatives, is also closely related to the counterparty credit risk management and measurement in terms of CVA (Credit Valuation Adjustment) that will be in detail discussed in the last section of this chapter.

5.1 Risk Management Responsibilities and Organization

The *market risk management* function's primary responsibility in a financial (or non-financial) institution is the identification, measurement, control, and the hedging of risks related to market factors.

In general, by *risk* we understand the dependence of a portfolio future's value, a financial strategy, or, generally, any business activity's future profit/loss, on external factors, whose future development is out of our control and, today, unpredictable. The risks can be generally classified into four broad categories: market, liquidity, credit, and operational. *Market risk* management focuses on the market factors such as interest rates, foreign exchange rates, stocks, and commodity prices. *Liquidity risk* is the risk that the institution will not be able to make cash payments when they become due, because of its inability to sell (liquidate) its assets on the financial markets. *Credit risk* management deals with the risk of losses due to the defaults of debtors, bond issuers, or counterparties. Finally, *operational risk* covers all the

© The Editor(s) (if applicable) and The Author(s), under exclusive license to Springer Nature Switzerland AG 2020

J. Witzany, *Derivatives*, Springer Texts in Business and Economics,
https://doi.org/10.1007/978-3-030-51751-9_5

remaining sources of risk, in particular losses caused by human failures (intentional or non-intentional), robberies, technology breakdowns, natural disasters, etc.

It should be emphasized that there is no clear-cut borderline between the four risk categories. For example, most of the derivative mishaps mentioned in Sect. 1.4 were caused by huge market exposures caused by a breach of internal rules by one or more responsible persons—i.e., there was a combination of market and operational risks. During the recent financial crisis, the market values of CDOs unexpectedly went down, causing large losses to many financial institutions. The fall of prices was, however, related to a dramatic increase in the US subprime mortgage default rates. Moreover, it turned out that the markets had trusted too much specific valuation models and the rating agencies that had assessed many of the structured papers with excessive optimism, possibly having a serious conflict of interest. The financial institutions which were going bankrupt additionally caused large losses to other financial market counterparties due to their large mutual OTC derivative exposures. The almost frozen liquidity of the financial markets has subsequently driven to bankruptcy other institutions that could have survived otherwise. Hence, in this case, the markets have experienced a devastating "synergy" of all the market, credit, operational, and liquidity risks, i.e., *systemic risk*.

The first task of a risk manager is to identify and map all the relevant risks endangering the activities under his management. In order to control the risks, the manager needs first to understand and measure them. Even experienced managers may forget about or strongly underestimate a source of risk. The case of LTCM is an example of insufficient risk identification. The hedge fund strategy was based on the convergence of prices of similar bonds at their maturity. The long position in one bond was assumed to be almost perfectly hedged by the short position in another bond. But the fund managers underestimated the risk of high margin requirements and limited funding possibilities in the case of a financial crisis, which, unfortunately later came, i.e., the liquidity risk. The seemingly hedged positions had to be liquidated before maturity, and the fund suffered huge losses.

5.1.1 Risk Management Organization

Even if the risks are measured correctly, risk managers need to have powers, or at least sufficient influence, to stop and prevent unacceptable risk taking. Figure 5.1 shows a good example of a risk management organization structure and its relationship to the other parts of a large commercial bank.

It is of key importance that the risk management function should be independent of the business units of the bank, otherwise it would be too easy to force "Risk" to close its eyes or even give explicit approval to speculative operations which might yield a high profit in the short-term horizon, but which would present an unacceptable level of risk, in particular from the long-term horizon perspective. Optimally, the *Chief Risk Officer* (CRO) should be a member of the *Executive Board* and should have no direct responsibility for the business activities of the bank. It is recommended that the CRO's position be strengthened by a direct link to the

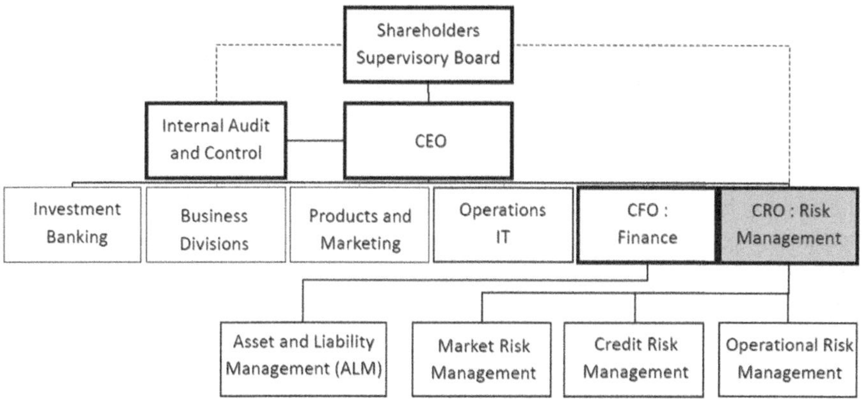

Fig. 5.1 A large bank risk management function organization

Supervisory Board. Regular risk reports should also be presented to the Supervisory Board, and the CRO should be appointed (or dismissed) only with the consent of the board. The three typical units reporting to the CRO are the Market (responsible also for liquidity risk), Credit, and Operational Risk departments. Their role consists mostly in identifying risk, measuring, limit setting, and controlling. The Risk departments must not be involved in normal business or trading. Sometimes, "Risk" may take over responsibility for a mismanaged portfolio or individual transactions. This might be, for example, a case of bad credits transferred to the recovery department within the risk organization. In the case of market risk, the risk managers could, exceptionally, be assigned a problematic trading portfolio to be hedged or liquidated. This would be a kind of forced administration in the case of dealers' serious failure.

On the other hand, the *Asset and Liability Management* (ALM) department's task is regularly to monitor, analyze, and actively hedge the banking book market and liquidity risks. Banks classify all assets, liabilities, and financial transactions in the *banking book* and the *trading book*. Classical banking products such as loans, guarantees, ordinary deposits, long-term security investments, ordinary client payments, and FX operations all belong to the banking book. The ALM must, first of all, aggregate and analyze all the banking book transactions, measure the FX, interest rate, liquidity, and other risks, and then implement appropriate hedging operations. Hedging instruments typically include money market deposits, short-term papers, FX forwards, FX swaps, interest rate swaps, cross-currency swaps, etc. Since the role of the ALM is rather active, although its task is mainly to hedge, it usually reports to the *Chief Finance Officer* (CFO), and not directly to the CRO. However, the overall hedging strategy is to be discussed and approved by an *Asset and Liability Management Committee* (ALCO), on which "Risk" is significantly represented. The ALM risks and activities should, in addition, be independently monitored by the Market Risk department.

The trading book contains transactions related to all *market making* activities and to *proprietary trading*, i.e., short-term investments (less than 1 year) and speculations. The book is under the active management of the *Investment Banking* unit. The role of risk management, with respect to the trading book, is rather supervisory and controlling. Not all banks are involved in active trading; small banks usually have just an ALM department that uses the market only to hedge the banking book positions. However, if ALM dealers start to trade actively, or even speculate, e.g., on foreign exchange rates or on interest rates, then the transactions must be monitored in a separate trading portfolio. The bank could suffer unexpected losses if ordinary hedging operations were mixed together with speculative transactions in the banking book.

Bigger banks have large *dealing* (or equivalent *trading*) *rooms* with many noisy dealers surrounded by a multitude of Reuters or Bloomberg screens and other ICT equipment. Transactions are usually entered into every minute or even second, and so the correct measurement and identification of risks in such a trading book require a perfect and up-to-date knowledge of all transactions, positions, and reliable market data. The first component, a complete database of transactions and positions, is of upmost importance. Highly sophisticated models are of no use if some transactions are missing from the databases, or if some notional amounts are wrong. Many banks have a dedicated *Middle Office* department, which stands somewhere between the *Front Office* (Trading Department) and the Risk department (see Fig. 5.2). The responsibility of the Middle Office team is to maintain the transaction database, to enter and control market, and counterparty limits, to perform the market revaluation of all market positions, and to prepare regular reports, etc. The Middle Office should be close to the Risk organization, but its responsibility is, at the same time, to support day-to-day front office operations. Indeed, it stands in the middle between the "Risk," the Back Office, and the Front Office. The core "Risk" responsibilities are valuation and risk measurement methodology, setting market limits, risk analysis, and in-depth reporting for the CRO, Executive and Supervisory Board. Any new derivative product valuation and risk measurement methodology must be approved by Risk, and properly implemented in the trading systems before authorization for

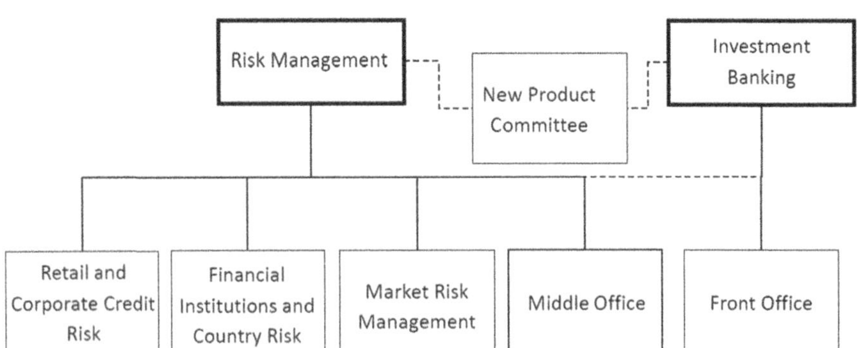

Fig. 5.2 A typical organization structure between the Risk and Investment Banking

trading. Active trading with a new derivative without those prerequisites could lead to risk positions and potential losses out of the control of Risk and even of the traders themselves. As it is mainly Investment Banking that is interested in new (hopefully profitable) product development and approval, there is usually a *New Product Committee*, where the proposals are discussed between the two sides.

Counterparty risk has become extremely important in the financial crisis and postcrisis period. It is the risk that a counterparty of an outstanding transaction (e.g., a forward or a swap) defaults before the final settlement of the contract. *Credit limits* on corporate or financial counterparties are set by the credit risk departments (Fig. 5.2), typically in terms of a loan equivalent. The market risk management must define and implement a viable methodology of limit utilization. In the case of an open OTC derivative position, the limit utilization is often based on the simple formula

$$\max(\text{market value}, 0) + x\% \times \text{notional amount}.$$

The second part of the expression (called the *add-on*) takes into account a possible increase in the market value of the position, and hence estimates the potential exposure if the counterparty defaults. A more sophisticated approach would model the distribution of possible market values over time until the contract maturity (see Sect. 5.5). Therefore, the task is, in fact, even more difficult than the valuation of derivatives themselves, or the calculation of basic market risk measures.

5.2 Classical Market Risk Measures

The goal of the simplest market risk measures is to characterize roughly the risk of individual and overall portfolio positions. The advantage of simple measures is their easy and unambiguous calculation. Their disadvantage is that they do not always make it possible to distinguish a higher risk from a lower risk position. In spite of their shortcomings, simple risk measures are often more appropriate for limit setting and management compared to more advanced, but less comprehensible and more difficult to evaluate measures like the Value at Risk (see Sect. 5.3).

An example of a very simple risk measure is the amount invested in a security or in a class of securities with similar risk characteristics. Limits set on the total value of individual stocks, bonds, or their classes are indeed often used as the basic limits for portfolio managers or security dealers. Nevertheless, the measure will be insufficient if forwards or futures on the securities are being traded. The initial invested amount in a forward contract is zero, but the risk can clearly be high—there is a leverage effect. The problem arises already with not-yet-settled spot operations. For a single security, the easy solution is to add up the amounts both of already held and of purchased but not-yet-settled securities calculated with the positive sign, and the amounts of sold but not-yet-settled securities with the negative sign. The resulting position might be *long* (positive sign) or *short* (negative sign).

Another question that arises is how to combine long and short positions in different securities. If the positions are to be netted, then we certainly have to add up the market value equivalents rather than the numbers of securities. Yet it has to be taken into account that the securities can be correlated only partially, and a loss in one position is not automatically offset by a profit in another position. A limit on the net position in two or more securities should be complemented with limits on individual securities, or with a limit on the sum of absolute positions (disregarding the signs).

5.2.1 Foreign Exchange Risk

A similar approach is used by central banks regulating the FX risk of commercial banks. For example, the Czech National Bank (CNB) defines the *FX position* in currency as the net sum of long and short positions (domestic currency equivalents) taking into account all on-balance sheet and off-balance sheet items in the currency. The CNB used to put a limit on the FX position in a currency (short or long), restricting it to 15% of the bank's capital.[1] The *overall short position* is defined as the sum of all FX positions in currencies where the individual positions are short. The *overall long FX position* is defined analogously. The *overall FX position*, which is defined as the maximum of the short and the long FX position, should not exceed 20% of capital (according to the older regulation). Note that the CNB allows the netting of positions in one currency, but not between different currencies, although there is often a positive correlation. Naturally, this approach tends to be rather conservative.

Example 5.1 Let us consider the following position of a CZK based trader: 1 million EUR, 1 million USD, 2 million GBP in cash, and an unsettled forward contract to sell 10 million EUR for 13 million USD. The limits on FX positions in individual currencies are 240 million CZK. The overall position limit is 320 million CZK. Let us check whether the limits are observed. The FX position in EUR is $1 - 10 = -9$ million EUR, i.e., -225 million CZK, since we have to take into account the unsettled forward contract representing 10 million EUR liability against 13 million USD receivable with $S_{\mathrm{EUR/CZK}} = 25$. Consequently, we can say that the limit is observed in the case of the EUR position. The other positions in USD and GBP are similarly calculated in Table 5.1. The limit in USD is violated, because we must combine the long cash position in USD and the long USD position of the forward contract. Moreover, the overall long FX position is $280 + 60 = 340$ million CZK and the short one is -225, and so the overall FX position, as defined by the CNB regulation, violates the limit of 320 million CZK, too.

The net FX position $340 - 225 = 115$ million CZK appears better, but it would be indeed incorrect to assume that losses on the long USD or GBP positions are always offset by profits on the EUR short position or vice versa. The exchange rates

[1]The CNB regulation 6/1995 has been replaced by the set of Basel regulatory documents.

Table 5.1 FX positions in individual currencies

Currency	FX position	Exchange rate (CZK)	FX position in CZK
EUR	−9 million EUR	25	−225 million
USD	14 million USD	20	280 million
GBP	2 million GBP	30	60 million

EUR/CZK, USD/CZK, and GBP/CZK are jointly influenced by CZK appreciation and depreciation (primarily on the EUR/CZK currency pair and automatically transferred to other exchange rates), but more importantly by the cross rates between EUR, USD, and GBP, which are quite volatile. For example, if EUR/USD goes up, then USD/CZK will go down and EUR/CZK probably will go up, and so there will be a loss on both positions. Therefore, it would be insufficient to limit and monitor only the net FX position. In reality, the correlation between the daily or weekly returns of EUR/CZK and USD/CZK has been around 50% (during 2017–2019), and we can say that there is some offsetting effect, but only a partial one, between the long position in USD and the short position in EUR.

5.2.2 Interest Rate Risk Measurement

In the case of (fix coupon) bond trading, riskiness depends on maturity—long-term bonds are more sensitive to interest rate movements than short-term ones. Interest rate sensitivity is typically measured by the modified (or Macaulay) duration defined in Sect. 3.2. The limits could be given, for example, on volume invested and on duration at the same time. However, this approach does not take into account that a large volume invested into bonds with low duration might have a similar risk compared to a lower volume invested into bonds with higher duration. The combined risk can be measured using the concept of Basis Point Value (BPV), also introduced in Sect. 3.2 and generally defined as

$$BPV = V(r - 0.01\%) - V(r), \tag{5.1}$$

where $V(r)$ is the value of the monitored portfolio as a function of interest rate (yield to maturity) r. Hence, if the interest rates go down by one basis point, the portfolio value will increase by the BPV; alternatively, if the interest rates go up one basis point then the portfolio value will go down (approximately) by the BPV. Generally, if the interest rates go up or down by x basis points, then the change in portfolio value can be approximated by $x \times BPV$. Note that this is indeed just an approximation since the function $V(r)$ is usually nonlinear; it is concave in the case of a single bond and might have different shapes for a general portfolio of interest rate instruments.

If a portfolio consists of a number of investments into different bonds with values $V_1(r_1), \ldots, V_n(r_n)$ depending on the yields corresponding to different maturities, then the expression (5.1) is interpreted as the change of value when the whole yield curve shifts one basis point down in parallel, i.e., it equals to the sum of the BPVs of the individual bonds in the portfolio.

$$BPV = \sum_{i=1}^{n} V_i(r_i - 0.01\%) - \sum_{i=1}^{n} V_i(r_i) = \sum_{i=1}^{n} BPV_i. \tag{5.2}$$

If D_i denotes the (modified) duration of bond i, then according to the definition of duration

$$BPV_i \approx D_i \times V_i \times 0.01\%, \text{ and so}$$

$$BPV \approx D \times V \times 0.01\%. \tag{5.3}$$

It is easy to show that the portfolio duration D used in (5.3) can be expressed as a weighted average of individual durations

$$D = \frac{1}{V} \sum_{i=1}^{n} D_i V_i. \tag{5.4}$$

The portfolio duration and the BPV characterize well the sensitivity to interest rates parallel shifts, but not to nonparallel shifts, for example, when short-term interest rates go down, but long-term interest rates go up. The concept of BPV can be easily refined when the portfolio value $V = V(r_1, \ldots, r_m)$ is represented as a function of interest rates with different maturities entering the valuation formulas. These might be the market rates (deposit rates, yields to maturity, or swap rates) entering the zero-coupon rate construction (see Sect. 3.1), or directly the zero-coupon rates corresponding to different maturities. Basis point value with respect to the rate r_i is then simply defined as

$$BPV^i = V(r_1, \ldots, r_i - 0.01\%, \ldots, r_n) - V(r_1, \ldots, r_n). \tag{5.5}$$

Note that the BPV^i is not necessarily the same as the BPV_i with respect to the i-th instrument in the portfolio. If the input rates are zero-coupon rates for different maturities, then all the bonds with maturity greater than or equal to the maturity corresponding to r will have, in general, non-zero sensitivity to this factor.

If V is a linear or even near-linear function of the input parameters $r_1, \ldots, r_m$ then its change, caused by the interest rates moving by $x_1, \ldots, x_m$ measured in basis points, can be approximated as a linear combination of the BPVs:

$$\Delta V \approx \sum_{i=1}^{m} x_i BPV^i. \tag{5.6}$$

Thus, the risk manager can put a limit not only on the parallel shift BPV, but also on the individual maturity BPVs, or on sums of BPVs corresponding, for example, to a parallel shift in short interest rates only, to a twist in the term structure of interest rates, and so on.[2] This approach is in particular appropriate for portfolios including interest rate swaps.

Example 5.2 A bond trader holds 500 one-year maturity 3% coupon bonds, 200 three-year maturity 4% coupon bonds, and 100 ten-year 5% coupon bonds. The nominal values of the bonds are 10,000 CZK and the market yields-to-maturity are 2.5%, 3%, and 4%, respectively. Let us analyze the interest rate risk of the portfolio. Table 5.2 shows the bond prices, durations, and BPVs calculated in Excel using the standard functions PRICE and MDURATION (modified duration).

The column "BPV" shows the numerically calculated basis point values for the individual bonds and the portfolio based on (5.1), while the last column "BPV_approx" gives the approximate values calculated using the durations according to (5.3). We can see that the results are almost identical and both approaches can be used. The total BPV means that if the yield curve goes up by 1%, i.e., 100 basis points, then the loss would exceed 191,500 CZK. Table 5.2 also shows that although the investment in the 10-year bonds is much lower than the investment into 1-year bonds, the interest rate risk measured by BPV is larger for the 10-year bonds. Note that the values shown in the tables are also BPVs according to (5.5) with respect to the yields of the 1-, 3-, and 10-year bonds. If the risk manager sets the limits as 2000 CZK on the total BPV and 700 CZK on BPVs with respect to 1, 2,. . ., 10, or longer maturity yields, then all the limits will be observed, with the exception of 10-year maturity.

Alternatively, valuation and sensitivity analysis could be performed with respect to the zero rates $r_1, \ldots, r_{10}$ in continuous compounding. The portfolio value can, in general, be expressed as the sum of discounted cash flows C_i paid in T_i years

$$V = \sum_{i=1}^{10} C_i e^{-r(T_i)\times T_i}. \tag{5.7}$$

For a maturity T_i, which is not an integer multiple of a year, the interest rate must be interpolated from the given rates $r_1, \ldots, r_{10}$ in the way outlined in Sect. 3.1. In this respect $V = V(r_1, \ldots, r_{10})$ is a function of the input rates and the sensitivities can be calculated according to (5.5). In the case of our simple portfolio, no interpolation is needed and the sensitivities are quite easy to calculate in an Excel spreadsheet. In fact, BPV for the year T depends only on the aggregate cash flow at that maturity. The maturities with the highest risk sensitivities, i.e., BPVs shown in Fig. 5.3 are again 1, 3, and 10 years when the bond principals are repaid, but there are also certain sensitivities in the remaining maturities due to coupon payments discounted by the zero rates according to (5.7).

The hedging approach based on the step-by-step elimination of sensitivities shown in Fig. 5.3 might seem cumbersome in the case of the simple three-bond

[2] The most important patterns of the term structure movements can be identified by the so-called *Principal Component Analysis* (see Sect. 7.2).

Table 5.2 Bond portfolio interest rate risk analysis

Maturity	Nominal	Coupon (%)	YTM (%)	Price (%)	Duration	Number	Value	BPV	BPV_approx
1	10,000	3.0	2.5	100.49	0.97	500	5,024,314	488.94	488.82
3	10,000	4.0	3.0	102.83	2.80	200	2,056,519	576.20	576.03
10	10,000	5.0	4.0	108.11	7.87	100	1,081,067	851.49	850.86
Total					2.35		8,161,900	1916.63	1915.70

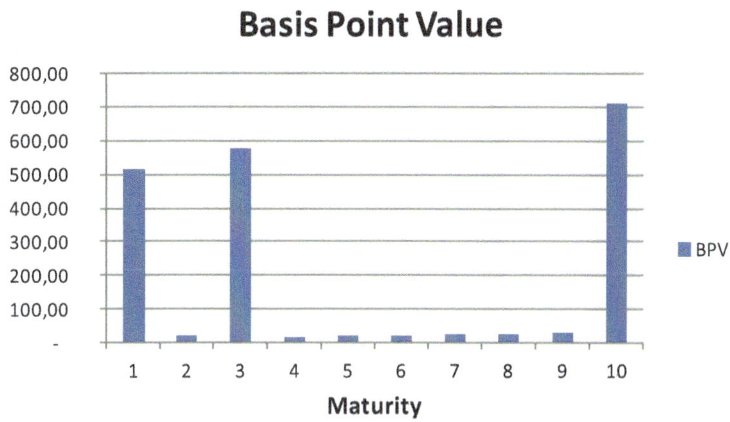

Fig. 5.3 Zero-coupon rate sensitivity for a range of maturities

portfolio, but it is a necessity if the portfolio contains many bonds with different maturities and other interest rate instruments such as interest rate swaps, FRA, etc. Interest rate risk (i.e., dependence on many different yields and rates) is then reduced to a set of sensitivities with respect to a scale of zero-coupon rates, or basic input interest rates. The risk sensitivity structure might be easily visualized as in Fig. 5.3. If a trader, or a risk manager, wants to limit or minimize (hedge) the interest rate risk, then a combination of interest swaps, FRA, and other instruments needs to be found. This can be done by going from the longer to shorter maturities: start with the longest maturity and enter an interest rate swap *neutralizing* (*immunizing*) BPV for the maturity, i.e., make it equal or close to zero; then go on to the next shortest maturity in the basic scale, and so on.

Example 5.3 Let us hedge the portfolio analyzed in Example 5.2 using interest swaps. According to Sect. 3.3, the market value of a fix-receiving IRS equals the market value of a long position in the corresponding fix-coupon bond and a short position of a floating-rate note (FRN) with the same principals. Hence, to hedge the 10-year BPV shown in Fig. 3.3 we need to enter into a fix-paying position with a 10-year BPV which is the opposite of the portfolio 10-year BPV which is 711 CZK. We have used the 10-year zero rate 3.9% in continuous compounding obtained from the bond prices by bootstrapping. The IRS notional can be roughly estimated at 1 million CZK (i.e., equal to the total principal of the 10-year bonds). More precisely, one can calculate the BPVs of, for example, 1 million CZK notional fix-paying IRS, and then adjust the notional to achieve the precise target BPV. The 10-year IRS is currently quoted at 3.5%, therefore

$$BPV_{10YIRS} = -1,035,000\left(e^{-3.89\%\times 10} - e^{-3.9\%\times 10}\right) = -701.1.$$

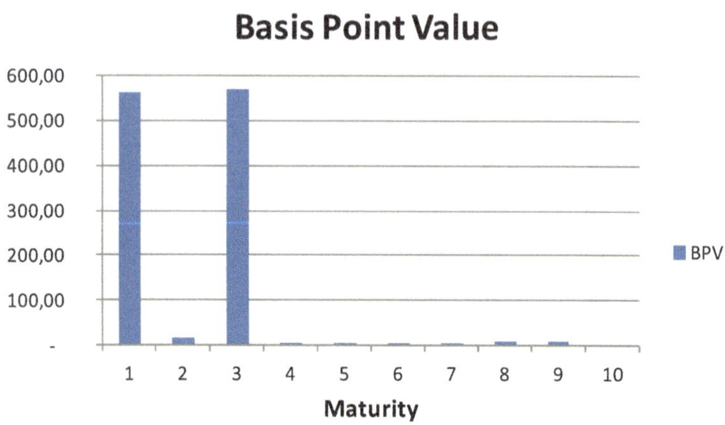

Fig. 5.4 Interest rate sensitivities of the bond after hedging with a 10Y IRS

The 10-year BPV of the target IRS with notional X million CZK is $X \times BPV_{10YIRS}$ and so we need to enter the fix-paying 10Y IRS with the notional equal to $711/701.1 = 1.01$ million CZK. Equivalently, we simply need to enter an IRS making the cash flow in 10 years equal exactly to zero.

Since the sensitivities for the maturities 4–10Y of the portfolio after the first hedging step, shown in Fig. 5.4, are negligible, we may proceed to the 3Y maturity by entering into a 3Y fix-paying swap with the notional approximately equal to 2 million CZK. The exact notional can be calculated similarly as above. The remaining 1Y maturity interest rate risk can be offset by a money market deposit, FRA, or 1Y IRS. In the case of the simple bond portfolio, the hedge IRS notional amounts corresponding to the three bonds can be easily estimated. In the case of a more complex portfolio, the precise BPV based calculation is necessary.

5.2.3 Interest Rate Gap Analysis

The duration and basis point value approach is based on the concept of market value calculated by discounting future cash flows to their present value. It is more appropriate for trading book portfolios, where changes in market values are directly accounted into the P/L (Profit/Loss) statement. This is not the case of banking book instruments with accrual accounting. Loans and deposits are not revalued based on their present value; they are accounted in the balance sheet according to their principal, and only the accrued interest enters the P/L statement. The present value (duration or BPV analysis) still makes sense, as it reflects the long-term value and risks of the book, but managers will usually focus on accrued interest accounting results at the end of the day. This is the reason why so-called *gap analysis* is so popular in ALM departments. The interest rate accrued on a loan or deposit is known today, but it may change later, either due to the reset of a floating interest rate, or due to the maturity of a product which is replaced with a new one. If the rates of assets and liabilities are reset in the same periods (in terms of the total principal of the reset assets compared to the total principal of the reset liabilities) then the net interest income should not be impacted. However, if there is a gap between the reset amounts (the assets and liabilities), then the future interest income is sensitive to future interest rates and may be impacted in a positive or negative direction by a change in rates that is unpredictable today.

Formally, let $A_1, \ldots, A_n$ denote the principals of assets (with the positive sign) and liabilities (with the negative sign) in our book and let $t_1, \ldots, t_n$ be the times of their interest rate reset or principal repayment. The time axis is usually divided into several subintervals (*baskets*) $[0, T_1)$, $[T_1, T_2)$, $\ldots$, $[T_{k-1}, T_k)$, $[T_k, \infty)$ and the gaps are calculated simply as

$$Gap(T_{i-1}, T_i) = \sum_{t_j \in [T_{i-1}, T_i)} A_j, \text{ for } i = 1, \ldots, k+1, \tag{5.8}$$

where $T_0 = 0$ and $T_{k+1} = \infty$. For example, $Gap(0, 0.5)$ calculates the difference between assets and liabilities, for which the rate is reset during the next 6 months. If the resets, in a worst-case scenario, take place today, and the rates move up by Δr (p.a.), then the monthly interest income will immediately change by approximately $Gap(0, 0.5) \times \Delta r/12$, meaning a reduction of income, if the gap is negative, and an increase, if the gap is positive. Similarly, $Gap(0.5, 1)$ indicates the possible change $Gap(0.5, 1) \times \Delta r/12$ of interest income after 6 months. But since we do not know the net interest income after 6 months and we may compare the future income only with respect to the actual interest income, we also need to take into account the resets that took place during the first 6 months. This is achieved by defining the *cumulative gap*

$$Gap(0, 1) = Gap(0, 0.5) + Gap(0.5, 1).$$

Generally, $Gap(0, T) \times \Delta r/12$ indicates possible changes in the monthly net interest income, compared to today, if all the rates reset until time T go up by Δr.

The gap analysis is a very rough, but quite a practical tool. The ALM department, in order to minimize the interest rate risk and to fulfill the limits set by ALCO, will try to balance the assets and liabilities to make the gaps corresponding to individual baskets close to zero. This can be achieved in a longer time horizon by the appropriate pricing of loans and deposits, and much faster with interest rate swaps or other interest rate derivatives.

In practice, the calculation outlined by Eq. (5.8) certainly needs to be elaborated for more complex products. Gradually repaid loan principals can be decomposed into a number of smaller principals repaid at different maturities. Interest rate swaps are decomposed into a fixed coupon bond and an opposite FRN, but more complex interest rate derivatives do not fit well into the framework. In the case of current accounts, the effective reset time is surprisingly infinite or very large, if the balance is assumed to be fixed—the current account interest rate is close to zero and it is (more-or-less) never reset. On the other hand, the possibility of withdrawing certain money from the current account and transferring it to a term deposit or saving account means that the interest can be reset upward. Interest rate gap analysis should not be confused with *liquidity gap* analysis, where the treatment of current accounts is even more difficult.

Example 5.4 The ALM department has a small portfolio of assets and liabilities given in Table 5.3. There is one fixed-rate mortgage, two loans with float interest rates linked to Pribor, a 1-year deposit, a short-term (approximately 1-month) deposit, and current accounts with the reset time being effectively infinite (set to 50 years in Table 5.3).

It should be sufficient to do the gap analysis for the time baskets corresponding to 0, 0.5, 1, 10, and infinity. Figure 5.5 shows that the −5 million first 6-month gap is offset by the +5 million gap between the 6-month and the 1-year maturity. Then there is a negative gap between 1- and 10-year maturities. The cumulative gap analysis shows that if there were a sudden single increase in interest rates, the net interest income would be negatively impacted during the following 6 months. The increase would then be offset during the next 6 months, but there would be a decline in income (compared to today's expectation) between year 1 and year 10. The problem is that the liabilities have (in the sense of the analysis) shorter effective maturities than the assets.

The remedy proposed by the ALM department could be to increase the rate on long-term deposits and cut the rate on short-term deposits. The bank clients would hopefully transfer more money to the long-term deposits. Similarly, the ALM could make the floating rate loans with short reset periods (e.g., 3 months) more attractive. This can be achieved through an *internal interest rate* (*Fund Transfer Pricing*—FTP policy), which is used by branches to offer and set loan interest rates. The reaction to these changes will take time and is uncertain. An immediate solution would be achieved by a long-term interest rate swap where the bank pays the fixed rate and receives the float rate. For example, a 10-year swap paying the fix on 5 million CZK equivalent to a 10-year effective maturity asset and 6-month effective maturity liability will hedge the gap between the 1-year and the 10-year maturity.

Interest rate risk has attracted more of the regulators' attention in recently issued regulatory documents (BCBS 2016b or EBA 2018a) focusing on interest rate risk in the banking book (IRRBB). Besides setting prudent governance standards, the regulation requires banks to calculate the impact of various interest rate stress scenarios not only on the *Net Interest Income* (ΔNII) but also on the *Economic Value of Equity* (ΔEVE), where the EVE is defined as the discounted value of all asset and liability cash flows (not including the bank's own capital) that are mapped to time buckets according to their interest rate sensitivity. The regulation defines a minimum set of stress interest rate scenarios (parallel shock up and down, steepener and flattener shock, short rates up and down shock) and encourages banks to use more scenarios specific to the market conditions and banks' balance sheet structure. Particular attention is paid to products with behavioral optionalities, i.e., to fixed rate loans subject to prepayment risk, fixed rate loan commitments, term-deposits subject to redemption risk, and non-maturity deposits (NMD). National legislation

Table 5.3 A simple banking portfolio of assets and liabilities

Description	Principal (million CZK)	Reset time (years)
Fixed rate mortgage	10	10
Float rate loan	5	0.75
Float rate loan	5	0.2
1-Y term deposit	−5	1
Short-term deposit	−10	0.1
Current accounts	−5	50

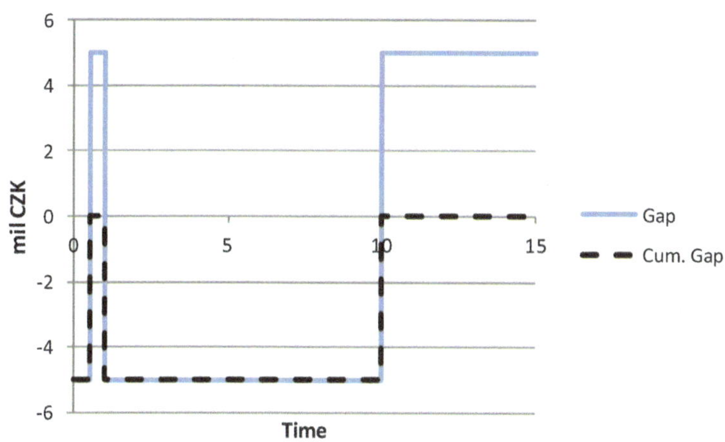

Fig. 5.5 Interest rate gap analysis

(including that of the Czech Republic) often provides significant protection to retail borrowers who can repay their mortgages or consumer loans with minimal or no penalties. But this means that the fixed rate loans subject to repayment cannot be 100% classified in the IR gap according to their maturity, and a repayment rate depending on the borrowers' behavioral characteristics and conditional on the interest rate scenario must be estimated and taken into account. In the case of non-maturity deposits, BCBS (2016b) offers a standard approach, where the NMD balances are split into the *core deposits* that are stable and unlikely to reprice due to significant changes in the interest rate environment, and the remaining *noncore deposits*. The analysis must consider the differences in behavioral characteristics between retail and wholesale (corporate) borrowers and the differences between the product characteristics of transactional (current) and non-transactional (saving) accounts. In the standard approach, the core deposits are classified into one or more longer maturity buckets, while the noncore deposits are classified only to the shortest (O/N) time bucket. BCBS (2016b) sets strict cap limits on the proportion of core deposits in the overall NMD balance in the range of 50–90% and with an average maturity of 4–5 years, depending on the segment/product group. The logic of this conservative approach is related to the fact that NMDs typically serve banks as the primary source of funds bearing no or minimum interest, and are used to finance long-term assets such as mortgages and other loans. If the overall NMD balance goes up with respect to the expectation, then the excess of liquidity can be invested in the inter-bank market with a nice unexpected additional interest rate profit, so it is not a risk. But if the NMD balance goes down, then it must be replaced by short-term financing on the inter-bank market with a certain unexpected interest rate cost. The recent global financial crisis has shown that this might be a fatal problem not only in terms interest rate risk but also in terms of liquidity risk. Therefore, for the purpose of prudent interest rate management it would not be correct to calculate only with the expected NMD balance development (which is usually increasing) but rather with a conservative NMD outflow scenario that can be related to increasing interest rates, market competition, the bank's reputational issues, etc. The estimated impact on capital (ΔEVE and ΔNII) is not currently used within the Pillar I capital requirement calculations, but it does belong to the Pillar II *Internal Capital Assessment Process* (ICAAP), which contributes to the regulatory *Overall Capital Requirements* (OCR).

5.2.4 General Delta Sensitivity Approach

The concept of delta sensitivity and delta hedging applied in Sect. 4.3 to a portfolio of options and forwards can be used for a general portfolio of securities and derivative contracts. Let us assume that a portfolio value $V = V(x_1, \ldots, x_m)$ depends on m market factors. The factors may be stock prices, stock index values, interest rates, or yields to maturity, exchange rates, investment commodity prices, or even implied volatilities, if the portfolio includes options. According to the Taylor's theorem, the change in the portfolio value caused by the market factor movements can be approximately expressed using the first-order partial derivatives of the function V as

$$\Delta V = V(x_1 + \Delta x_1, \ldots, x_m + \Delta x_m) - V(x_1, \ldots, x_m) \approx \sum_{i=1}^{m} \frac{\partial V}{\partial x_i} \Delta x_i. \tag{5.9}$$

The partial derivatives $\Delta_i = \frac{\partial V}{\partial x_i}$ can be called *deltas* or *sensitivities* with respect to the individual market factors. To control the risk, i.e., the potential variability of the portfolio value, the market risk manager needs only to control the deltas. The partial derivatives can be approximated numerically, e.g., as

$$\frac{\partial V}{\partial x_i} \approx \frac{V(x_1, \ldots, x_i + \Delta x_i, \ldots, x_m) - V(x_1, \ldots, x_m)}{\Delta x_i}$$

for small Δx_i. Alternatively, the portfolio value $V = \sum_j f_j$ and its partial derivatives can be decomposed into the market values and partial derivatives of the individual instruments,

$$\frac{\partial V}{\partial x_i} = \sum_j \frac{\partial f_j}{\partial x_i}.$$

If the deltas of the instruments in the portfolio can be expressed analytically, then the delta of the whole portfolio can be calculated analytically as well.

Notice that the deltas of simple instruments are identical or proportional to the classical risk measures. For example, the value of a position in N stocks is $f = N \times S$ where the stock value S is the market factor. And so, the delta is indeed just the number of stocks, $\Delta = \frac{\partial f}{\partial S} = N$. The delta of a forward contract to buy N non-dividend paying stocks is also N since the forward value is $f = N \times S - N \times Ke^{-r(T-t)}$. In the case of interest rate instruments, the sensitivity with respect to an interest rate r is approximately proportional to the BPV introduced above, since according to the definition of the partial derivative

$$\Delta = \frac{\partial V}{\partial r} \approx \frac{V(r) - V(r - 0.01\%)}{0.01\%} = -10,000 \times BPV. \tag{5.10}$$

In fact, the sensitivities and price differences in (5.9) can be appropriately rescaled so that we can work directly with the classical measures.

Example 5.5 Let us consider a portfolio of stocks A with the total current value 1 million CZK, of stocks B with the value 2 million CZK, and 5 million CZK of 10-year government bonds with the modified duration of 7 years. There are three main sources of risk—possible negative returns R_A, R_B on stocks A and B, and a possible increase in the 10-year government bond yield r_{10}. The numbers of stocks are not given, but if N_A and S_A denote the number and the price of stock A, respectively, then the factor in (5.9) corresponding to A can be written as

$$N_A \times \Delta S_A = N_A \times S_A \times (\Delta S_A / S_A) = V_A \times R_A,$$

where $V_A = 1$ million CZK is just the value of the investment in A and R_A is the percentage change in the stock A. The component of (5.9) corresponding to the yield r_{10} can, according to (5.10), be written as $BPV \times \Delta r_{bps}$ where $\Delta r_{bps} = -\Delta r \times 10,000$ is the downward change of the yield in basis points and

$$BPV \approx 7 \times 0.01\% \times 5,000,000 = 3500 \text{ CZK},$$

see (5.3). Hence, the change in the portfolio value can be decomposed into the three components as follows:

$$\Delta V \approx V_A \times R_A + V_B \times R_B + BPV \times \Delta r_{bps} \tag{5.11}$$

where $V_A = 1$ million CZK, $V_B = 2$ million CZK, and $BPV = 3500$ CZK. The approximation (5.11) can be used to analyze various scenarios; for example, in the stress scenario where the stocks go down by 10% and the yield goes up by 150 bps, the loss can be estimated as

$$\Delta V \approx -10^6 \times 10\% - 2 \times 10^6 \times 10\% - 3500 \times 150 = -825,000 \text{ CZK}.$$

If such a potential loss is not acceptable the investment should be adjusted accordingly.

To be precise, when analyzing changes in the portfolio value over a time interval from t to $t + \Delta t$ the time should also be one of the factors in (5.9). For example, in the case of a bond or other fixed cash flow instrument

$$V(r,t) = \sum \frac{CF_i}{(1+r)^{(T_i - t)}},$$

where T_i is the maturity date of cash flow CF_i and V is a function of the yield r and time t. Using the elementary rules of differentiation, we obtain

$$\frac{\partial V}{\partial t} = \ln(1+r) \times V = rV,$$

and so the time component of the change in V can be approximated by $rV\Delta t$ as expected. The deterministic time component is usually neglected when the time interval is short and large changes in the (true) market factors are considered.

The delta sensitivity approach is analogous to the Greeks' option portfolio analysis method (see Sect. 4.3), where we have used partial derivatives with respect to the underlying asset value (Delta), interest rate (Rho), implied volatility (Vega), and time (Theta). We have argued that in the case of the underlying asset price, where the dependence is strongly non-linear, it is important also to use the second-order partial derivative (Gamma). In that case, the Taylor's expansion (5.9) can be generally extended with the second-order terms:

$$\Delta V \approx \sum_{i=1}^{m} \frac{\partial V}{\partial x_i} \Delta x_i + \frac{1}{2} \sum_{i,j=1}^{m} \frac{\partial^2 V}{\partial x_i \partial x_j} \Delta x_i \Delta x_j. \tag{5.12}$$

The second-order derivatives with respect to the market factors can be called the *gammas*. The risk management and analysis of the portfolio value changes using (5.12) certainly become more complex, and so we try to include the gammas only if necessary. For simple (*linear*) products, like cash or asset holdings, the second-order derivative is zero or close to zero. For interest rate products, the second-order derivative is generally non-zero (recall the notion of convexity introduced in Sect. 3.2), but it can usually be neglected if the maturity is short and the analyzed interest rate changes are not too large. For longer maturity interest rate products, the second-order convexity component of (5.12) should be taken into account. The second-order derivatives (gammas) should also be included in the risk analysis, if option-like products are present in the portfolio under management.

5.3 Value at Risk

We have seen in Examples 5.4 and 5.5 that it is not obvious how to treat the joint effect of several market factors moving in different directions and causing a possible loss. The key question for a manager is: "How large a loss can we suffer on a portfolio (or business activity)? And what is the probability of such a loss?" Implicitly, the question assumes a time horizon in which the possible loss is analyzed. The concept of the *Value at Risk* (*VaR*) aims to give a simple answer to the simple question. It is very easy to interpret, but quite difficult to estimate (In practice, VaR is always just an estimation).

To define the concept more precisely, let $V = V(\mathbf{x})$ denote the value of a portfolio depending on a vector of market factors. The factors $\mathbf{x} = \mathbf{x}(t)$ are known today, at time t, but not in the future (with the exception of the time variable t itself, which could be generally included in the vector $\mathbf{x}$). Let us consider a time horizon Δt and define the change in the portfolio value $\Delta V = V(\mathbf{x}(t + \Delta t)) - V(\mathbf{x}(t))$ over an interval from t to $t + \Delta t$. In this expression $V(\mathbf{x}(t))$ is known, and so deterministic, while $V(\mathbf{x}(t + \Delta t))$ is unknown, hence it must be treated as a random variable. Optimally, we would like to know, for various time horizons, the full probability distribution of the random variable ΔV. If we knew the full probability distribution of ΔV, then given a probability level α (e.g., 95 or 99%) the question above could be perfectly answered by the $(1 - \alpha)$-quantile[3] $q_{1-\alpha} = q_{1-\alpha}[\Delta V]$ of the random variable ΔV, i.e., by the critical loss value $q_{1-\alpha}$ so that the loss could not be worse than $q_{1-\alpha}$ with the probability α. Since this quantile is normally negative, the (*absolute*) *Value at Risk* is defined as the opposite value, i.e.,

$$Var^{abs}(\Delta t, \alpha) = -q_{1-\alpha}. \tag{5.13}$$

Sometimes, in particular for longer time horizons, the *relative* Value at Risk is defined with respect to the expected profit (see Fig. 5.6), i.e.,

$$Var^{rel}(\Delta t, \alpha) = E[\Delta V] - q_{1-\alpha}. \tag{5.14}$$

For example, in the case of a portfolio investment, the expected profit covers the risk-free interest rate plus a risk margin accrued on the invested amount. The risk is a possible deviation from the expectation, and so we should analyze the relative rather than the absolute VaR. Since the relative VaR consistently takes into account the time value of money and $Var^{abs} = Var^{rel} - E[\Delta V]$, by Value at Risk we will mean rather the relative VaR, if not stated otherwise.

---

[3]If X is a random variable with the cumulative distribution function F and if $p \in (0,1)$ is a probability, then the least value q such that $F(q) \geq p$ is the *p-quantile of* X, i.e. $q = \inf\{x; F(x) \geq p\}$. If the function F is continuous and strictly increasing (which is usually the case if we are working with parametric distributions) then the quantile q is uniquely determined by the equation $F(q) = p$, i.e. $q = F^{-1}(p)$.

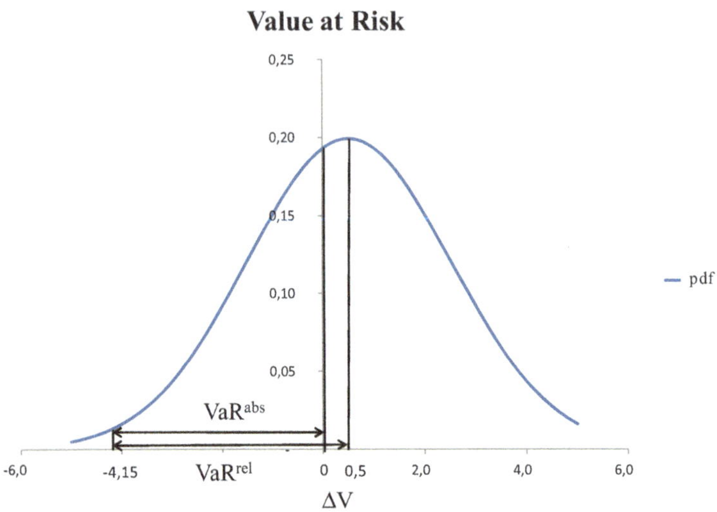

Fig. 5.6 Absolute and relative Value at Risk

Example 5.6 The distribution of future profits and losses of a US stock portfolio over a 1-month horizon shown in Fig. 5.6 has a normal distribution with the mean 0.5 million USD and standard deviation 2 million USD. The 1%-quantile of the distribution is approximately −4.15 million USD, and so according to (5.13) and (5.14) the absolute Value at Risk is 4.15 million USD while the relative VaR is 4.65 million USD.

The investors may suffer, with 1% probability, the absolute loss of at least 4.15 million USD with respect to today's portfolio value, and equivalently, the loss of 4.65 million USD with respect to the expected portfolio value reflecting the expected return on the invested amount.

Example 5.6 illustrates the so-called normal VaR when ΔV is assumed to have the normal distribution $N(\mu, \sigma)$ with a mean μ and a standard deviation σ. Since in this case $q_{1-\alpha}[\Delta V] = \mu + \sigma \times q_{1-\alpha}^N$, where $q_{1-\alpha}^N = -q_\alpha^N$ is the $(1-\alpha)$-quantile of the standard normal distribution $N(0,1)$, the Value at Risk at a given probability level α is simply

$$Var^{rel} = \mu - \left(\mu + \sigma \times q_{1-\alpha}^N\right) = -\sigma \times q_{1-\alpha}^N = \sigma \times q_\alpha^N. \tag{5.15}$$

Consequently, the risk is, in this case, fully characterized by the standard deviation σ. VaR on a probability level α is just a constant multiplied by σ, with the constant depending only on the confidence level α.

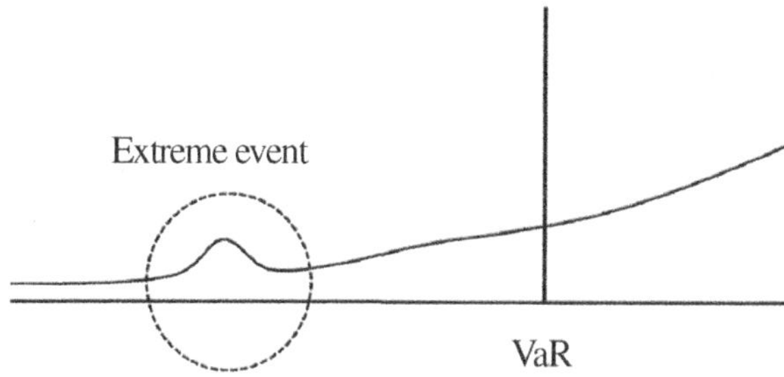

Fig. 5.7 Non-Gaussian distribution of extreme losses

5.3.1 Conditional Value at Risk

If the distribution of profits and losses ΔV is not normal, then the standard deviation or VaR at a probability level does not fully characterize the risk. The VaR tells us that a loss of a certain magnitude may happen, but we do not know how large the loss could be, i.e., what the distribution of losses on the left-hand side of the VaR value is. For example, Fig. 5.7 shows a non-Gaussian distribution of losses with a probability mass hump representing a possible extreme event—in other words, if there is a loss exceeding VaR, then it will probably be much larger than VaR itself, and even larger compared to what could be expected if the tail of the distribution.

The distribution of losses on the left-hand side of VaR can be partially characterized by the conditional mean of losses in an absolute value larger than VaR, or alternatively as the mean Var with confidence levels larger or equal than α. It answers the simple question "If things get bad, how bad can they be on the average?" Consequently, *Conditional Value at Risk* (*CVaR*), or equivalently *Expected shortfall* (*ES*), is defined by

$$CVar^{abs}(\alpha) = \frac{1}{1-\alpha}\int_{\alpha}^{1} Var^{abs}(q)dq \quad \text{and}$$

$$CVar^{rel}(\alpha) = CVar^{abs}(\alpha) + E[\Delta V] .$$

It is easy to show that $CVar^{\text{abs}} = E[\Delta V | \Delta V \leq -Var^{abs}]$ provided $\Pr[\Delta V = -Var^{abs}] = 0$, i.e., the two definitions are equivalent if the cumulative distribution function of ΔV is continuous at $x = -Var^{abs}$ (see Embrechts and Wang 2015). Another argument for conditional VaR is the issue of coherency. According to Artzner et al. (1999) a risk measure should have the following four *coherency* properties. By a risk measure (reflecting the potential magnitude of absolute losses), they understand a function $m(X)$ that assigns a nonnegative number to a random variable representing P/L of a portfolio over a fixed time horizon. The intuitively meaningful requirements are:

1. *Monotonicity*: If $X \leq Y$, i.e., if the P/L of the portfolio X is never better than the P/L of Y, then the risk of X is not lower than the risk of Y, $m(X) \geq m(Y)$.

2. *Translation invariance*: If a riskless amount K (deterministic profit or loss) is added to X, then the risk measure should be reduced by K, $m(X + K) = m(X) - K$.
3. *Homogeneity*: Multiplying the portfolio by the factor $\lambda > 0$ should result in the same risk measure multiplied by λ, $m(\lambda X) = \lambda m(X)$.
4. *Subadditivity*: The risk measure of two portfolios together should not be greater than the sum of the risk measures of the two separate portfolios, $m(X + Y) \leq m(X) + m(Y)$.

It is easy to see that the absolute VaR satisfies the first three conditions. The translation invariance and homogeneity properties are almost immediate, since $q_p(X + K) = q_p(X) + K$ and $q_p(\lambda X) = \lambda q_p(X)$ for any random variable X, and constants K, $\lambda > 0$. Regarding monotonicity, let X, Y be two random variables such that $X \leq Y$, i.e., the profit/loss given by X is always lower than the one given by Y. Then the cumulative distribution function (cdf) of X dominates the cdf of Y, i.e., $F_X(x) \geq F_Y(x)$ for any $x \in R$. Therefore, for any probability $p = 1 - \alpha$, the p-quantile of X is less or equal than the p-quantile of Y,

$$q_p(X) = \inf\{x; F_X(x) \geq p\} \leq \inf\{x; F_Y(x) \geq p\} = q_p(Y), \text{and so}$$

$$VaR(X, \alpha) = -q_p(X) \geq -q_p(Y) = VaR(Y, \alpha).$$

Unfortunately, neither the absolute nor the relative VaR satisfy the subadditivity property. The property is related to the effect of diversification: if we put two portfolios together, then there should be some benefit from diversification. If the distributions of X and Y were normal, then VaR would satisfy the last condition as well. Generally,

$$\sigma(X + Y) = \sqrt{\sigma(X)^2 + 2\rho\sigma(X)\sigma(Y) + \sigma(Y)^2} \leq \sigma(X) + \sigma(Y),$$

as the correlation $\rho \leq 1$. The normal absolute VaR on a probability level is, according to (5.15), just a constant multiple of the standard deviation plus the expected value, and so the subadditivity condition in this case holds. The subadditivity holds also for other distributions such as the Student t-distribution, logistic distribution, or more general distributions satisfying certain regularity tail conditions. However, for distribution with more exotic tail behavior, the subadditivity property can be violated (Danielsson et al. 2005).

Example 5.7 Suppose that we have two corporate bonds. The probability of default in a 1-year horizon for each of them is 0.9%, the probability that at least one of them defaults is 1.5%, and the probability of two defaults at the same time is 0.5%. In the case of default, the loss would be 50% of the bond value. Let us assume, for simplicity, that there is no interest rate or other risk, and the variability of future results is caused only by the possible defaults. Let X be the P/L variable of the portfolio with 1 million EUR invested only in the first bond, and let Y be the P/L variable corresponding to 1 million of EUR invested in the second bond. The value of X will be equal to zero with 99.1% probability, and there will be a loss of 0.5 million EUR with 0.9% probability. Consequently, the (absolute) $VaR(X)$ on the 99% confidence level is zero, and the same holds for $VaR(Y) = 0$. When we put the two portfolios together, there are three possible outcomes: no default, one default, and two defaults. The probability of at least one default is 1.5%, and so on the 99% confidence level

$$VaR[X + Y] \geq 0.5\,\text{mil} > VaR[X] + VaR[Y] = 0,$$

violating the subadditivity condition.

It seems that the violation of the subadditivity condition in the example above is related to the fact that the distributions are discrete but other counter-examples with continuous distributions can be constructed as well (for example, mixing a single large extreme loss distribution with a normally distributed variable as indicated in Fig. 5.7). Although VaR is not generally a coherent risk measure as demonstrated above, we will show that CVaR is a coherent risk measure (on the class of continuously distributed variables[4]).

Our elementary argument will be based on the fact that a continuously distributed can be arbitrarily well approximated by a discretely distributed variable with uniform probabilities. Let X be a random variable given by a set of not necessarily distinct values $x_1, \ldots, x_N$ with probabilities $1/N$ where N is a finite or even hyperfinite integer. Assume that N is large enough so that $k = pN$ is also an integer for a given probability $p = 1 - \alpha$ The VaR and CVaR measures can be then simply obtained by ordering the values[5] $x_{(1)} \leq x_{(2)} \leq \ldots \leq x_{(N)}$ and setting

$$VaR(X, \alpha) = -x_{(k)} \quad \text{and} \quad CVaR(X, \alpha) = -\frac{1}{k}\sum_{i=1}^{k} x_{(i)}.$$

The first three coherency conditions of VaR and CVaR follow quite easily from this representation. The translation invariance and homogeneity properties are again immediate. If $Y = X + K$ where K is constant then $y_i = x_i + K$ for $i = 1, \ldots, N$ is the discrete representation of Y. Therefore, $y_{(i)} = x_{(i)} + K$ for $i = 1, \ldots, N$, and so

$$CVaR(Y, \alpha) = -\frac{1}{k}\sum_{i=1}^{k}\left(x_{(i)} + K\right) = -\frac{1}{k}\sum_{i=1}^{k} x_{(i)} - K = CVaR(X, \alpha) - K.$$

Similarly, if $Y = \lambda X$ where λ is a positive constant, then we have $y_{(i)} = \lambda x_{(i)}$ for $i = 1, \ldots, N$, and

[4] X is continuosly distributed if its cdf $F_x(x)$ is continuous, or equaivalenetly if the probability of any single value $\Pr[X - x]$ equals to zero (or is infiniesimal in the nonstandard approach).

[5] It means that $x_{(1)} = x_{i1}$ where x_{i1} is the least element in the set $\{x_1, \ldots, x_N\}$, and so on.

$$CVaR(Y,\alpha) = -\frac{1}{k}\sum_{i=1}^{k}\left(\lambda x_{(i)}\right) = -\lambda\frac{1}{k}\sum_{i=1}^{k}x_{(i)} = \lambda CVaR(X,\alpha).$$

Regarding monotonicity, let another random variable $Y \geq X$ be represented by a set of values $y_1, \ldots, y_N$ with uniform probabilities on the same finite or hyperfinite probability space as the variable X. The assumption $X \leq Y$ is then equivalent to the condition $x_i \leq y_i$ for $i = 1, \ldots, N$. Therefore the least sum of the k elements (which is the sum of k least elements) from the set $\{x_1, \ldots, x_N\}$ must be less or equal than the least sum of k elements from the set $\{y_1, \ldots, y_N\}$. Since CVaR is the negative average of the sum of k least values for both X and Y, we have shown the desired inequality,

$$CVaR(X,\alpha) = -\frac{1}{k}\sum_{i=1}^{k}x_{(i)} \geq -\frac{1}{k}\sum_{j=1}^{k}y_{(j)} = CVaR(Y,\alpha).$$

To show subadditivity, let X and Y be two random variables represented by sets of values $\{x_1, \ldots, x_N\}$ and $\{y_1, \ldots, y_N\}$ with uniform probabilities, and their sum $Z = X + Y$ be represented by $z_i = x_i + y_i$ for $i = 1, \ldots, N$. We may assume, without loss of generality, that the three sets have the same number of elements N because repetitions are allowed. Note that the indices of the sequences $x_{(1)} \leq x_{(2)} \leq \ldots \leq x_{(N)}$ and $y_{(1)} \leq y_{(2)} \leq \ldots \leq y_{(N)}$, as well as of $z_{(1)} \leq z_{(2)} \leq \ldots \leq z_{(N)}$ might have completely different ordering. In particular, the k-th largest Z-value $z_{(k)}$ does not (typically) equal to $x_{(k)} + y_{(k)}$. In fact, it is easy to construct many examples of sequences where $z_{(k)} = x_{(i)} + y_{(j)}$ with both $i, j > k$, i.e., exactly violating the VaR subadditivity property, if $k = pN$ is as above. However, we can prove the subadditivity property of CVaR. Let $u(i)$ and $v(i)$ denote the indices of the X and Y components of $z_{(i)} = x_{u(i)} + y_{v(i)}$. Then we have

$$CVaR(Z,\alpha) = -\frac{1}{k}\sum_{i=1}^{k}z_{(i)} = -\frac{1}{k}\sum_{i=1}^{k}x_{u(i)} - \frac{1}{k}\sum_{i=1}^{k}y_{v(i)}$$

$$\leq -\frac{1}{k}\sum_{i=1}^{k}x_{(i)} - \frac{1}{k}\sum_{i=1}^{k}y_{(i)} = CVaR(X,\alpha) + CVaR(Y,\alpha).$$

The inequality follows from the fact that the sum $\sum_{i=1}^{k}x_{u(i)}$ is larger or equal than the sum of the k least values in the set $\{x_1, \ldots, x_N\}$, i.e., $\sum_{i=1}^{k}x_{(i)}$, and the same applies to the sum of Y-values.

We have shown that CVaR is coherent in the class of discrete distributions with uniform probabilities as defined above. Since any continuous distribution can be arbitrarily well approximated by a discrete distribution with uniform probabilities, the coherency properties must hold also for this more general class. There are many alternatives to this relatively elementary proof, for example, see Embrechts and Wang (2015) providing seven alternative proofs of CVaR subadditivity.

The estimation of CVaR is generally more difficult than in the case of VaR. For a normally distributed loss $X \sim N(\mu, \sigma)$, the relative CVaR on a probability level α can be also expressed as a constant multiple of the standard deviation σ

$$CVaR[X] = \frac{N'\left(q_\alpha^N\right)}{1-\alpha}\sigma, \tag{5.16}$$

where $N(x)$ is the cumulative distribution function and $N'(x)$ the probability density function of the standard normal distribution. Theoretically, CVaR as a risk measure does not, in this case, have any added value, compared to the standard deviation risk measure, but (5.16) gives us a feasible way to estimate CVaR if the P/L variable is approximately normal. The result provides an answer to senior managers asking what the magnitude of losses could be if things go wrong.

The difference between the normal and other distributions with fatter tails can be illustrated by the Student t-distribution, which is also quite popular in financial return modeling. The shape of the t-distribution is determined by its number of degrees of freedom ν. The t-distribution with 1 degree of freedom has the fattest tails, and it gets closer to the normal distribution when the number of degrees of freedom goes to infinity (see Fig. 5.8). The mean of the (central) t-distribution is 0 for $\nu \geq 2$, and its variance is $\nu/(\nu - 2)$ for $\nu \geq 3$, or infinite for $\nu = 2$ (the mean and variance of the Student distribution are undefined for $\nu = 1$).

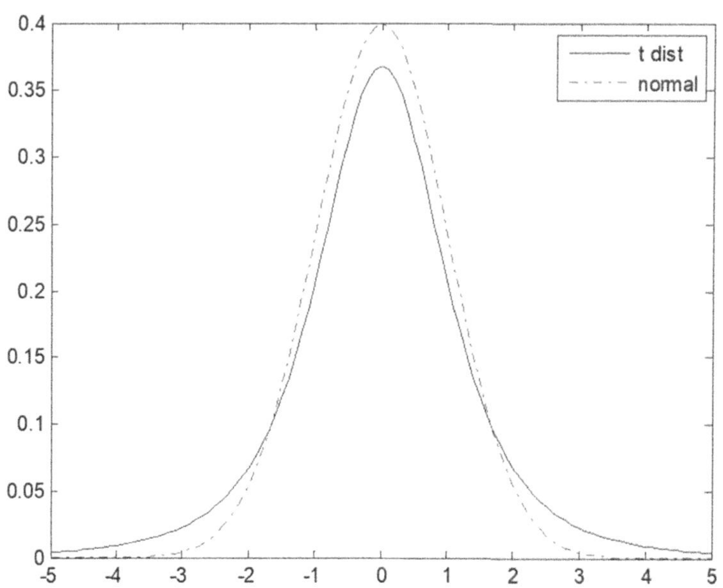

Fig. 5.8 Student t-distribution with three degrees of freedom versus the standard normal distribution

Table 5.4 Comparison of VaR and CVaR assuming $\sigma = 2$ and *t*-distributions with different degrees of freedom

Degrees of freedom	VaR (95%)	CVaR (95%)	VaR (99%)	CVaR (99%)
3	2.71	4.47	5.24	8.11
5	3.11	4.54	5.21	6.88
7	3.24	4.20	4.94	5.99
Infinite (normal)	3.29	4.12	4.65	5.33

We say that a variable X with mean μ and standard deviation σ has a *t-shaped* (non-central) distribution if the standardized variable $\sqrt{\frac{\nu}{\nu-2}}\frac{X-\mu}{\sigma}$ has the Student *t*-distribution with $\nu \geq 3$ degrees of freedom. If we estimate the mean and standard deviation of a portfolio change ΔV and assume that the distribution is t-shaped with a number of degrees of freedom, then VaR and CVaR may differ significantly from the normal VaR and CVaR values. The quantiles of the *t*-distribution are available, for example, in Excel and in other statistical or computational applications (Fig. 5.8 was created in Matlab). CVaR can be estimated numerically using the *t*-distribution pdf.

Example 5.8 Let us consider a portfolio with $\Delta V \sim N(0, 5, 2)$ and $VaR(1\,\text{year}, 0.99) = 4.65$ in million USD from Example 5.6. According to (5.16) the 99% relative CVaR is

$$CVaR = \frac{N'\left(q_{0.99}^{N}\right)}{0.01}2 = 5.33\,\text{million}\,\text{USD}.$$

Hence, with 1% probability the critical VaR loss of 4.65 million USD can be exceeded, and the loss we should expect in this case is, on average, 5.33 million, i.e., only around 15% higher compared to the VaR. Similarly, on the 95% probability level, $VaR = 3.29$ and $CVar = 4.12$ million USD. Table 5.4 compares the estimates if ΔV had, in fact, a t-shaped distribution with the same mean and standard deviation, but different degrees of freedom.

The last row corresponds to the normal VaR, since the *t*-distribution with infinite degrees of freedom is just the normal distribution. The estimated VaR and CVaR values do not differ too much on the 95% probability level and for VaR on the 99% probability level, but there are significant differences between the CVaR estimates on the 99% probability level. The difference becomes dramatic at 99.9% probability: normal CVaR amounts only to 6.73 million USD, while 3-degrees of freedom *t*-distribution VaR is around 23.73 million USD, i.e., more than three times the normal VaR. This illustrates that CVaR is an important complement of VaR characterizing the shape of the tail loss distribution.

5.3.2 Historical VaR

A universal and relatively simple approach to VaR estimation is based on the simple replication of returns that we have observed in the past. In order to sample the distribution of $\mathbf{x}(t + \Delta t)$ and $V(\mathbf{x}(t + \Delta t))$, we start from $\mathbf{x}(t)$ and move to $\mathbf{x}(t + \Delta t)$ using the returns of the market factors observed in the past. We have already argued in Sect. 4.2 that it is reasonable to assume that the relative returns distribution remains stable over time (rather than the distribution of absolute prices or their absolute changes). Recall that the model underlying the Black-Scholes valuation argument is the geometric Brownian motion

$$\frac{dS}{S} = \mu dt + \sigma dz,$$

where the return over a short time interval is assumed to have the normal distribution with the mean μdt and standard deviation $\sigma\sqrt{dt}$. The historical simulation approach does not assume that the distribution of returns is necessarily normal; we only need to assume that it is stable (*stationary*) over time. This assumption is reasonable for asset prices but less realistic in the case of interest rates (see Chap. 7 for a discussion of various interest rate models), but still applicable in practice. In the case of more market factors, the model implicitly assumes that the multivariate distribution of returns remains stable over time.

Let $\mathbf{x}^0, \ldots, \mathbf{x}^N$ be a sequence of vectors with m market factors observed in the past over time intervals of certain fixed length Δt (e.g., daily with $\Delta t = 1/252$). Calculate the historical arithmetic returns $\mathbf{r}^i = \langle r_1^i, \ldots, r_m^i \rangle$ setting

$$r_j^i = \frac{x_j^i - x_j^{i-1}}{x_j^{i-1}} \quad \text{for factors} \quad j = 1, \ldots, m \quad \text{and times} \quad i = 1, \ldots, N. \tag{5.17}$$

Since the equally weighted set of historical returns $\{\mathbf{r}^1, \ldots, \mathbf{r}^N\}$ is assumed to represent the multivariate distribution of returns between t and $t + \Delta t$ we can use it to sample the future values of $\mathbf{x}$ at $t + \Delta t$. Set $\mathbf{y}^i = \langle y_1^i, \ldots, y_m^i \rangle$ where

$$y_j^i = x_j(t)\left(1 + r_j^i\right) \quad \text{for} \quad j = 1, \ldots, m \quad \text{and scenario} \quad i = 1, \ldots, N.$$

Then, according to our assumption, the set $\{\mathbf{y}^1, \ldots, \mathbf{y}^N\}$ represents the empirical multivariate distribution of $\mathbf{x}(t + \Delta t)$, and so $\{V(\mathbf{y}^1), \ldots, V(\mathbf{y}^N)\}$ represents the empirical distribution of the variable $V(\mathbf{x}(t + \Delta t))$. Therefore, the quantiles of the distribution give us *historical* VaR estimates. By averaging the losses that are larger or equal than the VaR we may, in addition, easily obtain the CVaR estimates.

Example 5.9 Let us consider an investment into the US stock S&P 500 index ETF (Exchange Traded Fund tracking the index S&P 500 index) and into the European DJ Stoxx 600 index ETF from the perspective of a Czech investor. The price X_1 of

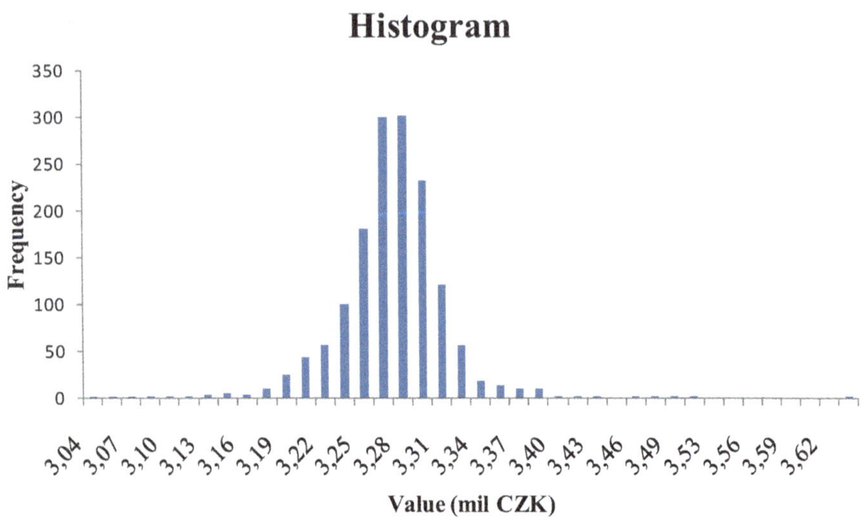

Fig. 5.9 Simulated portfolio values based on a series of historical returns

the S&P 500 ETF is denominated in USD while the price X_2 of the DJ Stoxx 600 ETF is denominated in EUR. Hence, the value of the investment into a units of the S&P 500 ETF and b units of the DJ Stoxx 600 ETF is a function of four market factors

$$V = aS_1X_1 + bS_2X_2, \tag{5.18}$$

where S_1 is the USD/CZK exchange rate and S_2 is the EUR/CZK exchange rate. Specifically, let $a = b = 1000$ and let us use the historical prices of the SPDR ETF issued by State Street, of DJ STOXX 600 (DE) issued by iShares, and of USD/CZK, EUR/CZK exchange rates starting on 1.1.2006 and continuing until 30.12.2012. The series gives 1523 return scenarios that can be applied to prices as of 10.1.2012 to generate possible portfolio values for the next (business) day. The calculations can be easily performed, e.g., in the Excel application, which is able to produce a histogram (see Fig. 5.9) and calculate the empirical quantiles.

The initial value of the portfolio, as well as the mean of the portfolio simulated values, is 3.266 million CZK, the 95% quantile is 3.199 million CZK, and the 99% quantile is 3.137 million CZK. Thus, the daily $VaR(95\%) = 0.067$ million CZK and $VaR(99\%) = 0.130$ million CZK. The empirical distribution appears to be close to the normal distribution, but it is not exactly normal. For example, the tail of the distribution on the left-hand side looks fatter than the tail on the right-hand side. Indeed, the normality of the distribution is rejected by the Jarque-Bera test on the 99.9% confidence level. In particular, the 99% normal VaR calculated from the sample standard deviation, which is 0.043 million CZK, turns out to be only 0.1 million CZK, i.e., significantly smaller than the historical VaR.

The historical VaR values are, again, only estimates of the "true" VaR values. A statistician also asks about the standard errors of the estimates and confidence intervals for the true values. Hull (2010) proposes the following two approaches. If q is the $(1-\alpha)$-quantile estimated from N simulated values, and if $f(q)$ is its empirical density, then the standard error can be estimated by the formula:

$$s=\frac{1}{f(q)}\sqrt{\frac{(1-\alpha)\alpha}{N}}. \tag{5.19}$$

For example, the estimation error of $VaR(99\%)=0.13$ from the case above is estimated at around 0.03 while the error of $VaR(95\%)=0.067$ is only around 0.003. The reason for the relatively large error of $VaR(99\%)$ is that the 1% loss quantile out of 1523 simulated values is only the 15th largest loss—there are only a few losses below the 1% quantile. According to (5.19) we would have to increase the number of simulated values 100 times in order to improve the standard error 10 times.

The second way to estimate the standard error suggested in Hull (2010) is based on the universal concept taken from statistics called *bootstrapping*. We assume that the dataset of returns $\Omega=\{\mathbf{r}^1,\ldots,\mathbf{r}^N\}$ represents the probability space of all observable returns and we sample a new sequence $\left\{\tilde{\mathbf{r}}^1,\ldots,\tilde{\mathbf{r}}^N\right\}$ drawing each $\tilde{\mathbf{r}}^i$ from Ω with possible repetitions. The new sequence is used to estimate a bootstrapped VaR value. The argument is that any such set could be possibly observed as a sequence of historical returns for which we can recalculate the VaR estimate. This bootstrapping estimation procedure is repeated many times (e.g., 1000 times using a code rather than a simple spreadsheet). Finally, the distribution of these estimates is used to estimate the confidence interval for VaR.

5.3.3 Linear VaR

The linear VaR estimation technique is based on the assumption that the change in a portfolio value can be expressed, or reasonably approximated, as a linear combination of absolute changes (or rather relative returns—see (5.9)) in the underlying market factors:

$$\Delta V\approx\sum_{j=1}^{m}\alpha_j r_j. \tag{5.20}$$

If we know the covariance matrix $\mathbf{Cov}=\left(\mathrm{cov}_{ij}\right)_{i,j=1}^{m}$ of the vector of random variables $\mathbf{r}=\langle r_1,\ldots,r_m\rangle$ then the variance of ΔV can be calculated by the standard probability formula

$$\sigma_V^2 = \boldsymbol{\alpha} \cdot \mathbf{Cov} \cdot \boldsymbol{\alpha}' = \sum_{i,j=1}^{m} \alpha_i \alpha_j \mathrm{cov}_{ij}, \tag{5.21}$$

where $\boldsymbol{\alpha} = \langle \alpha_1, \ldots, \alpha_m \rangle$. In addition, if ΔV is assumed to be normal (or to have another known parametric distribution, e.g., the Student distribution with known degrees of freedom) then VaR and CVaR can be calculated from the estimated standard deviation σ_V (see (5.15) and (5.16)). Note that there is no requirement for the distribution of $\mathbf{r} = \langle r_1, \ldots, r_m \rangle$, although it is usually assumed to be multivariate normal. In order to find the covariance matrix, it is enough to estimate the standard deviations of returns $\sigma_j = \sqrt{\mathrm{cov}_{jj}}$ and the matrix of correlations $\rho_{ij} = \mathrm{cov}_{ij}/(\sigma_i \sigma_j)$.

Example 5.10 The portfolio value (5.18) considered in Example 5.9 is not exactly linear, but its change can be relatively well approximated according to (5.9) as a linear combination of the four market factor returns:

$$\Delta V \approx \sum_{i=1}^{4} \frac{\partial V}{\partial x_i} \Delta x_i = \sum_{i=1}^{4} \alpha_i r_i, \text{ where } \alpha_i = \frac{\partial V}{\partial x_i} x_i \text{ and } r_i = \frac{\Delta x_i}{x_i}.$$

Hence,

$$\Delta V \approx aS_1X_1r(X_1) + aS_1X_1r(S_1) + bS_2X_2r(X_2) + bS_2X_2r(S_2),$$

where $a = b = 1000$, the exchange rates are $S_1 = 20.206$, $S_2 = 25.47$, and the ETF values are $X_1 = 129.1$, $X_2 = 25.47$. The standard deviations and correlations returns estimated in the usual way from the data set, used for the historical simulation in Example 5.9, are shown in Table 5.5. Note that the stock indices returns are positively correlated and the two exchange rates are positively correlated, but there are negative correlations between the stock and exchange rate returns.

The standard deviation of ΔV calculated according to (5.21) turns out to be 0.043 million CZK and the related normal VaR 0.070 million CZK on the 95% probability level (slightly higher than the historical VaR), and 0.100 million CZK on the 99% probability level (lower than the historical VaR estimate). It is also interesting to compare the linear normal CVaR on the 99% probability level, the 0.114 million CZK calculated according to (5.16) and the significantly higher historical CVaR, and

Table 5.5 Volatilities and correlations of DJ Stoxx 600, S&P 500 ETFs, and USD/CZK, EUR/CZK exchange rate returns over the period 1.1.2006 to 30.12.2012

Volatilities	$r_{\text{DJ Stoxx}}$	r_{SPY}	r_{USDCZK}	r_{EURCZK}
	1.52%	1.53%	0.97%	0.52%
Correlations	$r_{\text{DJ Stoxx}}$	r_{SPY}	r_{USDCZK}	r_{EURCZK}
$r_{\text{DJ Stoxx}}$	1.00	0.59	−0.32	−0.27
r_{SPY}	0.59	1.00	−0.42	−0.28
r_{USDCZK}	−0.32	−0.42	1.00	0.69
r_{EURCZK}	−0.27	−0.28	0.69	1.00

the 0.161 million CZK estimated as the average of the worst percentile of the simulated losses.

5.3.4 Analytical VaR with the Cornish–Fisher Expansion

The difference between the normal and historical VaR estimates illustrated in Example 5.10 may have two reasons. Firstly, the distributions of returns and losses are not normal—there are the "fat tails"; secondly, the dependence of the portfolio value on market factors is not linear and the error of the approximation (5.20) is too large. In this case, the expansion might be improved by including the second-order derivatives as in (5.12). In fact, the function (5.18) from Example 5.9 is not exactly linear in the four variables; the mixed second-order derivatives with respect to S_1, X_1 and with respect to S_2, X_2 are positive. The effect of the second-order derivatives, of convexity or concavity, is important in particular for portfolios including options (see Sect. 4.3).

Let us now assume that the change in the portfolio value can be expressed as a quadratic function in terms of the market factor returns

$$\Delta V \approx \sum_{j=1}^{m} \alpha_j r_j + \frac{1}{2} \sum_{i,j=1}^{m} \gamma_{ij} r_i r_j, \tag{5.22}$$

where $\alpha_j = \frac{\partial V}{\partial x_j} x_j$ and $\gamma_{ij} = \frac{\partial^2 V}{\partial x_i \partial x_j} x_i x_j$. The loss variable ΔV will not be normal even if the returns are multivariate normal. However, the quantiles of ΔV can be approximated by the Cornish–Fisher expansion that requires the first three moments of the random variable, i.e., $E[\Delta V]$, $E[(\Delta V)^2]$, and $E[(\Delta V)^3]$. Namely, let us start with the mean$\mu_V = E[\Delta V]$, the standard deviation $\sigma_V = E[(\Delta V - \mu_V)^2]$, and the skewness $\xi_V = \frac{1}{\sigma_V^3} E\left[(\Delta V - \mu_V)^3\right]$. According to the Cornish-Fisher expansion, the p-quantile of ΔV can be then approximated by

$$q_p[\Delta V] \approx \mu_V + w_p \sigma_V, \text{ where}$$
$$w_p = q_p^N + \frac{1}{6}\left(\left(q_p^N\right)^2 + 1\right)\xi_V,$$

and where q_p^N is the p-quantile of the standard normal distribution. It is not easy to work with the Eq. (5.22), but it can be used to calculate the first three moments provided the asset returns are multivariate normal.[6]

[6]Basically, taking the second or third power of (5.22), we need only to know the mixed moments of the type $E\left[r_i^{\mu} \cdot r_j^{\nu}\right]$. Since the density function of a multivariate normal distribution is determined by the covariance matrix, all the moments are known functions of the covariances. For example, $E\left[r_i^3 r_j\right] = 3\text{cov}_{ii}\text{cov}_{ij}$, etc.

5.3.5 Monte Carlo Simulation VaR

Complicated analytical calculations, like the Cornish–Fisher expansion, can be eliminated using modern computational power and the method of Monte Carlo simulations. The idea is to simulate, based on certain distributional assumptions, the future market factors $\mathbf{x}(t+\Delta t)$ and the portfolio values $V(\mathbf{x}(t+\Delta t))$. The empirical distribution is then used to determine the VaR and CVaR on a given probability level. For example, if the market factor returns $\mathbf{r} = \langle r_1, \ldots, r_m \rangle$ are assumed to be multivariate normal with a covariance matrix, then the Monte Carlo simulation algorithm will sample $\mathbf{r}$ from the multivariate normal distribution, calculate $\mathbf{x}(t+\Delta t)$, given $\mathbf{x}(t)$ and the returns $\mathbf{r}$, and revaluate the portfolio. One of the advantages of this approach is that it is relatively easy to implement alternative distributional assumptions for $\mathbf{r}$, for example, multivariate t-distribution, or returns generated by stochastic processes with jumps and stochastic volatility (see Chap. 8), etc. A disadvantage is that Monte Carlo simulations are not fast enough in certain situations, compared to analytic VaR estimates. This is the case if the portfolio is large with many complex instruments to be revaluated in each simulation run and the result needs to be calculated in a very short time.

5.3.6 Volatility and Correlation Estimates

Covariance matrix is a key input of analytic and Monte Carlo VaR calculations. Given a historical return series $\{\mathbf{r}^1, \ldots, \mathbf{r}^N\}$ defined according to (5.17) and assuming the mean return is zero (which is approximately true for shorter time horizons), the standard sample estimate of the covariances is

$$\mathrm{cov}^{jk} = \frac{1}{N}\sum_{i=1}^{N} r_j^i r_k^i, \text{ or in the matrix form } \mathbf{Cov} = \frac{1}{N}\sum_{i=1}^{N} (\mathbf{r}^i)'\mathbf{r}^i. \tag{5.23}$$

The covariance matrix used in Example 5.10 has been obtained in this way. It is based on a series of returns from 1.1.2006 to 30.12.2012. Similarly to Sect. 4.2, it can be argued that in (5.23) the recent periods should have larger weights than the periods in the more distant past. The EWMA (Exponentially Weighted Moving Average) volatility estimation approach can be applied to estimate the covariance matrix:

$$\mathbf{Cov} = \frac{1}{c}\sum_{i=1}^{N} \lambda^{N-i}(\mathbf{r}^i)'\mathbf{r}^i, \text{ where } c = \sum_{i=1}^{N} \lambda^{N-i} = \frac{1-\lambda^N}{1-\lambda}, \tag{5.24}$$

where $\lambda \in (0, 1)$ is the exponential weight, e.g., $\lambda = 0.97$. Both estimations (5.23) and (5.24) automatically ensure that the covariance matrix is symmetric and satisfies the positive semi-definiteness[7] consistency condition.

In Sect. 4.2, we saw that EWMA estimates might overreact to recent changes in the volatility historical pattern. Nevertheless, volatility has the *mean-reversion property*; it tends to go back to a long-term level after a period of unusually high or low values. The mean-reversion property is not incorporated in the EWMA model, but it is incorporated in the *Generalized Autoregressive Conditional Heteroscedasticity* GARCH(1,1) model proposed by Bollerslev (1986).

Let us first formulate the model in the one-dimensional case given a series of historical returns $u_1, \ldots, u_N$ assuming for simplicity that the mean of the returns is zero, i.e., $u_i = \ln S_i - \ln S_{i-1}$. The model starts with an initial variance estimate, e.g., $\sigma_1^2 = u_1^2$, and for $i > 1$ it recursively updates

$$\sigma_i^2 = \gamma V_L + \alpha u_{i=1}^2 + \beta \sigma_{i-1}^2, \tag{5.25}$$

where V_L is a long-term variance estimate and the weights sum to one, i.e., $\alpha + \beta + \gamma = 1$. Hence, the updated variance σ_i^2 is a weighted average of the long-term level V_L, previously estimated variance rate σ_{i-1}^2, and the last observed variance u_{i-1}^2 realized during the day $i - 1$. While in the case of EWMA, the exponential weight λ could be set expertly, or based on past experience, in the case of GARCH (1,1), the weights and, in particular, the long-term variance must be estimated specifically for the given historical series.

Unlike the equal-weighted or EWMA estimates, the next day volatility estimate cannot be used as our best estimate for all subsequent days. If u_N is the last observed return, then the expected variance for the following day according to the GARCH (1,1) model is

$$\sigma_{N+1}^2 = \gamma V_L + \alpha u_N^2 + \beta \sigma_N^2,$$

and so $E\left[u_{N+1}^2\right] = \sigma_{N+1}^2$. The Eq. (5.25) cannot be directly applied for $i = N + 2, \ldots$ but we can apply it recursively with the expectation operator starting from $i = N + 2$, $E\left[\sigma_{N+1}^2\right] = \sigma_{N+1}^2$, and $E\left[u_{N+1}^2\right] = \sigma_{N+1}^2$:

$$E\left[\sigma_i^2\right] = \gamma V_L + \alpha E\left[u_{i-1}^2\right] + \beta E\left[\sigma_{i-1}^2\right].$$

[7]We say that a matrix C is positive semidefinite if $w'Cw \geq 0$ for all vectors w. C is positive definite if the inequality is strict, i.e., $w'Cw > 0$ for all $w \neq 0$. The positive semidefiniteness condition follows from the fact that if C is a covariance matrix of returns and w is a vector of the weights of the market factors in a portfolio, then $w'Cw$ is the variance of the portfolio returns, and so it should be nonnegative. The variance might be zero if the factors are collinear and the weights are set up so that the portfolio is riskless.

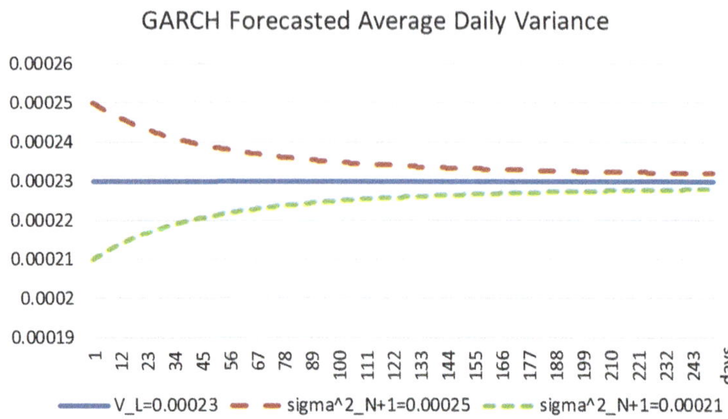

Fig. 5.10 GARCH forecasted average daily variance with the next day forecast above and below the long-term variance

Given $E\left[\sigma_i^2\right]$ we have $E\left[u_i^2\right]=E\left[\sigma_i^2\right]$ by the law of iterated expectations, and so we can evaluate $E\left[\sigma_{i+1}^2\right]$. Note that the recursive equation can be equivalently expressed in the form

$$E\left[\sigma_i^2\right]-V_L=(\alpha+\beta)\left(E\left[\sigma_{i-1}^2\right]-V_L\right), \text{and so by induction}$$

$$E\left[\sigma_{N+k}^2\right]-V_L=(\alpha+\beta)^{k-1}\left(\sigma_{N-1}^2-V_L\right)$$

Since $\alpha + \beta < 1$ the daily variance estimate converges to the long-term variance V_L. In order to estimate the total variance of returns in the time horizon $N + k$ from the perspective of day N we have to aggregate all the daily estimates:

$$\text{var}[\ln S_{N+k}-\ln S_n]=\sum_{i=1}^{k}E\left[\sigma_{N+i}^2\right]=kV_L+\frac{1-(\alpha+\beta)^k}{1-\alpha-\beta}\left(\sigma_{N+1}^2-V_L\right).$$

Therefore, by dividing the equation by k, we can see that the average expected daily variance over k days converges to V_L, or equivalently, the annualized volatility converges to the estimated long-term annualized volatility $\sqrt{252V_L}$. The convergence is from above if $\sigma_{N+1}^2 > V_L$, from below if $\sigma_{N+1}^2 < V_L$ (see Fig. 5.10), and the model has, indeed, the mean-reverting property as expected.

The standard statistical approach applicable in this case is the *Maximum Likelihood Estimation (MLE) method*. Generally, we are given certain data assumed to be generated by a model with unknown parameters $\boldsymbol{\theta}$. If we knew the parameters, then the observed data would have a certain likelihood, i.e., a probability density function value $f(\text{data}|\boldsymbol{\theta})$ aggregated over all observed datapoints, conditional on the model and its parameters. The method seeks the parameters $\boldsymbol{\theta}$ maximizing the likelihood function. The maximization problem can sometimes be solved analytically, but more often it can be solved only numerically.

Let us consider the GARCH(1,1) model where we need to estimate the three parameters: $\omega = \gamma V_L$, α, and β (the weight $\gamma = 1 - \alpha - \beta$ and the long-term variance $V_L = \omega/\gamma$). The weights are required to be positive, and so $\alpha > 0$, $\beta > 0$, and $\alpha + \beta < 1$. Given the historical returns $\mathbf{u} = \langle u_1, \ldots, u_N \rangle$ and the GARCH parameters, the volatility estimates $\sigma_1, \ldots, \sigma_N$ are recursively calculated according to (5.25). Since u_i should be a realization from the $N(0, \sigma_i^{\,2})$ distribution, its likelihood, conditional on the model and its parameters, is

$$\varphi(u_i; 0, \sigma_i^2) = \frac{1}{\sigma_i \sqrt{2\pi}} \exp\left(-\frac{u_i^2}{2\sigma_i^2}\right).$$

The returns $u_1, \ldots, u_N$ are assumed to have zero autocorrelation, hence the likelihood function can be written as

$$f(\mathbf{u}; \omega, \alpha, \beta) = \prod_{i=1}^{N} \frac{1}{\sigma_i \sqrt{2\pi}} \exp\left(-\frac{u_i^2}{2\sigma_i^2}\right). \tag{5.26}$$

The likelihood function depends on the parameters ω, α, β through the calculation of variances (5.25). It, or rather its logarithm, can be maximized to improve numerical efficiency in various statistical packages, or even in Excel using the Solver routine. Similarly, the maximum likelihood method can be used to find an optimal EWMA weight λ.

Example 5.11 Let us consider the series of the Czech stock index returns from Example 4.15. The example illustrates differences between the equal-weighted and EWMA volatilities. The GARCH coefficients estimated using the Excel Solver routine are $\omega = 6.2 \times 10^{-6}$, $\alpha = 0.139$, and $\beta = 0.835$. Figure 5.11 compares the 1-year window equal-weighted, EWMA, and GARCH 1-year volatility forecast of the PX index returns. The equal-weighted and EWMA 1-year forecasts are based only on the 1-day ahead forecasts multiplied by $\sqrt{252}$ while the GARCH 1-year forecasts incorporate the mean reversion property as explained above. Figure 5.11 indicates that the EWMA volatility forecasts are too jumpy, while the equal-weighted estimates react to changes on the market with a significant delay. On the other hand, the GARCH method is able to react fast, like at the beginning of the financial crisis in 2008 and at the same time, it reflects well the estimated long-term volatility level (24.2%) and the mean reverting property.

The GARCH(1,1) model can be generalized to estimate the covariance matrix of a multivariate return series in various ways—see Bauwens et al. (2006) for a survey. The simplest model could have the form

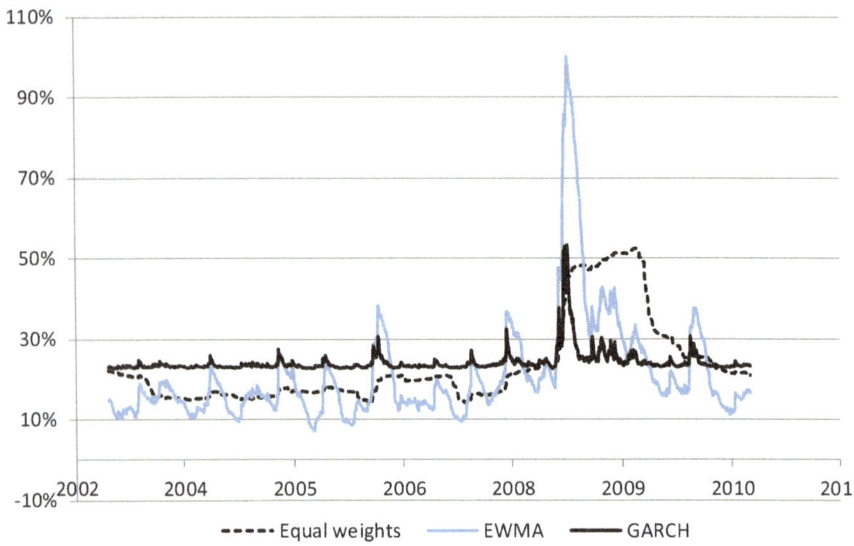

Fig. 5.11 Comparison of equal-weighted, EWMA, and GARCH one-year volatility forecasts

$$\mathbf{Cov}_i = \Omega + \alpha(\mathbf{r}_{i-1})'\mathbf{r}_{i-1} + \beta\mathbf{Cov}_{i-1}, \tag{5.27}$$

where α and β are scalars and Ω is a positive definite matrix. The disadvantage of this model is that the GARCH weights α and β must fit at the same time all the covariances. Engle and Kroner (1995) proposed the so-called BEKK model in the form

$$\mathbf{Cov}_i = C'C + A(\mathbf{r}_{i-1})'\mathbf{r}_{i-1}A' + B\mathbf{Cov}_{i-1}B', \tag{5.28}$$

where A, B, and C are $N \times N$ matrices, and C is lower triangular. The model is difficult to apply because there are $N(5N + 1)/2$ parameters to be estimated. If the time series is not long enough then there is a risk of overfitting. The model can be simplified requiring A and B to be diagonal. The model is then called the *diagonal* BEKK. The model (5.27) is also called the *scalar* BEKK.

A computationally feasible and quite popular compromise between the scalar BEKK and the full BEKK model is the DCC GARCH model proposed by Engle (2002). Firstly, the method estimates conditional variances using N separate univariate GARCH(1,1) models. Secondly, it normalizes the returns by the estimated conditional variances, and estimates correlations with the scalar model (5.27).

Specifically, let ${\sigma^i}_j$ be the univariate GARCH volatilities and define ${x^i}_j = {r^i}_j/{\sigma^i}_j$ for $j = 1, \ldots, m$ and $i = 1, \ldots, N$. The normalized returns $\mathbf{x}^i = \langle x_1^i, \ldots, x_m^i \rangle$ are then used in the scalar BEKK model

$$Q_i = \overline{Q} + \alpha(\mathbf{x}_{i-1})'\mathbf{x}_{i-1} + \beta Q_{i-1}$$

where α and β are scalars and $\overline{Q}$ is an unconditional variance matrix estimate ($\overline{Q}$ could be estimated separately, and so only the two scalar parameters remain to be estimated). The matrix Q_i is expected to be an approximation of the correlation matrix, but its diagonal elements may slightly differ from one. Consequently, we have to adjust Q_i to obtain a consistent correlation matrixR_i:

$$R_i(k,l) = Q_i(k,l)/\sqrt{Q_i(k,k)Q_i(l,l)}, \text{for } k,l = 1, \ldots, m.$$

Finally, set $cov_i^{kl} = R_i(k,l) \times \sigma_k^i \times \sigma_l^i$ and $\mathbf{Cov}_i = (cov_i^{kl})$.

5.3.7 Copula Correlations

We have argued that distributions with fatter tails, like the Student t-distribution, might be more appropriate for modeling financial returns. Similarly, correlations appear to increase in extreme market conditions. This relationship is not well captured by the ordinary correlation concept, which reflects a linear dependence between two given random variables. For example, let X and Y be independent normal variables with the exception of the lower and upper 0.5% quantile, where $X = Y$. Figure 5.12 shows a scatter plot of the two variables. The linear correlation

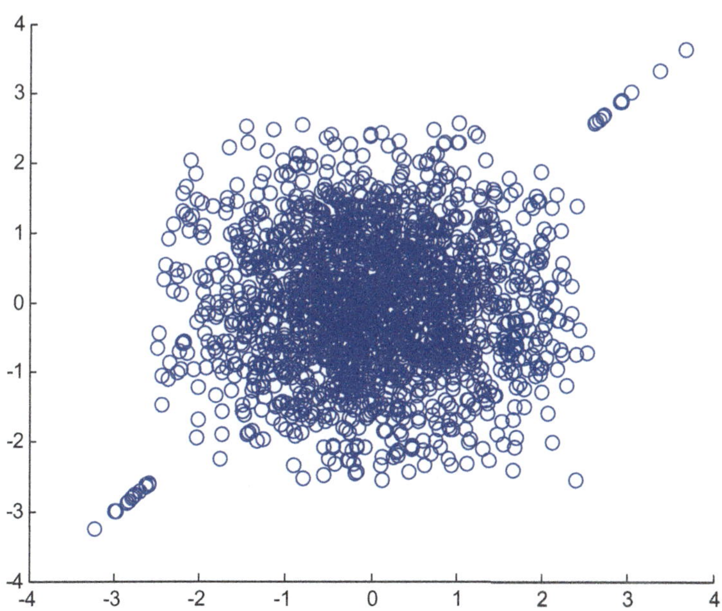

Fig. 5.12 Scatter plot of two random variables with a high tail correlation (2000 samples)

turns out to be less than 10%, but for a risk manager it is almost 100%, because when things go wrong, the variables move in exactly the same way.

The dependence between two or more variables can, in a more flexible way, be modeled using copulas. The idea is to transform two or more given random variables $X_i, \ldots, X_m$ with cumulative distribution functions $F_1, \ldots, F_m$ to a certain type of standard parametric distribution (normal, uniform, *t*-distribution, etc.) and specify the multivariate correlation structure of the transformed variables. The copula distributions can be relatively easily sampled, e.g., in a Monte Carlo simulation. The copula parameters can be estimated from historical data using the maximum likelihood method, but it is, in general, more difficult to handle the dynamics of these dependencies (correlations) analytically.

Specifically, we say that $X_1, \ldots, X_m$ have a *Gaussian copula* correlation structure with the correlation matrix Ω, if the vector of transformed standard normal distributions $U_1, \ldots, U_m$ is multivariate normal with the correlation matrix Ω. Here, $U_i = N^{-1}(F_i(X_i))$ is the quantile-to-quantile transformation of X_i to U_i. The correlation structure is defined as the *Student t-copula* with v degrees of freedom, and a correlation matrix, if the transformed variables have a multivariate Student *t*-distribution, i.e., $U_i = T^{-1}(F_i(X_i))$, where T is v degrees of freedom *t*-distribution cumulative distribution function.

A *general copula* can be defined as a correlation structure on the vector of uniform distributions $U_i = F_i(X_i)$. According to the Sklar's Theorem, any function $C : [0,1]^m \to [0,1]$ satisfying certain elementary consistency conditions corresponds to the distribution function of a multivariate uniform distribution.[8] There is a wide class of popular *Archimedean copulas* that can be still handled parametrically. Generally, an Archimedean copula function has the form

$$C_A(u_1, \ldots, u_n) = \varphi^{-1}(\varphi(u_1) + \ldots + \varphi(u_n))$$

where φ is a strictly decreasing and convex function from $[0, 1]$ onto $[0, +\infty]$. For example, if $\varphi(u) = (-\ln(u))^\alpha$, $\alpha > 1$, then the copula is called *Gumbel's*. The parameter $\alpha > 1$ that determines the level of tail dependence can also be estimated from historical data by the MLE method.

5.3.8 Extreme Value Theory VaR

Applying the normal, or another analytic VaR modeling approach, we try to fit a certain parametric distribution to the whole set of observed returns. However, when estimating VaR we are interested in extreme losses only. The idea of Extreme Value Theory (EVT) VaR estimation is to fit an appropriate parametric distribution to a tail of historical observations. The estimated extreme loss distribution then yields the

[8]Specifically, $C(u_1, \ldots, u_1, 0, u_{i+2}, \ldots, u_m) = 0$, $C(1, \ldots, 1, u_i, 1, \ldots, 1) = u_i$, and the probability mass of any subrectangle of $[0, 1]^m$ given by$C$is nonnegative.

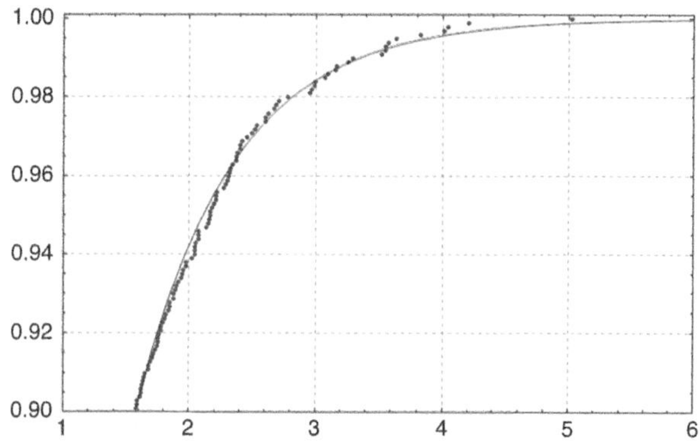

Fig. 5.13 Generalized cumulative Pareto distribution (solid line) fit to 100 tail losses (blue dots) with $u = 1.57$ and the parameter estimates $\xi = 0.01,\ \beta = 0.77$ (Source: Baran and Witzany 2012)

required quantiles (VaR) and expected shortfalls (CVaR) at various probability levels.

Let X be a loss variable (negative return) with the distribution function F, and define the conditional distribution function of *excess losses* over a threshold u by

$$F_u(y) = P[X - u \leq y | X > u] = \frac{F(u+y) - F(u)}{1 - F(u)}, \quad (y > 0). \tag{5.29}$$

According to Gnedenko (1943), for a wide class of distributions F, the excess function $F_u(y)$ converges to the generalized Pareto distribution as the threshold u increases. The *generalized Pareto cumulative distribution* (GPD) has the parametric form:

$$G_{\xi,\beta} = \begin{cases} 1 - (1 + \xi x/\beta)^{-1/\xi}, & \xi \neq 0, \\ 1 - \exp(-x/\beta), & \xi = 0. \end{cases} \tag{5.30}$$

Given a series of historical returns, we need first to choose an appropriate threshold, for example, the 90%-quantile of losses, making sure that we have a sufficient number of observations. Then the parameters ξ and β can be estimated by the maximum likelihood method. In practice, we require $\xi > 0$ with values not expected to be larger than 0.5. For example, Fig. 5.13 shows the empirical excess loss function of the 100 worst historical losses of a CZK portfolio investment in the DJ Stoxx 50, Prague stock index PX, and a EUR/CZK put option (see Baran and Witzany 2012). In order to apply the maximum likelihood method, we need to use the density of the generalized Pareto distribution function (i.e., the derivative of (5.30)),

$$g_{\xi,\beta}(x) = \frac{1}{\beta}\left(1 + \frac{\xi}{\beta}x\right)^{-1-1/\xi},$$

to express and maximize the log-likelihood function.

When the (true) loss function is approximated by $F_u(y) \approx G_{\xi,\ \beta(u)}(y)$ we can go backward, express the underlying distribution function from (5.29)

$$F(x) \approx (1 - F(u))G_{\xi,\beta(u)}(x - u) + F(u),$$

and calculate the loss quantiles, VaR, and CVaR. The probability that $X > u + y$, conditional on $X > u$ is $1 - G_{\xi,\ \beta(u)}(y)$, and the probability of $X > u$ is $1 - F(u)$. Hence, if u is the n_u-th largest loss out of n observations, then $1 - F(u)$ can be estimated by n_u/n and for $x > u$

$$\Pr[X > x] \approx (1 - F(u))(1 - G_{\xi,\beta(u)}(x - u)) \approx \frac{n_u}{n}(1 + \xi\frac{x - u}{\beta})^{-1/\xi}. \tag{5.31}$$

To find absolute VaR on a confidence level α we need only to solve the equation $\Pr[X > VaR] = 1 - \alpha$. Therefore, from (5.31) we easily get

$$VaR = u + \frac{\beta}{\xi}\left[\left(\frac{n}{n_u}(1 - \alpha)\right)^{-\xi} - 1\right].$$

It can also be shown (e.g., see Hull 2010) that the expected shortfall can be calculated by the simple formula from the VaR and the GPD parameters:

$$CVaR = \frac{VaR + \beta - \xi u}{1 - \xi}.$$

We have applied the EVT VaR estimation method to a univariate series of returns generated as in the case of the historical VaR. It is also possible, but computationally more difficult, to use EVT to model the tails of more market factors at the same time. The marginal distributions of the individual factors can then be "glued" together with an appropriate copula function (see Baran and Vořìšek 2011).

5.3.9 Operational Risk Modeling

The EVT technique is also popular for operational loss modeling. Operational risk is one of the three categories, besides market and credit risk, covered by the Basel regulatory Pillar I capital. This covers losses resulting from inadequate or failed internal processes, people and systems, or from external events, especially losses caused by internal and external fraud, human and IT failures, natural disasters, etc. As we have seen in the examples of derivative scandals, there may be an overlap with market risk. The main goal of operational risk modeling is, based on historical data, to estimate the distribution of losses, VaR and CVaR over a future period, usually one year. It turns out that a typical empirical operational loss distribution is very skewed with relatively few very large losses and many relatively small recorded losses. Hence, the popular Loss Distribution Approach (LDA) is based on the separate modeling of low severity high frequency (LSHF) and high severity low frequency (HSLF) losses (see, e.g., Chernobai et al. 2008). Since there is a random number of loss events observed over a period, the modeling is, in addition, split into frequency distribution modeling using, e.g., the Poisson or Negative Binomial distributions, and loss severity modeling using, e.g., truncated Lognormal, Weibull, or the GPD distribution for high severity losses. Once the parameters are estimated, the distribution of losses over the next year can be obtained by the Monte Carlo simulation. It turns out that high operational loss quantile estimates such as 99 or 99.9% are extremely sensitive to the good choice and robust fit of the high severity loss distribution as well as to the choice of the threshold between low and high severity losses similarly to EVT.

5.3.10 Choice of the Time Horizon and of the Confidence Level

The choice of the time horizon and the confidence level for VaR/CVaR estimations depends on the purpose the estimate. From a general perspective, risk managers should analyze the possible losses of a bank or financial institution over a 1-year horizon and compare them to the available capital. If a bank wants to maintain a rating with a theoretical probability of default p then the capital should equal at least the total bank's VaR on the $\alpha = 1 - p$ confidence level. In fact, this concept has been implicitly adopted by the Basel regulation as a guiding principle for the calculation methodology of the regulatory capital with $p = 0.1\%$, i.e., with the high confidence level $\alpha = 99.9\%$.

For the purpose of daily ongoing risk management, especially of a trading portfolio, VaR is considered in a short-time horizon, typically 1 day, 1 week, or 2 weeks. A trading portfolio is continuously changing, and, in any case, it would not make much sense to estimate VaR in a longer time horizon based on the portfolio's actual characteristics. The confidence level is also lower, corresponding to extreme losses that could occur once or a few times during a year, e.g., 99.5, 99, or 95%.

The choices of the time horizon and confidence level also have a practical estimation aspect. In order to estimate VaR directly over a time horizon Δt, we need to have a series of observations over time intervals of length Δt. The number of the observations should be in the hundreds, or, even better, in the thousands. This is possible on a daily or maybe weekly level, but hardly for the 1-year horizon. Moreover, to analyze extreme losses, we need to have some historical observations of the extreme events, so the confidence level should not be too high. For example, if $\alpha = 99.9\%$, then only one observation in 1000 belongs to the loss tail corresponding to the confidence level.

Hence, a typical choice of time horizon is one day. The VaR is then recalculated to a longer horizon using the *square-root-of-time rule*, if needed. Recall that for a geometric Brownian motion, if $\sigma^2\Delta t$ is the variance of log-returns over a Δt horizon, then $n\sigma^2\Delta t$ is the variance of log returns over a time horizon of length $n\Delta t$. If our portfolio results are defined in terms of log returns, and if we consider the normal VaR, then

$$VaR(\Delta t, \alpha) = \sigma\sqrt{\Delta t}q_{\alpha}^{N}.$$

Since the volatility sigma remains constant in the geometrical Brownian motion model, VaR can be easily rescaled to any other time horizon $n\Delta t$ and to any probability level $\widetilde{\alpha}$:

$$VaR(n\Delta t, \widetilde{\alpha}) = \sigma\sqrt{n\Delta t}q_{\widetilde{\alpha}}^{N} = \sqrt{n}\frac{q_{\widetilde{\alpha}}^{N}}{q_{\alpha}^{N}}\sigma\sqrt{\Delta t}q_{\alpha}^{N} = \sqrt{n}\frac{q_{\widetilde{\alpha}}^{N}}{q_{\alpha}^{N}}VaR(\Delta t, \alpha). \tag{5.32}$$

Normal conditional VaR can be rescaled similarly based on (5.16):

$$CVaR(n\Delta t, \widetilde{\alpha}) = \sqrt{n}\frac{N'\left(q_{\widetilde{\alpha}}^{N}\right)}{N'\left(q_{\alpha}^{N}\right)} \times \frac{1-\alpha}{1-\widetilde{\alpha}}CVaR(\Delta t, \alpha). \tag{5.33}$$

Example 5.12 In Example 5.10, we estimated the daily normal VaR on the 99% probability level to be 0.100 million CZK. Let us recalculate, applying (5.32), the VaR to the 1-year horizon and to the 99.9% probability level:

$$\begin{aligned} VaR(1\text{ year}, 0.999) &= \sqrt{252}\frac{q_{0.999}^{N}}{q_{0.99}^{N}}VaR(1\text{ day}, 0.99) = \sqrt{252}\frac{3.09}{2.33}0.1 \\ &= 2.099\,\text{million CZK}. \end{aligned}$$

The rescaled 1-year conditional CVaR on the 99.9% probability level turns out to be 2.287 million CZK. Regarding the number of days used, 1 year has 365 (or 366) calendar days, but trading activity happens only during the business days. The number of business days is around 252, and so we have set $N = 252$. Again, this is a simplifying assumption that could be reconsidered for some assets like global foreign currencies where trading takes place almost continuously.

The scaling rule (5.32) is often used even if the VaR estimate on the short time horizon is not based on the normality assumption. However, it should be kept in mind that the result might underestimate the true risk. For example, it is known that days with a high volatility of returns are, with a higher probability, followed by days with high volatility again. The volatility appears to be stochastic and autocorrelated. Thus, fat tails and the possible autocorrelation of daily returns need to be taken into account in more advanced models.

5.3.11 VaR Backtesting

We have introduced a variety of VaR estimation methods, but how should we decide, which one is the best? The hypothesis according to which the future loss is greater or equal than $VaR(\Delta t, \alpha)$ with probability $1 - \alpha$, can be tested statistically on historical data, as well as on an ongoing basis. We cannot test the validity of today's VaR estimate, but we can test a specific method by repeating the estimation procedure using all the historical data available until the time of estimation, and then compare the VaR with the loss realized over the following period. The days when the actual loss exceeds VaR are referred to as *exceptions*. The relative number of exceptions over a large number of observations should tend to $p = 1 - \alpha$, but it could differ slightly from the target proportion over a shorter time horizon, even if the estimation procedure were perfect. An appropriate statistic and its p-value can tell us how far the observed number of exceptions is from its target value.

This could be done, for example, using the *Binomial test*. Generally, if n is the number of independent trials with outcome 0 or 1, where the probability of 1 is p, then the probability of exactly d ones and $n - d$ zeros out of n trials is given by the *binomial distribution:*

$$B(d; n, p) = \binom{n}{d} p^d (1 - p)^{n-d}.$$

Similarly, we can express the probability of at most or of at least d ones:

$$B(\leq d; n, p) = \sum_{j=0}^{d} \binom{n}{j} p^j (1 - p)^{n-j},$$

$$B(\geq d; n, p) = 1 - B(\leq d - 1; n, p) = \sum_{j=d}^{n} \binom{n}{j} p^j (1 - p)^{n-j}.$$

Testing the performance of a VaR estimation method, our hypothesis is that the true probability of an exception is $p = 1 - \alpha$, and we expect to see $d_0 = pn$ exceptions. If we observe $d < d_0$, then we should calculate the probability $B(\leq d; n, p)$ of seeing d or fewer exceptions. If this probability is too low, lower than our statistical test confidence level (e.g., 5%), then we reject the one-sided hypothesis that the true exception probability is at least p. If $d > d_0$ and $B(\geq d; n, p)$ is too small, then we reject the one-sided hypothesis that the true exception probability is at most p.

The binomial tests proposed above are one-tailed tests. *Kupiec* (1995) has proposed a relatively powerful two-tailed test. If the true probability of an exception is p and d exceptions are observed in n trials, then the statistic

Table 5.6 Binomial and Kupiec test p-values

VaR method	A	B	C
d	6	13	17
$B(\leq d; 1000, 1\%)$	0.129	0.866	0.986
$B(\geq d; 1000, 1\%)$	0.781	0.082	0.007
K	1.886	0.831	4.091
$\Pr[\chi > K]$	0.170	0.362	0.043

$$K = -2\ln\left((1-p)^{n-d}p^d\right) + 2\ln\left((1-d/n)^{n-d}(d/n)^d\right) \tag{5.34}$$

should have a chi-squared distribution with one degree of freedom. If the statistic K is too large, e.g., larger than the 95% quantile of the chi-distribution, then the hypothesis is rejected on the corresponding probability level (the hypothesis states that the true probability of an exception equals to p).

Example 5.13 Let us assume that we test three methods, say A, B, and C, to estimate daily VaR at the 99% probability level. We back-test the methods on 1000 days using available historical data, count 5 exceptions for the method A, 13 exceptions for the method B, and 17 for C. The precise number of expected exceptions is 10. Can any of the three estimation methods be accepted at the 5% confidence level, and which one performed best? Table 5.6 shows the binomial distribution probabilities, Kupiec test statistics calculated according to (5.34), and the corresponding chi-distribution probability. All the calculations can be easily executed, for example, in Excel (using the BINOMDIST and CHIDIST functions).

Looking at the binomial test results in Table 5.6, methods A and B cannot be rejected at the 5% confidence level, but method C is rejected, since the one-sided p-value is only 0.7% (precisely speaking, the hypothesis is that the true probability of a VaR_C exception is at most 1%). The same conclusion is confirmed by the Kupiec test. Method A seems to perform best according to the binomial test (since the p-value is larger than in the case of B), but not according to the Kupiec test. Nevertheless, method A would probably be recommended because, according to the backtest, it appears to be slightly more conservative than method B.

5.3.12 Christofferesen Test

The binomial and Kupiec tests implicitly assume that the exceptions are independent, in particular, that the probability of an exception on day t does not depend on an exception taking or not taking place on day $t-1$. However, the market volatility turns out to be volatile, but at the same time quite persistent (see Sect. 8.4), i.e., days with high volatility are usually followed by days of high volatility, and so, if the VaR predictions are based on a static long-term volatility estimate, the exceptions might be clustered in periods of higher volatility meaning that these exception events are not in fact independent. The problem of stochastic volatility and autocorrelated exceptions might be solved by GARCH based VaR estimates aiming to utilize fully the day $t-1$ information in the VaR prediction for day t. Nevertheless, we still need a backtest that is correct even when the exceptions are possibly dependent. This problem is solved, for example, by the *Christofferesen* (1998) *test*.

The test introduces the transition rates $\pi_{ij} = \Pr[I_t = j | I_{t-1} = i]$ where $i, j \in \{0, 1\}$ and $I_t \in \{0, 1\}$ is the exception indicator variable; hence $\pi_{00} + \pi_{01} = 1$ and $\pi_{10} + \pi_{01} = 1$. The test assumes that the process is Markov—characterized by the transition probability matrix

$$\Pi_1 = \begin{bmatrix} \pi_{00} & \pi_{01} \\ \pi_{10} & \pi_{11} \end{bmatrix}$$

where the state I_t may depend on state I_{t-1} the day before but not on the states in the preceding days. If we observe a sequence of realizations $I_1, \ldots, I_T$ then the likelihood function for this process is

$$L(\Pi_1; I_1, \ldots, I_T) = \pi_{00}{}^{n_{00}} \pi_{01}{}^{n1} \pi_{10}{}^{n_{10}} \pi_{10}{}^{n_{10}}$$

where n_{ij} counts the number of transitions from state i to j (for $i, j \in \{0, 1\}$). The likelihood function can be easily maximized to obtain the standard estimates

$$\widehat{\pi}_{ij} = \frac{n_{ij}}{n_{i0} + n_{i1}}. \tag{5.35}$$

This relatively simple framework allows to test the hypothesis that the exceptions are independent and $E[I_t] = p$ (coverage) against the alternative hypothesis by comparing the likelihoods. Under the null hypothesis, the likelihood of the observed realizations $I_1, \ldots, I_T$ is

$$L(p; I_1, \ldots, I_T) = (1-p)^{n_0} p^{n_1}$$

where n_i is the number of observations with $I_t = i$, while the likelihood under the alternative hypothesis with the MLE parameters given by (5.35) is

$$L\left(\widehat{\Pi}_1; I_1, \ldots, I_T\right) = \widehat{\pi}_{00}{}^{n_{00}} \widehat{\pi}_{01}{}^{n1} \widehat{\pi}_{10}{}^{n_{10}} \widehat{\pi}_{10}{}^{n_{10}}.$$

Christofferesen (1998) shows that the standard likelihood ratio

$$\begin{aligned} LR_{cc} &= -2 \log \frac{L(p; I_1, \ldots, I_T)}{L\left(\widehat{\Pi}_1; I_1, \ldots, I_T\right)} \\ &= 2 \log L\left(\widehat{\Pi}_1; I_1, \ldots, I_T\right) - 2 \log L(p; I_1, \ldots, I_T) \end{aligned}$$

is asymptotically chi-squared distributed with two degrees of freedom. Therefore, if the likelihood under the alternative hypothesis is much larger than the likelihood under the null hypothesis (the exceptions independent and $E[I_t] = p$) as indicated by the p-value of the statistic and the chi-squared distribution, then the VaR model does not pass the test.

The test statistic can be decomposed into two parts allowing to test separately for independence and unconditional coverage:

$$LR_{cc} = LR_{ind} + LR_{uc}$$

where

$$LR_{ind} = -2\log\frac{L(\widehat{\pi}; I_1, \ldots, I_T)}{L\left(\widehat{\Pi}_1; I_1, \ldots, I_T\right)}$$
$$= 2\log L\left(\widehat{\Pi}_1; I_1, \ldots, I_T\right) - 2\log L(\widehat{\pi}; I_1, \ldots, I_T),$$

$$LR_{uc} = -2\log\frac{L(p; I_1, \ldots, I_T)}{L(\widehat{\pi}; I_1, \ldots, I_T)} = 2\log L(\widehat{\pi};; I_1, \ldots, I_T) - 2\log L(p; I_1, \ldots, I_T),$$

and $\widehat{\pi} = \frac{n_1}{n_0+n_1}$ is the observed proportion of exceptions. Notice that LR_{uc} is just the Kupiec statistic based on the assumption that the exceptions are independent. Independence itself (not including coverage) can be tested by the second statistic LR_{ind}. Christoffersen (1998) shows that both statistics are asymptotically chi-squared distributed with one degree of freedom. In an empirical study, Christoffersen (1998) demonstrates that the LR_{ind} statistic correctly rejects static VaR forecasts for a GARCH generated process. He also compares the static VaR, EWMA (RiskMetrics) based VaR, and GARCH-t VaR forecasts for a set of currency pairs return series using the three tests with the conclusion that the GARCH-t model performs the best passing the independence test and unconditional coverage test in majority of cases (unlike the static VaR model and to a certain extent also the EWMA VaR model).

5.3.13 CVaR Backtesting

Since CVaR is defined as the tail conditional mean of losses above VaR, the first step in back-testing of CVaR is, of course, to test VaR itself. However, this simplistic approach does not reflect how well the CVaR model captures the tail distribution. Backtesting of CVaR gained on importance with an advanced CVaR in practice and especially with the new Basel regulation (BCBS 2016a) requiring banks to use CVaR instead of VaR in the Internal Model calculations.

Tasche (2013) proposes a simple CVAR backtest based on an approximation of the expected shortfall by a four-term numerical integral:

$$CVaR(\alpha) = \frac{1}{4}(VaR(\alpha) + VaR(\alpha + 0.25p) + VaR(\alpha + 0.5p) + VaR(\alpha + 0.75p))$$

where $p = 1 - \alpha$. Therefore, instead of direct backtesting of CVaR, Tasche recommends just to backtest VaR on the four confidence levels.

Another nonparametric backtest of the Expected Shortfall was proposed by Acerbi and Szekely (2014). In this case, we need to observe not only exceptions I_t but also realized losses X_t. The goal of a VaR/CVaR model is to estimate

$$VaR_t^F(\alpha) = -F_t^{-1}(1 - \alpha) \text{ and } CVaR_t^F(\alpha) = E_F[X_t | X_t < VaR_t(\alpha)]$$

where the F_t is the profit/loss X_t true distribution that generally depends on t. The index t also indicates that the VaR/CVaR predictions should be dynamic, i.e., conditional on the information at time $t - 1$, rather than static, i.e., unconditional. The upper index F distinguishes the true (unknown) values of VaR and CVaR from the predictions $VaR_t(\alpha)$ and $CVaR_t(\alpha)$ obtained by the model. It is assumed that the model uses a predictive distribution P_t, and so the null hypothesis is that the $(1 - \alpha) -$ tail distributions of F_t and P_t coincide (or in the weaker form that $CVaR_t^F(\alpha) = CVaR_t(\alpha)$ and $VaR_t^F(\alpha) = VaR_t(\alpha)$). The first test proposed by Acerbi and Szekely (2014) assumes that the VaR predictions are correct (the VaR model has been already tested), i.e., $VaR_t^F(\alpha) = VaR_t(\alpha)$, and the CVaR statistic is calculated as follows:

$$Z_1 = \frac{1}{N_T} \sum_{t=1}^{T} \frac{X_t I_t}{CVaR_t(\alpha)} + 1,$$

provided $N_T > 0$, where $N_T = \sum_{t=1}^{T} I_t$. Since $CVaR_t^F(\alpha) = -E[X_t I_t]/(1 - \alpha)$, the value of the statistic is expected to be close to zero, i.e., $E[Z_1] = 0$. If our alternative (on-sided) hypothesis is that the predicted CVaR underestimates the true CVaR, $CVaR_t^F(\alpha) \geq CVaR_t(\alpha)$ with a strict inequality for some t, then a significantly negative value of the statistic indicates a problem (rejection of the null hypothesis). Unfortunately, the critical values of the statistic need to be calculated empirically by a Monte-Carlo simulation based on the predictive distributions P_t.

The first test statistic can be slightly modified to test VaR and CVaR at the same time:

$$Z_2 = \frac{1}{(1 - \alpha)T} \sum_{t=1}^{T} \frac{X_t I_t}{CVaR_t(\alpha)} + 1.$$

Again, the expected value of the statistic is zero under the null hypothesis and negative under the alternative hypothesis against which we are testing. Acerbi and Szekely (2014) show that the critical levels for Z_2 are quite stable across different distribution types (Student t-distributions with different degrees of freedom) and suggest to use the pre-computed critical levels. This approach is used, for example, by the function `esbacktest` implemented in Matlab.

The third test is based on the idea that the observed ranks $U_t = P_t(X_t)$ are i.i.d. uniformly distributed, $U(0, 1)$, if the model distribution is correct. This test is quite general, but its disadvantage is that the null hypothesis involves the entire distributions ($H_0 : F_t = P_t$) and the predictive distributions again need to be stored in order to simulate the critical levels of the test statistics (for details, see Acerbi and Szekely 2014).

A more practical test has been recently proposed by Du and Escanciano (2017). The test uses the rank values $H_t = \frac{1}{p}(p - P_t(X_t))I_t$, where $p = 1 - \alpha$, that need to be recorded only for the days with exceptions. If the predicted tail distribution is the true one, then $E[H_t] = \alpha/2$, and so the proposed test in the basic form assuming independent exceptions can be just a t-test. Simple calculations show that $E[H_t^2] = p/3$ and $\mathrm{var}[H_t] = p(1/3 - p/4)$. Hence, the t-statistic is as follows:

$$U_{ES} = \frac{\sqrt{T}(\overline{H} - p/2)}{\sqrt{p(1/3 - p/4)}}, \text{where } \overline{H} = \frac{1}{T}\sum_{t=1}^{T} H_t.$$

It can be shown, under certain conditions, that the distribution of the statistic is asymptotic normal, $N(0, 1)$. The assumptions can be relaxed in a more robust version of the test. The test can be, in fact, compared to the Kupies unconditional test. In addition, Du and Escanciano (2017) propose a conditional test that takes into account possible dependence in an analogy to the Christofferesen test.

5.3.14 VaR/CVaR Model Selection

The backtesting methods outlined above allow to select the most appropriate method from a number of possibilities. The choice will of course depend on the portfolio, return characteristics, or business process for which the VaR/CVaR models need to be developed. In any case, there are many research papers focusing on the comparison of the various VaR/CVaR estimation methods. For example, Baran and Witzany (2012) compare historical, linear, quadratic (Cornish-Fisher expansion), EVT, and EVT combined with GARCH value at Risk estimation methods. Based on similar tests, they conclude that EVT-GARCH outperforms the other methods.

5.4 Risk-Return Optimization, Economic, and Regulatory Capital

So far, we have been concerned mainly with market risk measurement and control. In addition, for any asset management activity, or in any business activity, the general target should be risk-return optimization, i.e., given an overall risk limit, a rational investor wants to maximize the return of his/her investment or business over

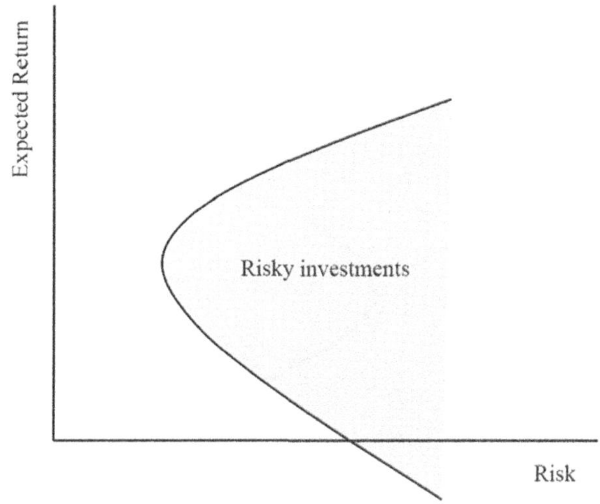

Fig. 5.14 A set of alternative risky investments

a time horizon. Or vice versa, given a targeted return, an investor wants to minimize the risk. Figure 5.14 shows the typical shape of a set of alternative risky investments in a plane with the risk measured on the x-axis and the expected return on the y-axis. Given the risk level, investors will seek an investment corresponding to the upper part frontier of the set (called the *efficient frontier*). Given a targeted expected return level, investors will minimize the risk on the left frontier of the set. Some investors might also want to optimize certain risk-return ratio without the limitation of any of the two indicators.

The optimization problem certainly depends on the risk measure used. Markowitz (1952) in his well-known theory of portfolio selection, uses the standard deviation of returns as the risk measure.

Let A and B be two possible investments (e.g., stocks) with expected returns $ER_A = E[R_A]$, $ER_B = E[R_B]$, standard deviations of returns σ_A, σ_B, and correlation ρ_{AB}. Let us consider a two-asset investment where A has the weight $\lambda \in [0, 1]$ and B has the weight $1 - \lambda$. The expected return and standard deviation of this diversified investment depending on the weight λ are:

$$\begin{aligned} ER_\lambda &= \lambda ER_A + (1 - \lambda)ER_B, \\ \sigma_\lambda &= \sqrt{\lambda^2\sigma_A^2 + 2\lambda(1 - \lambda)\rho_{AB}\sigma_A\sigma_B + (1 - \lambda)^2\sigma_B^2}. \end{aligned} \tag{5.36}$$

While ER_λ is just a weighted average of the expected returns of A and B, the standard deviation σ_λ is a nonlinear function of λ (provided $|\rho_{AB}| < 1$). Figure 5.15 shows an example of the set of mixed portfolios (5.36) in the risk-return coordinates with $\lambda \in [0, 1]$. It shows that for an appropriate choice of λ, the combined risk (standard deviation) will be even lower than the stand-alone risk of A or just B. In any case, the mixed portfolio allows for the significant improvement of the risk-return profile—there is a *diversification effect.*

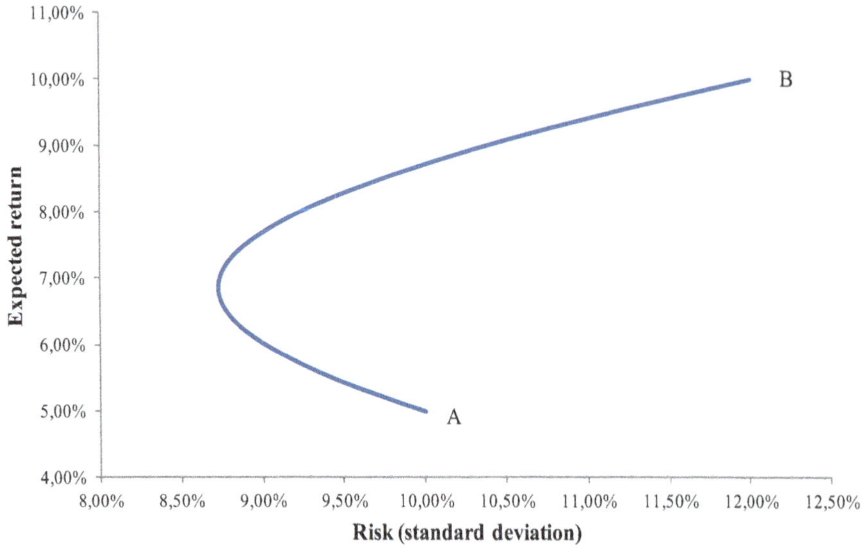

Fig. 5.15 Risk-return profile of a mix of two assets A and B with 30% correlation

The set of investments shown in Fig. 5.14 in the context of the Markowitz model corresponds to all the possible portfolios composed of a set of individual risky assets with given expected returns, positive standard deviations, and the correlations of returns. The typical shape of the universe of all investments is illustrated by the two-asset portfolio analysis shown in Fig. 5.15.

Let us now add a risk-free investment to our set of possible investments. The return of the investment in all scenarios equals the risk-free interest rateR_f, i.e., the standard deviation is zero, $\sigma_f = 0$. The beauty of the model with the standard deviation measuring the risk is that its algebraic property (5.36) and the elementary assumptions made can be used to show the existence of an *Efficient Market Portfolio* and the validity of the well-known *Capital Asset Pricing Model* (CAPM).

Consider a portfolio where λ (proportion of a total unit amount) is invested into a risky asset (or portfolio) X and $1 - \lambda$ into a risk free asset. Then, according to(5.36), its expected return and standard deviation are:

$$ER_\lambda = \lambda ER_X + (1 - \lambda)R_f,$$

$$\sigma_\lambda = \sqrt{\lambda^2 \sigma_X^2} = \lambda \sigma_X.$$

In this case, σ_λ is a linear function of λ, since $\sigma_f = 0$. We also allow $\lambda > 1$, in other words, we allow investments into risky assets to be partially financed by a short position in the risk-free asset, or equivalently borrowing $\lambda - 1$ at the risk-free rate R_f. Therefore, the set of all investments corresponds in the risk-return coordinates to the line starting at the point given by the risk-free asset and going through the point X.

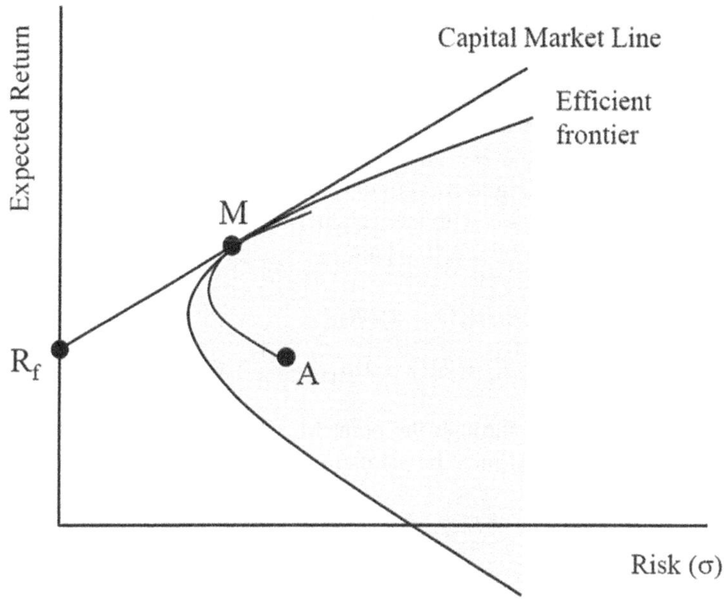

Fig. 5.16 The Markowitz Efficient frontier and Capital Market Line (CML)

An investor maximizing his/her return given a risk level will always seek the line that has the highest slope, i.e., the one that is tangent to the set of all risky investments (see Fig. 5.16). This line is called the *Capital Market Line* (CML). It can be shown (using (5.36)) that the set of risky investments is strictly convex (if the investment universe is non-trivial), and so the tangent line intersects the set of risky investments at one unique point corresponding to a portfolio of risky assets called the *Efficient Market Portfolio* (point M in Fig. 5.16). This portfolio is called efficient, because it maximizes the slope of the line, also called the Sharpe's ratio

$$S = \frac{ER_M - R_f}{\sigma_M}, \tag{5.37}$$

where the numerator can be interpreted as the *risk premium* over the risk-free rate R_f and the denominator measures the risk. Hence, with this interpretation, the portfolio M is the most efficient in terms of the risk-return trade-off.

The Capital Asset Pricing Model (CAPM) claims that if A is an arbitrary risky asset (or portfolio of risky assets), then its risk premium depends only on the systematic risk measured as the linear (regression) dependence of R_A on the efficient market portfolio return R_M. Specifically,

$$ER_A = R_f + \beta(ER_M - R_f), \text{where } \beta = \rho_{AM}\frac{\sigma_A}{\sigma_M}. \tag{5.38}$$

It is easy to see that β defined by (5.38) is exactly the *OLS* (*Ordinary Least Squares*) coefficient from the regression equation $R_A = \alpha + \beta R_M + \varepsilon$. The key message of the CAPM is that $\alpha = R_f(1 - \beta)$. In order to prove (5.38) let us consider the parametric curve starting at *A* and going through *M* corresponding to a mixed investment into the two assets with the weight $1 - \lambda$ put on A and λ put on M. The coordinates parameterized by $\lambda \in [0, 1]$ are

$$\begin{aligned} ER_\lambda &= \lambda ER_M + (1-\lambda)ER_A, \\ \sigma_\lambda &= \sqrt{\lambda^2\sigma_M^2 + 2\lambda(1-\lambda)\rho_{AM}\sigma_M\sigma_A + (1-\lambda)^2\sigma_A^2}. \end{aligned} \tag{5.39}$$

Since the curve passes through the point M, but does not cross the CML (see Fig. 5.16), its gradient at M must be parallel to the CML, i.e.,

$$\frac{\partial ER_\lambda}{\partial \lambda} \Big/ \frac{\partial \sigma_\lambda}{\partial \lambda} = \frac{ER_M - R_f}{\sigma_M}. \tag{5.40}$$

where the partial derivatives are evaluated at $\lambda = 1$. Differentiating (5.39) at $\lambda = 1$, we obtain

$$\frac{\partial ER_\lambda}{\partial \lambda} = ER_M - ER_A \text{ and}$$

$$\left.\frac{\partial \sigma_\lambda}{\partial \lambda}\right|_{\lambda=1} = \frac{1}{2\sigma_\lambda}\left(2\lambda\sigma_M^2 + 2(1-2\lambda)\rho_{AM}\sigma_M\sigma_A + 2\lambda\sigma_A^2\right)\Big|_{\lambda=1} = \sigma_M - \rho_{AM}\sigma_A.$$

Consequently, according to (5.40)

$$\frac{ER_M - ER_A}{\sigma_M - \rho_{AM}\sigma_A} = \frac{ER_M - R_f}{\sigma_M}.$$

Finally, expressing ER_A in terms of the other parameters, we have proved the CAPM:

$$ER_A = R_f + \rho_{AM}\frac{\sigma_A}{\sigma_M}(ER_M - R_f) = R_f + \beta(ER_M - R_f).$$

The CAPM Eq. (5.38) can be also written without the expectation operator in the form

$$R_A = R_f + \beta(R_M - R_f) + \varepsilon_A, \tag{5.41}$$

decomposing the (random) return of A into its risk-free part R_f, the systematic risk part $\beta(R_M - R_f)$, and an idiosyncratic part ε_A. While the systematic part has a positive expected value (there is a risk premium for the systematic risk), the idiosyncratic risk has mean zero and there is no reward for the idiosyncratic risk. According to the CAPM, the Efficient Market Portfolio diversifies the risk most efficiently—any other risky asset or portfolio has some residual idiosyncratic risk, which does not bring any premium.

5.4.1 Economic Capital Allocation

The risk-return optimization problem can also be formulated by defining the risk through VaR or CVaR. In fact, we have seen in Sect. 5.3 that normal VaR and CVaR are proportional to the standard deviation of returns, and so the conclusions of the Markowitz theory would remain unchanged. Things become more complex if the general concept of VaR or CVaR is used to define the risk without the normality of returns assumption. The advantage of VaR is that it can be meaningfully related to the capital available to cover the potential losses.

The role of a bank's capital is two-fold. Firstly, it can be invested into government or similar bonds to bring a risk-free return. The second and most important role of the capital is to provide a cushion against possible losses, so that the bank maintains high credibility in the eyes of its customers, counterparties, and investors. *Economic (or risk) capital* is a bank's own estimate of the capital needed to absorb losses over a certain time horizon and at a certain confidence level. We have already argued in the previous section that the 1-year time horizon and a confidence level corresponding to the target rating probability of default (e.g., AA and 0.1%) are a natural choice. Hence, $VaR(1 \text{ year}, 99.9\%)$ covering all the risks across a bank's activities could reasonably define the total economic capital. However, banks also want to allocate economic capital to individual business units, portfolios, activities, and risk categories in order to calculate and compare *risk-adjusted performance measures* (RAPM) like, for example, RAROC (*risk-adjusted return on capital*) defined by the formula:

$$\text{RAROC} = \frac{\text{Revenues} - \text{Costs} - \text{Expected losses}}{\text{Economic capital}}. \tag{5.42}$$

RAROC, in a way, generalizes Sharpe's ratio (5.37). The costs in the numerator include the cost of funds, i.e., the risk-free rate; hence the numerator represents the risk premium, while the economic capital (VaR) in the denominator measures the risk instead of the standard deviation of returns in the Sharpe ratio. The allocation of economic capital and the evaluation of RAROC for various units help to decide which activities should get a higher priority and which should be reduced or even shut down.

The economic capital can be allocated in a *top-down* or *bottom-up* approach. In the bottom-up approach VaR is estimated for different types of risk and for different units. It is then aggregated on the top level. A simple and conservative approach would be to define the total economic capital as the sum of the individual components

$$E_{sum} = \sum_{i=1}^{n} E_i.$$

This method, however, significantly overestimates the risk, if there are too many components and their correlations are low or negative. Generally, a more sophisticated aggregation should take the diversification effect into account. This can be achieved by adopting certain simplifying assumptions. For example, Hull (2010) recommends a *hybrid approach* estimating the total economic capital (VaR) by

$$E_{total} = \sqrt{\sum_{i,j=1}^{n} \rho_{ij} E_i E_j}, \tag{5.43}$$

where ρ_{ij} is the correlation between the risk components i and j. If the distributions were normal, then this approach would be exactly correct. If the distributions are not normal, then this is only an approximation that tries to take into account possible fat tails of the individual distributions.

If the aggregated capital is estimated according to (5.43), or in a similar way, then there is a positive diversification effect defined by the difference $E_{sum} - E_{total}$. The economic capital E_i allocated to an activity or business unit i must be rescaled, in a way, to $\widetilde{E}_i$ so that the total sum equals to the total diversified capital, $E_{total} = \sum_{i=1}^{n} \widetilde{E}_i$. A simple approach is to allocate the total capital proportionately to *the stand-alone* capital, $\widetilde{E}_i = E_i \frac{E_{total}}{E_{sum}}$. A more sophisticated approach is to allocate the economic capital based on the *incremental* VaR. If a new investment opportunity or activity is analyzed, then the bank should look at its incremental rather than the stand-alone VaR, i.e., at the difference between the VaR of the total portfolio including the new investment and the VaR of the portfolio without the investment. In this approach, $\widetilde{E}_i = E_i^{incr} \frac{E_{total}}{E_{sum}^{incr}}$, where E_i^{incr} stands for the incremental VaR, and E_{sum}^{incr} is the sum of all the incremental VaRs.

5.4.2 Regulatory Principles, the Standardized and Internal Model Approach

It might be tempting to say that there is no need for banking regulation. One could argue that the banks and their shareholders are adequately motivated to maintain high-risk management standards and sufficient capital levels. Unfortunately, not all banks and shareholders stick to these principles. Bank failures cause large losses to the whole economy and to the taxpayers, who in the end often pay the bill. Some banks are simply *too-big-to-fail*, and the governments are, in a sense, obliged to help.

There is a moral hazard when the banks are tempted to take excessive risks relying implicitly on government support in the case of large losses. This is the main reason for banking regulation that sets minimum risk management standards and capital requirements.

Due to the increasing globalization of banks, the supervisory authorities of the G-10 countries (Basel Committee for Banking Supervision, BCBS) approved in 1988 the first Capital Accord entitled "International Convergence of Capital Measurement and Capital Standards" (BCBS 1988) known as Basel I, or just "The Accord." Banks were required to hold a minimum regulatory capital calculated simply as 8% of their *risk-weighted assets*. Equivalently, the *capital adequacy (Cooke) ratio* defined as the eligible capital divided by the risk weighted assets must be at least 8%. The regulators have gradually adopted the Value at Risk philosophy—the required regulatory capital can be viewed as an approximation of Value at Risk in a 1-year horizon and at the 99.9% confidence level. The Basel I regulatory capital is an approximation of the VaR, although very simplified and related essentially only to credit risk, which is considered to be the most important banking risk component. The Accord was amended to incorporate market risks in 1996 (BCBS 1996). The Amendment explicitly allowed for the use of the Value at Risk concept. Later, in 2004, BCBS approved a new regulatory framework called Basel II or The New Capital Accord (BCBS 2004). The New Accord did not change the market risk regulatory capital methodology, but significantly improved the credit risk capital calculation standards, explicitly using the $VaR(1\text{ year}, 99.9\%)$ concept, and introduced the operational risk capital requirement. The 1996 market risk methodology was incorporated into the comprehensive BCBS regulatory document (2006).

The market risk regulatory capital can be calculated in two ways. The *standardized approach* assigns the capital separately to different types of securities and risks (foreign exchange, interest rate, equity, and commodity risk) and multiplies them by certain regulatory coefficients. There is a special treatment of options. It does not consider the internal estimates of the correlations between the assets. More advanced banks may apply for regulatory approval to use an *internal model-based approach*. The internal model should estimate VaR in a 10-day horizon and at the 99% probability level. The VaR could be estimated in the 1-day horizon and rescaled to the 10-day horizon as discussed in Sect. 5.3. The capital requirement is then defined as

$$\text{Regulatory capital} = \max\left(Var_{t-1}, k \times VaR_{avg}\right) + \text{SRC}$$

where $Var_{t-1} = VaR(10\text{ days}, 99\%)$ is the previous day value-at-risk estimate, VaR_{avg} an average of the estimates over the preceding 60 business days, the coefficient $k \geq 3$ is set by the regulator depending on the internal model quality, and where SRC stands for the *specific risk charge* applicable to certain instruments with significant idiosyncratic risk, such as corporate bonds. The regulators set a number of qualitative requirements in order to approve and maintain an internal model, including independent validation, back-testing, and regular stress testing. The internal model does not necessarily reduce the market risk overall capital requirement, but its regulatory approval is a positive sign of high-quality risk management standards, and many banks seek it as a reputational target.

5.4.3 Basel 2.5 and Basel III: A Lesson Learned from the Crisis

The 2007–2008 global financial crisis showed a number of weaknesses in the global financial system and its regulatory framework. The aim of the Basel III reform is to improve the banking sector's ability to absorb shocks arising from financial and economic stress, to improve further risk management and governance, and to strengthen banks' transparency and disclosures. The first part of the new regulatory package, also called "Basel 2.5," was approved as early as 2009 (BCBS 2009a, b) and the second part, called "Basel III," in 2010 (BCBS 2010a, b). Basel 2.5, which revises mainly the market risk framework, was introduced in 2011. Basel III introduces, in particular, stricter capital requirements, new liquidity standards, a leverage ratio limit, etc. (see Fig. 5.17 for an overview). Its implementation was phased in during the period 2013–2019.

According to Basel 2.5, banks are required to calculate, in the internal approach, not only the "ordinary" VaR but also the *stressed* VaR based on historical data covering the financial crisis period. The total capital requirement now is

$$\text{Regulatory capital} = \max\left(Var_{t-1}, k_c \times VaR_{avg}\right) + \max\left(sVar_{t-1}, k_s \times sVaR_{avg}\right) + \text{IRC}$$

where $sVaR$ is the stressed value-at-risk and $k_c, k_s \geq 3$ regulatory constants set by the regulator. In addition, banks are asked to estimate their unexpected credit risk and the Incremental Risk Charge (IRC), of their security trading portfolios. The banks may use, for example, CreditMetrics or a similar approach (see Witzany 2010). The estimation necessarily involves the integration of the market and credit risk modeling techniques.

Notice that the Basel 2.5 market risk capital requirements are more than twice the capital required by Basel II. Not surprisingly, the new regulation also brings much greater regulatory requirements on securitization positions.

The financial crisis has shown that many banks have weak capital bases. Basel III aims to improve the quality of capital by setting higher limits on common equity and Tier 1 capital. It completely eliminates the concept of Tier 3 capital. Moreover, it tries to mitigate the criticized procyclicality effect of the Basel II capital requirement. The market, credit, and operational capital requirements, faithfully reflecting the actual economic conditions, tend to go down during economic expansions and up during recessions (see Fig. 5.18). Therefore, the banks are motivated to give more credit during the expansionary periods and less credit during economic contractions, amplifying the procyclical effect. Basel III does not change the Value at Risk philosophy in calculating the capital requirement, but introduces new conservation and countercyclical capital buffers that are supposed to mitigate the procyclical

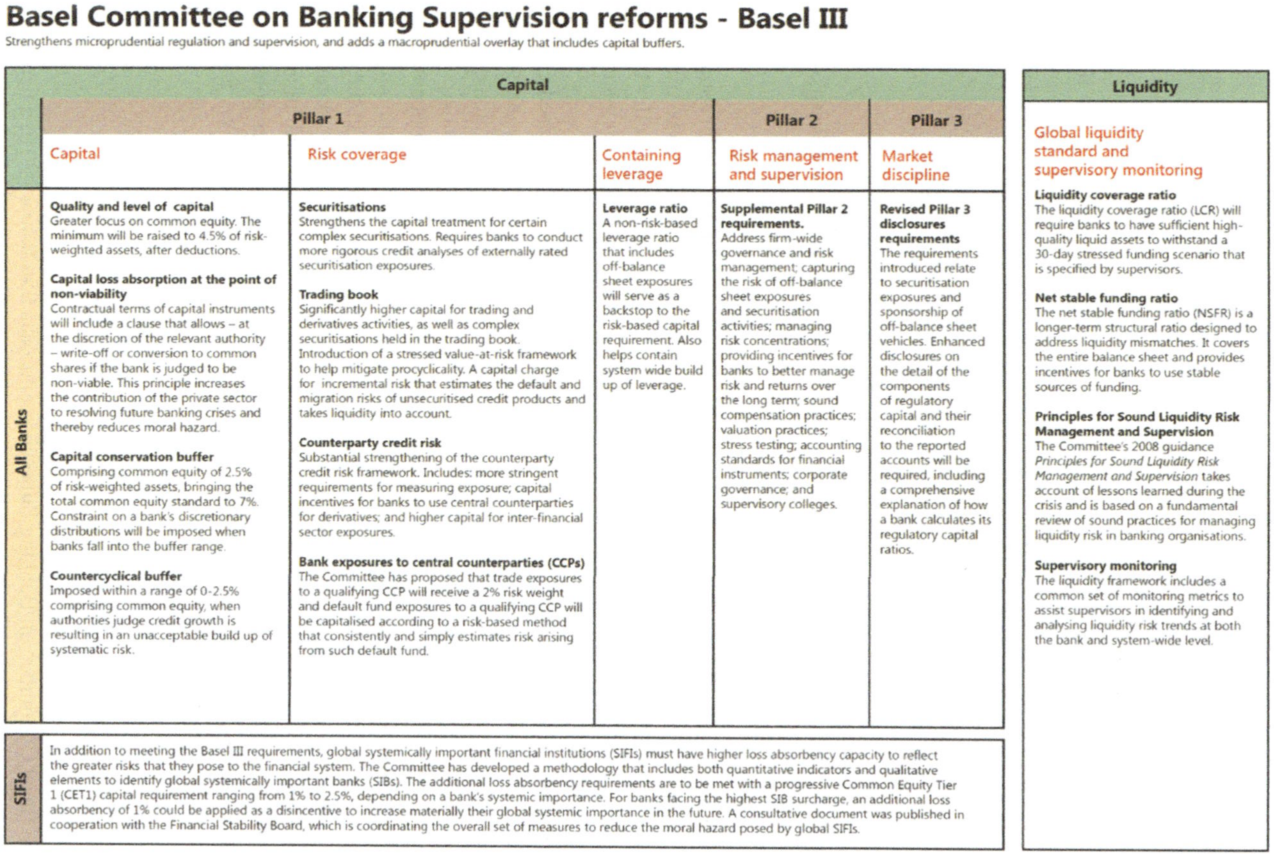

Fig. 5.17 Basel III overview (Source: BCBS)

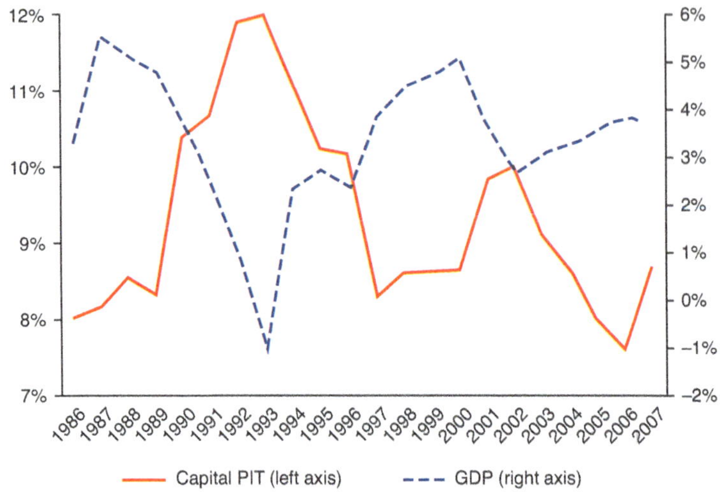

Fig. 5.18 Basel II capital requirement and GDP growth, Spain 1986–2007 (Source: Repullo et al. 2009)

effect. Banks are asked to accumulate (by limiting the payout of dividends) common equity capital in good times (up to 2.5% of risk weighted assets) that can be used to cover the losses in bad times. National regulators may even increase the buffer by an additional 2.5%. Supplementary requirements (1–2.5%) apply to systematically important financial institutions (SIFIs).

During the crisis, many institutions suffered large losses due to increased counterparty credit risk. It turned out that many of them suffered losses not only due to the bankruptcies of their counterparties, but also due to the accounting revaluation of the *Credit Value Adjustment* (CVA) when the credit quality of their counterparties deteriorated (BCBS 2010b). The CVA is a derivative market value component that is supposed to cover the potential losses of a financial derivative. It is minimal or negligible for exchange traded derivatives, but it has become significant in the case of OTC derivatives. Its incorporation into the balance sheet and the P/L statement has become a standard, and it is normally required by auditors and financial regulators. The idea of the new Basel III capital requirement is to cover potential losses due to CVA revaluation. We are going to discuss the concept of CVA and its stressing in the following section.

5.4.4 Fundamental Review of the Trading Book

The Basel Committee on Banking Supervision published in January 2016 a new set of Pillar I standards on "Minimum requirements for market risk" (BCBS 2016a) commonly called the Fundamental Review of the Trading Book (FRTB). The standards were revised in January 2019 and should become effective in January 2022 (BCBS 2019). This new reform aims to strengthen the financial system after the experiences of the Global financial crisis and is sometimes referred to as "Basel IV."

The main change of the Internal Market Model (IMM) Approach compared to Basel II and 2.5 is that the concept of Value at Risk (VaR) over a 10-day horizon is replaced by the Expected Shortfall (ES) over a 1-year horizon, but at the same time differentiates instruments according to their liquidity. This new philosophy also impacts the standardized approach. Other changes include a more granular approval process when IMM is approved at Trading Desk level (not at the Bank level as in the current approach), the Incremental Risk Charge (IRC) is replaced by the Default Risk Charge (DRC), which prevents the double counting of credit migration risk, and there are stricter constraints on the effects of portfolio hedging and diversification, etc.

Focusing on the IMM Approach, the capital charge IMCC is calculated only for modelable risk factors and a Stressed Capital add-on—stresses expected shortfall (SES) is added for the other non-modellable risk factors. The IMCC is based on a weighted average of the constrained and unconstrained expected shortfalls,

$$IMCC = \rho \cdot ES_{\text{unconstrained}} + (1 - \rho) \cdot ES_{\text{constrained}}$$

where $ES_{\text{unconstrained}}$ is the Expected Shortfall for all risk factors allowing diversification, while the constrained Expected Shortfall allows no diversification between the broad regulatory risk classes (interest rate risk, equity risk, foreign exchange risk, commodity risk, and credit spread risk),

$$ES_{\text{constrained}} = ES_{\text{IR}} + ES_{EQ} + ES_{FX} + ES_{CM} + ES_{CR}.$$

The weight ρ is assigned to the internal model by the regulator and its default value is 0.5. It can be interpreted as a regulatory correlation between the asset classes. The Expected Shortfall at the portfolio level or an asset class level is calculated by the bank using the 97.5% confidence level, $T = 10$ days as the base time-horizon with adjustments for instruments having longer liquidity time-horizons,

$$ES(P) = \sqrt{(ES_T(P))^2 + \sum_{j=2}^{5} (ES_T(P,j))^2 \frac{LH_j - LH_{j-1}}{T}}.$$

The instruments are mapped to risk factors, which are assigned regulatory liquidity time-horizons $LH_j = 10, 20, 40, 60, 120$ days for $j = 1, \ldots, 5$; for example, major currencies are assigned $LH_1 = 10$, small cap equity $LH_2 = 20$, and commodities (other than precious metals and energy) $LH_5 = 120$. The Expected Shortfall for a set of positions (portfolio) over the base time horizon is denoted $ES_T(P)$ while $ES_T(P,j)$ denotes the Expected Shortfall over the base horizon T only with respect to the risk factors that have liquidity horizon LH_j or longer. Note that at the level of one factor the formula is consistent with the square-root-of-time VaR/CVaR adjustment rule.

In addition, the ES measure must be calibrated to a period of stress. A bank must identify a 12-month stress period in which the portfolio experiences the largest loss. The observation horizon for determining the most stressful period must stretch back to and include 2007. Since not all factors might have a sufficient price history, the set of factors can be reduced, but it should include at least 75% of the factors in the full model. The final rescaled Expected Shortfall

$$ES = ES_{R,S} \times \frac{ES_{F,C}}{ES_{R,C}}$$

is based on the stressed $ES_{R,\ S}$ with the reduced set of risk factors multiplied by the ratio $ES_{F,\ C}/ES_{R,\ C}$ (floored at 1) between the Expected Shortfall, with the full set of factors and a reduced set of factors calculated on the current (most recent) 12-month observation period.

Regarding the non-modellable risk factors (NMRF) and Capital add-on SES, a bank needs to estimate the capital requirement for each NMRF separately using a stress scenario that is calibrated to be at least as prudent as the Expected Shortfall at the 97.5% confidence level. The stress liquidity horizon must be at least 20 days and the horizon assigned to the modellable risk factor. The individual requirements are then aggregated by the following regulatory formula:

$$SES = \sqrt{\rho^2 \left(\sum SES_{NM,k}\right)^2 + (1-\rho^2) \sum SES_{NM,k}^2} + \sqrt{\sum ISES_{NM,j}^2} + \sqrt{\sum ISES_{NM,i}^2}$$

where $SES_{NK,\ k}$ is the stress scenario capital requirement for (systematic) risk factor k, $ISES_{NK,\ j}$ the capital requirement for the idiosyncratic equity risk factor j, $ISES_{NK,\ i}$ the capital requirement for the idiosyncratic credit spread risk factor i, and the weight (correlation) is $\rho = 0.6$. Practically, it means that the volatilities and the factor stress movements need to be estimated at individual systematic or idiosyncratic factor level, but the correlations are taken care of by the regulatory formula.

Next, the Default Risk Charge (DRC) in the IMM approach should be measured using a VaR (1 year, 99.9%) model reflecting the risk of direct and indirect losses due to obligors' defaults. The default correlations must be based on credit spreads or equity prices covering a period of at least 10 years and including a period of stress. Note that this capital charge (unlike IRC) covers only the unexpected losses due to the events of default and not credit migrations. The price risk related to credit migrations should already be incorporated into IMCC or SES.

Finally, the capital requirement for a trading desk is calculated as DRC plus the maximum of the actual (previous day) $IMCC + SES$ estimation and the 60-day average of *IMCC* scaled by a regulatory multiplier ($m_C \geq 1.5$) plus the 60-day average of *SES*, i.e.,

$$IMA_{G,A} = \max\left\{IMCC_{t-1} + SES_{t-1}; m_C \cdot IMCC_{avg} + SES_{avg}\right\} + DRC.$$

The aggregate capital requirements for market risk are then equal to the sum of capital requirements for IMM approved trading desks and of capital requirements of non-IMM-approved trading desks that have been calculated by the standardized approach. Therefore, we can see that, although the ES (and VaR for DRC) internal models lie at the core of the IMM approach, there are many limitations and standardized calculation steps limiting the flexibility of the approach.

It is not obvious whether the Basel IV (FRTB) Expected Shortfall based calculations will always lead to higher capital requirements compared to the Basel 2.5 stressed Value-at-Risk concept or not. However, it turns out that the banks applying the IMM approach will probably face a significant increase in capital requirements under FRTB. According to several BIS quantitative impact studies (see, e.g., BCBS 2018), the participating banks reported average increased capital requirements of over 50%.

5.4.5 Internal Capital Assessment Process

The idea of Internal Capital Assessment Process (ICAAP) set up by Basel Pillar II (BCBS 2006) is to require banks to assess their overall capital adequacy in relation to their risk profile using their internal models independently of Pillar I calculations, which might be too simplified, especially in the standardized approach, and do not necessarily cover all possible sources of risk. Besides the market, credit, and operational risks covered by Pillar I, the banks should also assess the interest rate risk in the banking book (IRRBB), liquidity risk, reputational and strategic risk, and other risks. The ICAAP results are then used to set-up a capital buffer in excess of the minimum Pillar I requirements.

After Basel II implementation, in particular, following the Financial Crisis, the supervisors started to put gradually more and more emphasis on ICAAP. There is a long list of ICAAP guidance documents that have been issued by BCBS and the national regulators including ECB and EBA (European Banking Authority), see, e.g., ECB (2018) or EBA (2018b). So, on one hand, the risk managers are supposed to have more freedom to assess the risk within ICAAP, but the regulators at the end often tend to check the benchmarks and recommendations formulated in their guidance documents making the ICAAP modeling process sometimes even more challenging in terms of regulatory compliance.

The first decision a bank needs to make is on the confidence level, time-horizon, and modeling concept. The most common choice is the VaR at 99.9% or higher confidence level, or ES on a slightly lower confidence level, and the 1-year modeling time horizon. The ICAAP capital requirement then corresponds to the "*gone concern approach*," since the realization of losses of this magnitude would mean that the bank does not have any additional capital and is "gone." The *Overall Capital Requirement* (OCR) can then be defined as the maximum of Pillar I and of the ICAAP requirements. On the other hand, the "*going concern approach*" aims to estimate the capital buffer protecting the bank from breaching the minimum Pillar I regulatory requirements (allowing the bank to go on without regulatory interventions). The confidence level is then typically lower, e.g., 95%. The result of the ICAAP calculations then directly corresponds to the buffer that should be added to the Pillar I requirement to obtain the OCR.

The ICAAP modeling should be more advanced than the Pillar I calculations, which are subject to many regulatory restrictions. For example, if a bank uses the standardized approach for Pillar I market risk, then it might apply an internal VaR/ES model within ICAAP. The experiences with the ICAAP model can be used later to prepare an application for IMM approval.

The regulator requires banks to perform not only loss distribution VaR/ES modeling but also so-called stress testing. The stress testing exercise calculates losses conditional on several identified stress scenarios. The economic stress scenarios should be severe but plausible, and their specification is not an easy task. The scenarios should also be bank specific, reflecting the specific positions and exposures of the bank, and so general economic stress scenarios cannot always be used. The forward-looking risk factors then need to be forecasted conditionally on the stress scenarios, which are typically defined in terms of one or more macroeconomic indicators. The stress modeling then leads to several stress loss values over different time horizons (typically 1, 2, and 3 years), which should be analyzed by the management and also contribute to the regulatory Overall Capital Requirements. EBA has recently implemented several EU-wide stress tests in order to assess the resilience of financial institutions to adverse market developments based on a standardized stress testing methodology, and plans a new one in 2020 (EBA 2019). The EBA stress test methodology can be used as useful guidance for ICAAP stress testing.

5.5 Counterparty Credit Risk

Counterparty credit risk (CCR) is a specific form of risk arising from the possibility of a default of a counterparty before the maturity of a financial transaction (Witzany 2017). The products subject to counterparty credit risk are mainly OTC derivatives and securities financing transactions (e.g., repo operations). Exchange-traded derivatives or derivatives settled with a centralized counterparty are only theoretically subject to counterparty credit risk. Even an exchange or its clearinghouse may go bankrupt. However, the probability of default of an OTC counterparty is generally much higher.

Let us consider a financial contract, for example, a forward or swap transaction, between a financial institution and a counterparty. If the counterparty defaults at time τ before the contract's maturity T, then the institution may (but does not have to) suffer a loss depending on the legal set-up and the value of the transaction at the time of default. Under standard (ISDA) legal documentation the market value f_τ (from the perspective of the institution) will be frozen at the time of default, remaining cash flows will be canceled, and the amount f_τ will be either payable by the institution to the counterparty, if negative, or by the counterparty to the institution, if positive. Hence, if $f_\tau \leq 0$ then the institution does not suffer any loss due to the counterparty's default, since the transaction is simply closed out before maturity with a cash settlement based on its market value. However, if $f_\tau > 0$ then the institution's exposure with respect to the counterparty will probably be paid back only partially or not at all. If l denotes the fractional Loss Given Default coefficient (LGD), then the loss will be $l \times f_\tau$. The loss could be much higher in a nonstandard legal situation when the institution is obliged to fulfill all the future payments under the contract, but the counterparty's payments become part of bankruptcy claims. On the other hand, the loss on collateralized transactions, where the market value is secured by cash or other high-quality collateral, may be completely covered by the collateral in the case of the counterparty's default. The situation becomes more complicated when all losses and profits with respect to a single defaulting counterparty can be mutually netted.

In any case, in market terms the Credit Valuation Adjustment (CVA) can be defined as the difference between the market value of a transaction with respect to a theoretically risk-free counterparty and the market value of the identical transaction with respect to a specific risky counterparty. In other words, CVA is specified as the theoretical cost of insurance against counterparty credit losses, and therefore, under the risk-neutral valuation principle, CVA can be expressed as the expected discounted loss caused by a possible counterparty default event. Focusing on the standard set-up, we can define it theoretically as

$$CVA = E[\text{discounted CCR loss}] = E[e^{-r\tau} \times \max(f_\tau, 0) \times l \times I(\tau \leq T)], \quad (5.44)$$

where $I(\tau \leq T)$ denotes the indicator function, i.e., it is 1 if default takes place before maturity, $\tau \leq T$, and 0 otherwise, r is the risk-free rate in continuous compounding, and E is the expectation operator under the risk-neutral probability measure. Given CVA, we can adjust the derivative market value as $f_{def} = f_{nd} - CVA$ where f_{nd} is the no-default valuation and f_{def} the market value including the counterparty credit risk. It is clear that the CVA definition involves several uncertainties: the time and probability of default, and the derivative transaction value at the time of default. Even the loss given default and the discount rate, depending on the time of default τ, should be considered stochastic. Moreover, all these variables might be mutually correlated. Therefore, CVA modeling is generally even more challenging than the valuation of complex derivatives without considering the counterparty credit risk.

5.5.1 Expected Exposure and Credit Value Adjustment

There are many approaches to the CVA valuation. The most general method is based on a Monte Carlo simulation of future market factors, exposures, intensities of default, and times to default. On the other hand, a simplified practical "*add-on*" approach tries to express CVA as an expected exposure with respect to the counterparty, multiplied by the probability of counterparty default, and by the LGD. The expected exposure itself is, in general, estimated by a Monte Carlo simulation (that is usually much simpler compared to a full-scale simulation of all the factors). For some products, the expected exposure can be calculated analytically using an option valuation formula.

To come up with a realistic and practical formula, one needs to make a number of simplifying assumptions. First, let us assume independence between the time to default and exposure; moreover, assume that the loss rate l is constant, and that the discount rate $r(t)$ and the counterparty default (forward unconditional) intensity[9] $q(t)$ are deterministic functions of time. Dividing the time interval into subintervals $0 = t_0 < \ldots < t_m = T$, the simplified CVA expression (see, e.g., Gregory 2010) can be written as:

$$CVA \approx \sum_{j=1}^{m} e^{-r(t_j)\times t_j} EE(t_j) \cdot q(t_j) \cdot \Delta t_j, \tag{5.45}$$

where $\Delta t_j = t_j - t_{j-1}$ and the *expected exposure* $EE(t) = E[\max(f_t, 0)]$ is defined independently of the other factors. The expected exposure can be, in general, estimated by the Monte Carlo simulation method. For some products, like forwards or swaps, it can be expressed in an analytic or semianalytic form. On the other hand, if the derivative value f_t itself is not analytical, then we have to cope with a numerically difficult "double" Monte Carlo simulation embedded into another Monte Carlo simulation.

Example 5.14 Let us consider a simple outstanding 1-year forward to buy a non-dividend stock for $K = 101$ with the current spot price $X_0 = 100$. Assume that the risk-free rate is $r = 1\%$, that the counterparty risk-neutral intensity of default $q = 4\%$ and the LGD $l = 60\%$ are constant. The forward value at time t is $f_t = X_t - e^{-r(T-t)}K$, conditional on the spot price X_t, and so the time t expected exposure

$$EE(t) = E\left[\max\left(X_t - e^{-r(T-t)}K, 0\right)\right] = c\left(t, e^{-r(T-t)}K\right)e^{rt}$$

is just the market value of the European call option with maturity t, and the strike price $e^{-r(T-t)}K$, which can be analytically expressed using the Black-Scholes formula and re-discounted forward to the time t (i.e., multiplied by the factor e^{rt}). The adjustment can be then calculated simply by integrating the call option value from the time 0 to T

[9]That is $q(t)\Delta t$ is the probability of default over the period $[t, t + \Delta t)$.

$$CVA = \int_0^T l \times c\left(t, e^{-r(T-t)}K\right) \times q \times dt$$
$$= \sum_{j=1}^{m} l \times q \times c\left(t_j, e^{-r(T-t_j)}K\right) \Delta t_j. \qquad (5.46)$$

Table 5.7 shows the results of the calculation according to (5.46) dividing the time to maturity into 10 subintervals. We can see that in this case, the expected exposure of the forward transaction is an increasing function of time. The resulting CVA presents more than 10 bps of the forward notional amount. We have assumed, for simplicity, that the interest rates and forward default intensities are constant. The calculation would remain essentially the same, if we were given a term structure of interest rates and of the default intensities.

The calculation of the expected exposure for option-like products turns out to be relatively simple since the signs of the market values do not change. An option seller exposure would always be zero (once the premium was paid) while an option buyer exposure would always be just the option value. For a European option, the expected exposure would be, according to the principle of iterated expectations, simply the current option value re-discounted forward to the time t, i.e., $EE(t) = c_0 e^{rt}$.

For an outstanding interest rate swap, the expected exposure would be zero at the start date when there is no market value volatility, and at maturity when all the cash flows have been settled; its maximum would be attained somewhere in the middle between the start date and the maturity (see Fig. 5.19 for an illustration). In fact, the expected exposure turns out to be just a swaption value that can be evaluated using a generalized Black-Scholes formula (see Sect. 6.3).

Further simplification can be achieved by introducing the notion of *expected positive exposure* defined as an average of $EE(t)$ over time:

Table 5.7 Long forward CVA calculation

t	EE(t)	CVA contr.
0.1	1.89	0.005
0.2	2.73	0.007
0.3	3.38	0.008
0.4	3.93	0.009
0.5	4.42	0.011
0.6	4.88	0.012
0.7	5.29	0.013
0.8	5.69	0.014
0.9	6.06	0.014
1.0	6.41	0.015
	CVA	**0.107**

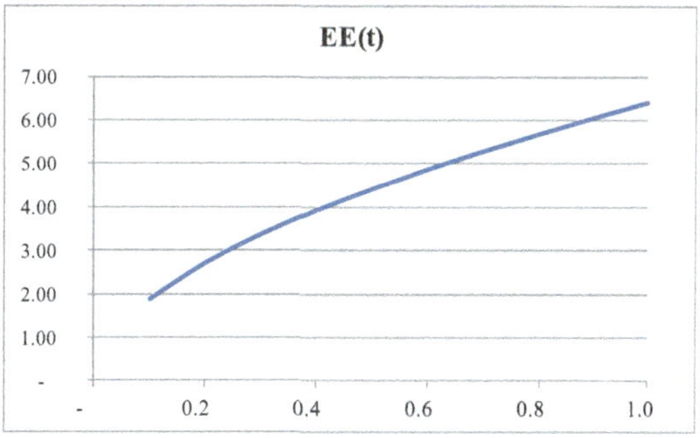

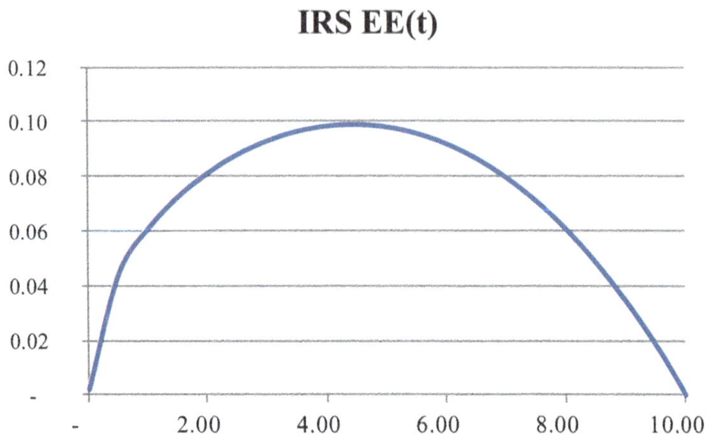

Fig. 5.19 Expected exposure of a 10-year interest rate swap

$$EPE = \frac{1}{T}\int_0^T EE(t)dt \approx \frac{1}{T}\sum_{j=1}^{m} EE(t_j)\cdot \Delta t_j. \tag{5.47}$$

If we assume that the intensity of default is constant or independent of the expected exposure, then CVA can be approximately written as

$$CVA \approx \overline{q} \times l \times EPE \times A(0) \tag{5.48}$$

where $\overline{q}$ is the average intensity of default, and

$$A(0) = \sum_{j=1}^{m} e^{-r(t_j)\times t_j}\Delta t_j \tag{5.49}$$

is the risk-free annuity value.

Example 5.15 The expected positive exposure of the forward contract from Example 5.14 can be estimated as an average of the values from the second column in Table 5.7: $EPE = 4.47$. Similarly, the annuity (5.49) can be estimated as an average of the discount factors applied: $A(0) = 0.99$. Therefore, we have a simple calculation

$$CVA \approx 0.04 \times 0.6 \times 4.47 \times 0.99 \approx 0.107$$

which gives us exactly the same result as in Table 5.7.

The concept of EPE is also useful in a connection with *Credit Default Swap* (CDS) quotations. A credit default swap is an OTC contract where one counterparty (*credit protection seller*) regularly receives a spread X on a notional amount L while the other counterparty (*credit protection buyer*) receives a payoff only if the CDS *reference entity* defaults. The payoff can be settled physically through a bond issued by the reference entity or in cash as $l \times L$ where l is a fixed LGD rate. The payoff takes place and the spread payments are terminated once the default takes place. Let us assume that the payoff is settled in cash and that default can take place only in times $t_1, \ldots, t_m$ when the spread is being paid. If $S(t)$ denotes the reference entity (risk-neutral) survival probability function,[10] then the spread must satisfy the classical insurance equivalence relation:

$$\sum_{j=1}^{m} e^{-r(t_j)\times t_j} \times X \times \Delta t_j \times S(t_j) = \sum_{j=1}^{m} e^{-r(t_j)\times t_j} \times l \times \Delta S(t_j), \tag{5.50}$$

where $\Delta S(t_j) = S(t_{j-1}) - S(t_j)$ is the probability of default during the time interval $[t_{j-1}, t_j)$, i.e. $\Delta S(t_j) = q(t_j)\Delta t_j$ using the concept of default intensity. The left-hand side of (5.50) corresponds to the expected discounted premium income, while the right hand side corresponds to the expected discounted payoff.

Now, if X is a market CDS spread quotation for the reference entity which is equal to our counterparty and for the maturity T, and if we replace $EE(t_j)$ by the constant EPE in (5.45), then

$$\begin{aligned} CVA = EPE \sum_{j=1}^{m} e^{-r(t_j)\times t_j} \times l \times \Delta S(t_j) = \\ = EPE \sum_{j=1}^{m} e^{-r(t_j)\times t_j} \times X \times \Delta t_j S(t_j) = X \times EPE \times A_{CDS}(0). \end{aligned} \tag{5.51}$$

Consequently, CVA can be simply approximated as the CDS spread multiplied by EPE and the risky (CDS) annuity

$$A_{CDS}(0) = \sum_{j=1}^{m} e^{-r(t_j)\times t_j} \times \Delta t_j \times S(t_j).$$

For swap-like products, the CVA is often expressed as a spread X_{CVA} that can be added to the periodic fixed or float payments. Since the payments are terminated in the case of default, we need to solve the equation

$$CVA = X_{CVA} \times L \times \sum_{j=1}^{m} e^{-r(t_j)\times t_j} \times \Delta t_j \times S(t_j), \tag{5.52}$$

where L is the swap notional. Therefore, combining (5.51) and (5.52) we get a nice and simple approximation:

[10] $S(t) = \Pr[\tau > t]$ is defined as the probability that default does not take place until time t.

$$X_{\mathrm{CVA}} \approx X_{\mathrm{CDS}} \frac{EPE}{L}. \tag{5.53}$$

Example 5.16 Let us consider a 10-year 1 billion CZK notional interest rate swap where we should receive the fixed coupon with the expected exposure profile as shown in Fig. 5.19. Without considering the counterparty credit risk, the IRS market rate would be 4%. The expected positive exposure (based on 40% volatility of the IRS rates) is 71 million CZK. Assume that the CDS spread quoted for our counterparty equals 250 bps. According to (5.53), the CVA spread comes out approximately at 18 bps. Consequently, the adjusted fixed payment paid by the counterparty should be 4.18%, which significantly differs from the 4% rate without the CVA adjustment.

The decompositions (5.48) or (5.51) allow for certain stressing of the CVA. Besides the probability of default (intensity of the CDS spread), the exposure can be stressed by introducing the *potential future exposure* (PFE) as a quantile of the average exposure (depending on paths from 0 to T). The Basel II internal approach also introduces the notion of *effective EE*, which is defined as EE, but with the additional requirement of being non-decreasing. *Effective EPE* is then defined as the average of effective EE. This approach is related rather to a netted portfolio of transactions with respect to a counterparty, where maturing transactions are expected to be replaced with new ones.

5.5.2 Collateralization and Netting

The counterparty credit risk of an OTC derivative can be mitigated in a way similar to the margin mechanism applied to exchange-traded products if the OTC counterparties agree to post collateral, usually in cash, covering the derivative market value. In practice, this can be achieved by signing the *Credit Support Annex* (CSA) of the ISDA master agreement. The collateralization can be two-way or one-way. For example, a bank would require a corporate counterparty to post collateral covering the exposure $\max(f_t, 0)$, but no collateral would be sent by the bank if f_t became negative. Two-way collateralization has recently become quite common between banking counterparties.

If the collateral were recalculated and posted on a continuous basis, then the CCR would be virtually eliminated. In practice, there is a standard remargining period or a minimum threshold, and so there is still a residual counterparty risk. For example, if the remargining period were 1 day, then there still should be a CVA corresponding to the 1-day horizon market value variation.

Another way to mitigate CCR with respect to a counterparty is a netting agreement allowing the netting of positive and negative market values in the case of counterparty default, i.e., the exposure needs to be defined and monitored on the portfolio basis as $E(t) = \max(V(t), 0)$ where $V(t) = \sum f_i(t)$ is the sum of the outstanding transactions' market values with respect to the single counterparty. For a more complex portfolio, there is little chance of finding a precise analytical formula for the expected netted exposure. Nevertheless, if we can assume that the portfolio value $V(t)$ depends linearly on the market factors and that it approximately follows the generalized Wiener process, i.e., if

$$V(t) = V(0) + \mu t + \sigma Z\sqrt{t} \quad \text{where} \quad Z{\sim}N(0,1),$$

then we can relatively easily find analytical formulas for *EE*, *EPE*, or *PFE* (Gregory 2010). In particular, if $V(0) = 0$and $\mu = 0$, then $EE(t) = \sigma\sqrt{t}\varphi(0)$ and

$$EPE = \frac{\sigma\varphi(0)}{T}\int_0^T \sqrt{t}dt = \frac{2}{3}\sigma\varphi(0)\sqrt{T} \approx 0.27\sigma\sqrt{T}.$$

For more complex non-linear portfolios, where we cannot assume normality, Monte Carlo simulations need to be used. Notice that the problem is technically like the VaR estimation. We need to model future exposure probability distribution, focusing in this case on the positive rather than on negative values of the portfolio. However, the time dimension makes the task even more challenging.

5.5.3 Wrong-Way Risk

We have already underscored that the simplified CVA formula (5.45) is based on the assumption that the exposure and the event of default are independent. The formula should not be used if there is an evidence of a link between default, or the intensity of default, and of the exposure. The formula can, in fact, be easily fixed by defining the expected exposure conditional on default $EE^*(t) = E[\max(f_t, 0) | t = \tau]$. Then the analogous formula

$$CVA = l\sum_{j=1}^{m} e^{-r(t_j)\times t_j} EE^*\left(t_j\right) \cdot q\left(t_j\right) \cdot \Delta t_j \tag{5.54}$$

becomes consistent with (5.44). In terms of causality, there is usually a common driver of the exposure and the event of default, and so we cannot say that one event causes the other or vice versa. For example, if a company is sensitive to currency devaluation and if the exchange rate impacts the exposure then there could be either a *wrong-way risk*, $EE^*(t) > EE(t)$, with the exposure increasing in the case of devaluation, or a *right-way risk*, $EE^*(t) < EE(t)$, with the exposure going down in the case of revaluation. Both wrong- and right-way risks exist similarly for interest rate products, but in the case of CDS exposures the risk is almost always in the wrong-way direction from the perspective of the credit protection buyer. If the systematic risk increases, e.g., during a financial crisis, then the CDS exposure goes up and the counterparty credit risk generally increases as well.

One way to express the conditional expected exposure analytically, in some cases, is to use the standard Gaussian copula model where the time to default $\tau = S^{-1}(\Phi(X))$ is driven by a normally distributed variable $X{\sim}N(0, 1)$ transformed using the inverse survival function. Similarly, let us assume that the derivative or portfolio value $V(t) = G(Z)$ at time t is driven by a normal variable $Z \sim N(0, 1)$ through an increasing function G. Now, the exposure-default correlation can be captured by a correlation ρ between the normal variables X and Z. Since a high value of X is translated into a low value of τ, as the survival function is decreasing, a positive correlation $\rho > 0$ corresponds to the wrong-way risk while a negative correlation $\rho < 0$ corresponds to the right-way risk. The conditional expected exposure can be then written as

$$EE^*(t) = E\left[G\left(\sqrt{1-\rho^2}Y + \rho X\right)^+ | X = \Phi^{-1}(S(t))\right] = \\ = \int_{-\infty}^{\infty} G\left(\sqrt{1-\rho^2}y + \rho\Phi^{-1}(S(t))\right)^+ \varphi(y)dy, \tag{5.55}$$

decomposing $Z = \sqrt{1-\rho^2}Y + \rho X$ where $X, Y \sim N(0, 1)$ are independent. For example, if $V(t) = \mu t + \sigma\sqrt{t}Z$ just follows the generalized Wiener process then

$$EE^*(t) = \int_{-a/b}^{\infty} (a + by)\varphi(y)dy = a\Phi\left(\frac{a}{b}\right) + b\varphi\left(\frac{a}{b}\right),$$

where $a = \mu t + \rho\sigma\sqrt{t}\Phi^{-1}(S(t))$and $b = \sqrt{1-\rho^2}\sigma\sqrt{t}$.

The principle (5.55) can be used to express the expected loss of a forward or to price an option with the wrong-way risk by a semi-analytic formula (Gregory 2010). In a similar fashion, Černý and Witzany (2013) obtained and tested a semi-analytical formula to price the CVA of interest rate swaps with the wrong-way risk. In general, Monte Carlo simulations for underlying market factors and the counterparty time of default need to be run (see, e.g., Brigo and Pallavicini 2008).

Example 5.17 Let us consider the outstanding 1-year forward from Example 5.14 and let us recalculate the CVA with the wrong-way risk Gaussian correlation $\rho = 0.5$. The stock price is lognormally distributed and can be written as $X_t(Z) = X_0 \exp\left((r - \sigma^2/2)t + \sigma\sqrt{t}Z\right)$. If the (unconditional) intensity of default $q = 4\%$ is constant then the survival function is linear, $S(t) = 1 - qt$. Therefore, in line with (5.55) the conditional expected exposure can be written as:

$$EE^*(t) = \int_{-\infty}^{\infty} \left(X_t\left(\sqrt{1-\rho^2}y + \rho\Phi^{-1}(S(t))\right) - e^{-r(T-t)}K\right)^+ \varphi(y)dy.$$

This integral could be solved analytically, as mentioned above, but we can also evaluate it numerically for $t = 0.1, 0.2, \ldots, 1$ similarly to Table 5.7. The conditional $EE^*(t)$ values, their average $EPE^* = 11.21$, and the corresponding $CVA = 0.282$ come out more than twice as high than without the wrong-way risk!

5.5.4 Bilateral Counterparty Credit Risk

So far, we have assumed that there is a default-free institution that has a counterparty with a positive probability of default. In reality, both counterparties may default. If the institution defaults at time t and if the outstanding derivative market value $f_t < 0$ is negative then the counterparty will lose and the institution will "save" the amount $-l_I \times f_t$ where l_I is the institution's LGD ratio. In this sense, the institution has an "option to default" with a potential positive payoff. *Bilateral credit value adjustment* (BCVA) takes into account both the effects of the potential loss due to counterparty's default and the potential "profit" due to the institution's own default. Let τ_I denote the institution's time of default and let τ_C denote the counterparty's time of default. Then the BCVA can be decomposed into two parts[11]

$$BCVA = CVA_C - CVA_I$$

where CVA_C covers the counterparty's default, provided that the institution has not defaulted sooner

$$CVA_C = E[e^{-r\tau_C} \times \max(f_{\tau_C}, 0) \times l_C \times I(\tau_C \leq T \& \tau_C < \tau_I)],$$

and analogously

$$CVA_I = E[e^{-r\tau_I} \times \max(-f_{\tau_I}, 0) \times l_I \times I(\tau_I \leq T \& \tau_I < \tau_C)].$$

The CVA_I is also sometimes called the *debit value adjustment* (DVA). If we assume that the institution and the counterparty cannot both default before T (or that the probability of this event is negligible) then CVA_C and CVA_I are just the "one-way" CVAs we have discussed so far from the opposite perspectives. The probability of joint default, i.e., $\tau_C, \tau_I \leq T$, is negligible if the defaults are independent and their probabilities are low. Otherwise, this possibility should be taken into account in the context of a correlation model.

The advantage of the concept of BCVA is that it makes derivatives valuation symmetric again. Note that with the "one-way" CVA the institution's market value is $f_I = f_{nd} - CVA_C$ while the counterparty's market value is $f_C = -f_{nd} - CVA_I \neq -f_I$. On the other hand, with the bilateral adjustment we have $f_I = f_{nd} - BCVA$ and $f_C = -f_{nd} + BCVA = -f_I$.

The accounting of CVA, DVA, and BCVA has gradually become a market standard, and according to IFRS 13 (IASB 2011) it has been mandatory since 1/2013. It should be noted that BCVA has the strange effect that the deterioration of the institution's own credit quality is translated into its accounting profit. This can be compared to marking down liabilities due to their own credit downgrade. During the crisis, this has indeed happened, when a large investment bank reported a 1 billion USD profit based on this effect. Such a situation is not acceptable to the

[11]Implicitly assuming that $\Pr[\tau_C = \tau_I] = 0$.

regulators, who prefer conservative accounting principles and tend to require banks to account just the CVA rather than BCVA.

5.5.5 What Are the Risk-Free Rates?

With the emergence of ever-present counterparty credit risk, the markets started to reconsider the classical approach to the construction of risk-free rates from the government bond yields or from interest rate swap rates. Figure 5.20 shows the German government 5-year CDS spread development (approximating the government bond spreads over the risk-free rate), which went as high as 100 bps during the Eurozone crisis 2009–2012. The risk of the German government is considered to be almost minimal compared to other countries, where the CDS spreads went up even to hundreds of basis points.

This market reality leads to a preference for zero-coupon curves built from interest rate swap rates, where the counterparty risk is much smaller (the expected exposure is usually just a fraction of the swap notional amount). A fixed IRS rate is the cost of the rolled-over financing of a reference rated (e.g., AA) bank, where the bank can be periodically replaced by another one, in the case of its credit deterioration. Nevertheless, we have to keep in mind that even a highly rated entity can default during the reset time horizon (3M or 6M), and so short rates do incorporate certain credit spreads that are, consequently, reflected in the IRS rates. Figure 5.21 shows the development of the 3M USD Libor and the US Treasury 3M bill rate (which is close to the ideal risk-free rate). The spread between the two rates went as high as more than 400 bps during the Financial crisis and remains relatively high to this day.

Fig. 5.20 Germany 5Y CDS spread (Source: Thomson Reuters)

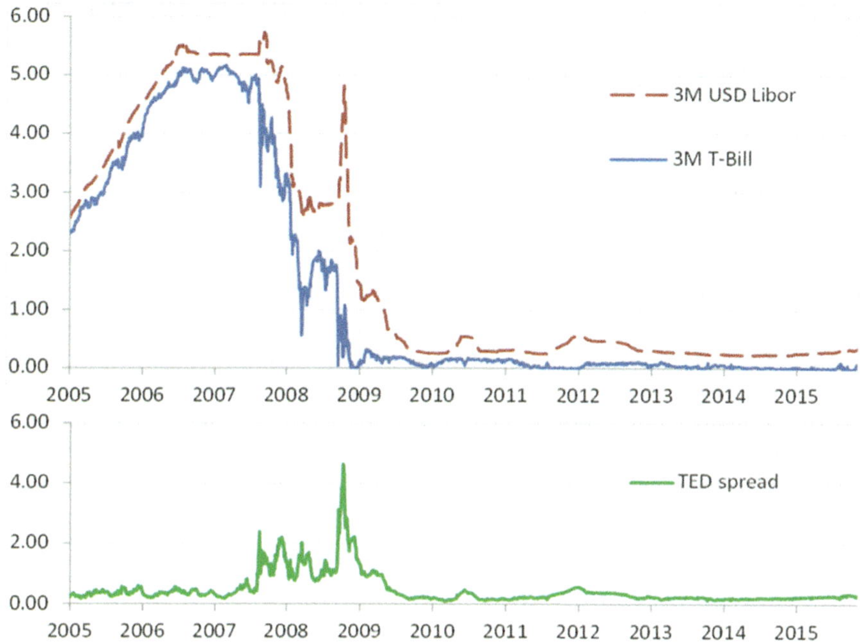

Fig. 5.21 The spread between 3M USD Libor and 3M US T-bill rates (Source: Bloomberg)

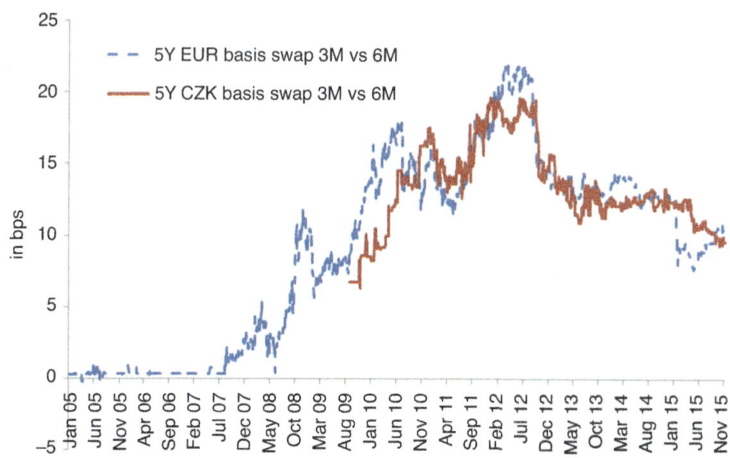

Fig. 5.22 The 5Y basis spreads for 3M/6M Euribor and 3M/6M Pribor (Source: Bloomberg)

Another argument against the IRS rates is the emergence of significant basis swap spreads. Figure 5.22 shows that the markets have recently perceived the 6M Euribor credit premium to be at least 10 bps higher than the 3M Euribor financing premium.

This phenomenon indicates that, in spite of the still relatively high rating of the financial market counterparties, the intensities of default are viewed as non-negligible and increasing in time to maturity. Practically, it means that the zero-coupon curve based on swaps with 6M float payments differs from the curve based on swaps with 3M periodicity, and so on. Using the swaps, there would be a multitude of risk-free curves, completely changing the paradigm of one single risk-free curve. The situation would become even more complicated if we considered cross-currency swap basis spreads (see Baran and Witzany 2013 or Baran 2016).

The current solution generally accepted by the market is to use the *Overnight Index Swap* (OIS) rates in order to construct a risk-free curve, since the 1-day horizon risk (default intensity) is considered minimal. An OIS is similar to plain vanilla IRS with the difference that the float rate is calculated daily (every business day) as an official overnight rate (the *Effective Federal Funds Rate* for USD, the *Euro Overnight Index Average* (EONIA) for EUR, the *Sterling Overnight Index Average* (SONIA) for GBP, etc.). To simplify the settlement, the O/N rates are compounded over longer periods, e.g., 3 months or 1 year,

$$r = \left(\prod_{t=1}^{n_b}\left(1 + \frac{r_t \times n_t}{360}\right) - 1\right)\frac{360}{n},$$

where r_t is the O/N rate, n_t is the number of calendar days in the O/N period (normally 1 day, but it can also be 3 days for a weekend), n_b is the number of business days in the compounding period, and n is the total number of days. OIS swaps tend to be short lived, often only 3 months or less (see, e.g., Fig. 5.23). For swaps of 1 year or less, there is only a single payment at maturity defined as the difference between the fixed rate and the compounded OIS rate. For longer swaps, the payments are made quarterly or annually.

A fixed OIS rate again represents the cost of the rolled-over financing of a reference rated bank, where the bank can be replaced by another one in the case of credit deterioration. In this case, the roll-over periods are only one business day and the probability of a full default of a reference rated entity (e.g., AA rated) during one single business day is considered almost negligible (there is usually a sequence of downgrades before an AA bank ends up in default). Indeed, Fig. 5.24 shows that the spread between the 3M USD Libor and the 3M OIS rates approximates well to the TED[12] spread (compare Fig. 5.21). Therefore, if there is a liquid market for the OIS swaps (which is the case for USD and EUR) then the rates can be used to construct an almost ideal risk-free zero-coupon curve. For currencies with a limited or no OIS market, an approximation needs to be used (see Baran and Witzany 2013, for a discussion).

Figure 5.23 shows an example of USD and CZK OIS quotes. The USD OIS market turns out to be quite liquid, with quotes available also for longer maturities,

[12]The TED spread is the difference between the interest rates on the interbank loans (USD Libor) and on short-term US government debt ("T-bills").

USDOIS — USD OIS FOCUS — LINKED DISPLAYS — MONEY

<USDVIEW> <USDIRS> <USDOIS> <USDFRA> <0#ED:> <USD/1>

	USD	FedFund Effectve			DEALING	
1M	1.9850	2.0050	BROKER	GFX	30AUG19	23:00
2M	1.9050	1.9250	BROKER	GFX	30AUG19	22:51
3M	1.8400	1.8600	BROKER	GFX	30AUG19	23:00
4M	1.7770	1.7970	BROKER	GFX	30AUG19	22:59
5M	1.7290	1.7490	BROKER	GFX	30AUG19	23:00
6M	1.6800	1.7000	BROKER	GFX	30AUG19	23:00
7M	1.6330	1.6530	BROKER	GFX	30AUG19	23:00
8M	1.5980	1.6180	BROKER	GFX	30AUG19	23:00
9M	1.5610	1.5810	BROKER	GFX	30AUG19	23:00
10M	1.5250	1.5450	BROKER	GFX	30AUG19	23:00
11M	1.4960	1.5160	BROKER	GFX	30AUG19	23:00
1Y	1.4660	1.4860	BROKER	GFX	30AUG19	22:59
2Y	1.2090	1.2290	BROKER	GFX	30AUG19	23:00

31.8.2019 11:55:02 (GMT+1)

CZKOIS — CZK OIS FOCUS — LINKED DISPLAYS — OIS/1

<CZKVIEW> <CZKIRS> <CZKOIS> <CZKFRA> <CZK/1>

	CZK	CZK EONIA			DEALING	
SW	1.4900	1.8900	KOMERCNI BKA	PRG KOBB	31AUG19	12:00
2W	1.4900	1.8900	KOMERCNI BKA	PRG KOBB	31AUG19	12:00
1M	1.4900	1.8900	KOMERCNI BKA	PRG KOBB	31AUG19	12:00
2M	1.4900	1.8900	KOMERCNI BKA	PRG KOBB	31AUG19	12:00
3M	1.4900	1.8900	KOMERCNI BKA	PRG KOBB	31AUG19	12:00
6M	1.4630	1.8630	KOMERCNI BKA	PRG KOBB	31AUG19	12:00
9M	1.4000	1.8000	KOMERCNI BKA	PRG KOBB	31AUG19	12:00
1Y	1.4100	1.6600	42 FIN SRV	PRG FTFS	31AUG19	12:00

31.8.2019 12:00:31 (GMT+1)

Fig. 5.23 USD and CZK OIS quotations (Source: Thomson Reuters, 31.8.2019)

but the CZK OIS market, unfortunately, provides quotes for only up to 1 year. The quotes can be compared with Libor, Pribor, or IRS rates, which, indeed, turn out to be at least 20–50 bps higher than the respective OIS rates

The "true" risk-free rates are a key input to derivative valuation models (Hull and White 2012a). Derivatives should be first valued assuming there are no defaults, and then adjusted for the (bilateral) counterparty credit risk $f = f_{nd} - BCVA$. It is not

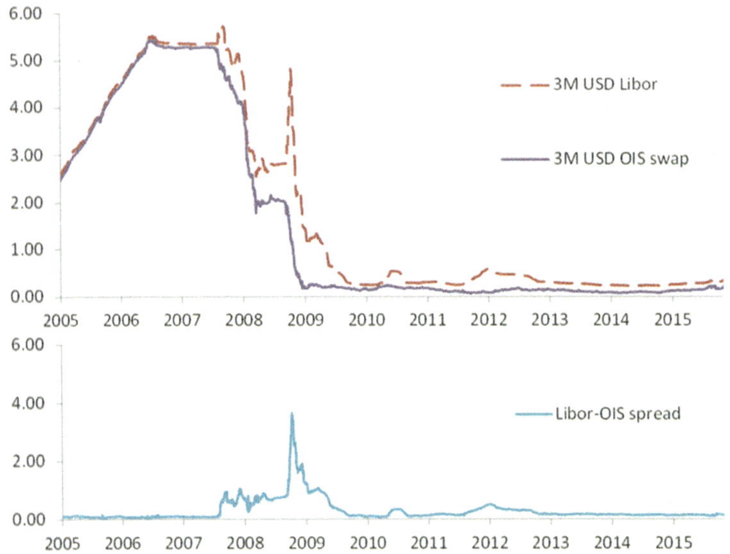

Fig. 5.24 The spread between 3M USD Libor and 3M USD OIS rates (Source: Bloomberg)

correct to use the interest rate reflecting the counterparty's cost of financing as an input to the derivative valuation model. For example, for a long European call option, it is correct to write $f = e^{-r_C T}\widehat{E}\left[(S_T - K)^+\right]$, where r_C is the counterparty's cost of financing (assuming no credit risk of the institution), but the risk-neutral expected value is based on the assumption that the drift of S_t is the risk-free rate r_0, not the risky rate $r_C > r_0$. In this particular case, the no-default option value can, in fact, be adjusted as $f = e^{-(r_C - r_0)T} f_{nd}$ using the counterparty's credit spread. Nevertheless, this formula is not applicable to derivatives like forwards or swaps, where the cash flows can be both positive and negative.

The same discussion applies to collateralized derivative transactions. If there is a two-way continuous collateralization, then the discounting rate can be effectively replaced by the rate r_M accrued on the margin account, yet the drift of the asset prices is still the risk-free rate (in the risk-neutral world). In this case, we can use the multiplicative adjustment $f = e^{-(r_M - r_0)T} f_{nd}$ for all types of derivatives. If $r_M > r_0$ then the collateral interest brings an advantage to the counterparty receiving a positive payoff, and vice versa if $r_M < r_0$.

5.5.6 Basel III CVA Capital Charge

The accounting of CVA and DVA has improved both the awareness and management of the counterparty credit risk. For large institutions, the total P/L impact of the seemingly small adjustments might be in billions of USD. The new BCVA accounting practice has, at the same time, highlighted the existence of a new price risk category: movements of counterparty credit risk (and of own credit risk), and of exposures, causing changes in the total BCVA with a positive or negative impact on the institution's P/L. BCBS (2010b) notes that while the Basel II standard covers the risk of counterparty default, it does not address the similar CVA risk, which was, during the financial crisis, a greater source of losses than those arising from outright defaults. Therefore, the Basel III regulation (BCBS 2010b) introduces a new CVA capital charge to cover the risk of mark-to-market losses on CVA for the OTC derivatives. Banks are not required to calculate this capital charge for transactions with respect to a *central counterparty* (CCP) or securities financing transactions. Note that the regulator does not, in contrast to IFRS 13, consider the bilateral BCVA, which could increase the total market value due to institutions' own credit deterioration.

In principle, the regulator wants banks, in the internal market model (IMM) approach, to calculate the VaR of their portfolio market value by incorporating the credit value adjustments

$$MV = \sum_i \left(f_{nd,i} - CVA_i \right)$$

depending not only on the market factors but also on the credit spreads of their individual counterparties. More precisely speaking, the CVA capital charge should be calculated separately from the pure market risk capital charge, i.e., considering counterparty risk as the only source of losses (but at the same time simulating future exposures depending on the underlying market factors). The regulatory CVA formula according to BCBS (2010b) is essentially the same as (5.45), i.e.,

$$CVA = LGD_{MKT} \sum_{i=1}^{T} \max\left(0, \exp\left(-\frac{s_{i-1} \cdot t_{i-1}}{LGD_{MKT}}\right) - \exp\left(-\frac{s_i \cdot t_i}{LGD_{MKT}}\right)\right) \times \left(\frac{EE_{i-1} \cdot D_{i+1} + EE_i \cdot D_i}{2}\right)$$

where s_i is the credit spread corresponding to maturity t_i. The banks should use market CDS spreads whenever available. If a CDS spread is not available, the banks are supposed to use a proxy. Similarly, LGD_{MKT} is the loss given default of the counterparty based on a market instrument, if available, and on a proxy otherwise. Finally, EE_i is the expected exposure and D_i the discount factor, both corresponding to the revaluation time t_i. Note that the term

$$\exp\left(-\frac{s_i \cdot t_i}{LGD_{MKT}}\right)$$

estimates the survival probability until time t_i implied by the loss given default LGD_{MKT} and the spread s_i (with an implicit assumption of a constant default hazard rate). Therefore, the first factor in the summation approximates the probability of the counterparty's default between t_{i-1} and t_i, while the second factor stands for an

Table 5.8 Regulatory CVA weights (BCBS 2010b, par. 104)

Rating	Weight w_i (%)
AAA	0.7
AA	0.7
A	0.8
BBB	1.0
BB	2.0
B	3
CCC	10.0

average discounted exposure during this time interval. According to the regulation, any internal model, based on direct CVA revaluation, or on spread sensitivities, must use this formula. Likewise, the Basel III market risk capital charge and the CVA capital charge must be calculated as the sum of non-stressed and stressed VaR components. The non-stressed VaR components use the expected exposures and spread variations corresponding to normal market conditions, while the stressed VaR must use stressed future exposures and spread variations corresponding to a financial crisis.

Looking at the Internal Market Model (IMM) requirements, it is not surprising that the majority of banks will opt, regarding the total capital charge, for a relatively simple standardized formula where the capital charge is calculated as a percentage of each exposure depending on the counterparty's rating, transaction maturity, and possible (counterparty) credit risk protection. The individual capital charges are then aggregated at the portfolio level:

$$K = 2.33\sqrt{\begin{array}{l}\left(\sum_i 0.5w_i\left(M_i EAD_i^{\text{total}} - M_i^{\text{hedge}}B_i\right) - \sum_{\text{ind}} w_{\text{ind}}M_{\text{ind}}B_{\text{ind}}\right)^2 + \\ +\sum_i 0.75w_i^2\left(M_i EAD_i^{\text{total}} - M_i^{\text{hedge}}B_i\right)^2\end{array}} \tag{5.55}$$

where w_i is the weight corresponding to counterparty i based on its rating and Table 5.8 (if there is no external rating, then the bank must, subject to regulatory approval, map its internal rating to an external rating); EAD_i^{total} is the total exposure with respect to the counterparty i (with or without netting) including the effect of collateral; M_i is the effective weighted maturity (duration) of the transaction with respect to the counterparty i; B_i is the notional of purchased single-name CDS hedges (with respect to the reference entity i) with maturity M_i^{hedge}; and, finally, B_{ind} is the notional of a purchased index CDS hedge with maturity M_{ind} and weight w_{ind}.

To get some intuition about the formula, note that w_i looks like a regulatory estimate of a "standard" annualized CVA change as a percentage of the duration and hedge adjusted exposure $M_i EAD_i^{total} - M_i^{hedge}B_i$. Therefore, the first part under the square root of (5.55) corresponds to an estimate of the total portfolio CVA standard deviation assuming that the counterparties are perfectly correlated, but allowing for a systematic CDS index hedge, while the second part under the square root corresponds to the standard deviation estimate assuming the counterparties are independent. The two estimates correspond to decomposition into single-factor systematic and idiosyncratic factors with the weight $\rho^2 = 0.5^2 = 0.25$ for the systematic variance and the weight $1 - \rho^2 = 0.75$ for the idiosyncratic variance. Therefore, the square root stands for a conservative portfolio CVA standard deviation estimate and the multiplier 2.33 is just the standard normal 99% quantile. So, indeed, the result of (5.55) appears to estimate the 99% CVA VaR in a 1-year horizon (for a more detailed explanation see, for example, Pykhtin 2012).

5.5.7 FVA, KVA, MVA, and Other XVAs

Besides the CVA, DVA, and BCVA, which have become more-or-less standard accounting and regulatory concepts, there are other more controversial valuation adjustments like FVA, LVA, KVA, or MVA, denoted collectively as XVAs (Gregory 2015).

To explain the reasoning behind the *Funding Value Adjustment* (FVA), let us consider as an example a non-collateralized derivative position with a positive market value, e.g., a long option. The derivative position is an asset that has been acquired by paying a premium, and the cost is funded internally by a rate corresponding to the institution's cost of financing. On the other hand, the market value accrues only the risk-free rate used for discounting the derivatives' expected cash flow, e.g., the OIS rate. Consequently, there is a difference between the institution's financing rate and the risk-free (OIS) rate, which calls for an additional valuation adjustment. The same, but positive, effect is applicable if a derivative market value is negative, i.e., a liability. In this case, the interest cost of the liability equals the OIS rate, but the interest revenue is the funding rate. The funding cost spread FS_C and the funding benefit spread FS_B could be generally different, so we should calculate separately the two $FVA = FCA + FBA$ components, i.e., the *Funding Cost Adjustment* (FCA) and the *Funding Benefit Adjustment* (FBA). Mathematically,

$$FVA = E\left[\int_0^T e^{-rt} \max(f_t, 0) \cdot FS_C(t) \cdot S(t) dt\right] - E\left[\int_0^T e^{-rt} \max(-f_t, 0) \cdot FS_B(t) \cdot S(t) dt\right],$$

where FS_C is the funding (cost) spread on the asset side, FS_B is the funding (benefit) spread on the liability side, and $S(t)$ the transaction survival probability, i.e., the joint survival probability for both counterparties.

The adjusted derivative market value should now be

$$f_0^{adj} = f_0^{nd} - CVA + DVA - FVA.$$

It should be noted that the concept of FVA remains controversial. According to Hull, White (2012b), FVA should not be reflected in the valuation of derivatives—the standard valuation argument says that derivative value should be equal to the risk-neutral expectation of the cash-flows discounted by the risk-free rate, not by any entity-specific funding rate. Moreover, if the funding costs are applied, then the valuation of derivatives will be asymmetric and arbitrage opportunities might exist. In spite of the continuing academic discussion, according to a market survey (Gregory 2015), the majority of large global banks do account for FVA with a total impact in billions of USD.

Regarding practical calculations, the usual assumption is that the exposure and funding spreads are independent. Moreover, depending on close-out assumptions (Gregory 2015), the survival probability can be neglected. Then, after a standard discretization, we have the following relatively simple formula:

$$FVA \approx \sum_{j=1}^{m} e^{-rt_j} EE(t_j) \cdot \bar{FS}_C(t_j)\Delta t_j - \sum_{j=i}^{m} e^{-rt_j} ENE(t_j) \cdot \bar{FS}_B(t_j)\Delta t_j \quad (5.56)$$

where $ENE(t)$ is the expected negative exposure at time t, $\overline{FS}_C(t)$ the expected (or forward) funding cost spread, and $\overline{FS}_B(t)$ the expected funding benefit spread.

However, defining the FVA component more precisely, it has become obvious that there is an overlap with the concept of DVA. The institution's funding spread should be theoretically equal to the product of the default probability and the institution's LGD, i.e. $\overline{FS}_B(t_j) = l \cdot q(t_j)$, hence

$$FBA \approx -\sum_{j=i}^{m} e^{-rt_j} ENE(t_j) \cdot \bar{FS}_B(t_j)\Delta t_j = -l\sum_{j=1}^{m} e^{-rt_j} ENE(t_j) \cdot q(t_j)\Delta t_j$$
$$= DVA.$$

Note that this argument does not apply to CVA and FCA, since CVA uses the counterparty's default probabilities, while the FCA institution's funding spread depends on its own default probabilities. One simple solution to this finding is to apply either FBA or DVA, but not both at the same time, i.e., the total adjustment would be either the *"CVA and symmetric funding"* $CVA + FVA$ or *"bilateral CVA and asymmetric funding"* $BCVA + FCA$. According to Gregory (2015), the market practice prefers the asymmetric funding approach.

Another consistent solution to the CVA/FVA overlap is to define the funding benefit spread only as the liquidity spread above the risk-free rate, plus a standard institution's credit spread, i.e., we should define rather the liquidity (benefit) spread rather as the difference between the real market spread and the theoretical spread: $LS_B(t_j) = MS_B(t_j) - l \cdot q(t_j)$. Some authors (Gregory 2015; Hull and White 2012b, c, or Crépey et al. 2014) argue that the same principle applies to FCA on the asset side as well. The reasoning is that the asset quality influences an institution's overall credit quality, and so the cost of funding should depend on the asset credit risk. For example, an investment into treasury bonds has no (negative) contribution to the institution's credit quality, and so should be funded by the risk-free rate, possibly plus an institution-specific liquidity spread. Therefore, we obtain an alternative definition of FVA, denoted rather as *LVA—Liquidity Valuation Adjustment* (Crépey et al. 2014), where the funding spread is replaced by the liquidity spread $LS_B(t_j) = MS_B(t_j) - l \cdot q(t_j)$ only. In market terms, the liquidity spread can be estimated as the difference between the institution's bond yield spread and the CDS spread. It seems that the concept of LVA resolves the academic controversy, and is accepted, for example, by Hull and White (2014).

Another XVA to be mentioned is the *Margin Valuation Adjustment—MVA*. While FVA is related to uncollateralized transactions, MVA arises due to standard overcollateralization requirements, mainly due to initial margin posting. Organized exchange derivatives positions or OTC positions cleared by a CCP (central counterparty) involve initial margin and maintenance margin management. The requirement is defined not to collateralize exactly the actual counterparty's losses, but in order to cover potential losses over a short-time horizon (1–10 days) and on a high confidence level (e.g., 99%). The excess margin balance usually exists even for bilateral transactions depending on the margin mechanism. In any case, the margin balance earns a return R_{IM}, which will at most be equal to the OIS rate, and its financing at the same time represents a funding cost FC that will be larger than the OIS rate. Therefore, the MVA can be defined mathematically as follows:

$$MVA = E\left[\int_0^T e^{-rt} IM(t) \cdot (FC(t) - R_{IM}(t)) \cdot S(t)dt\right]$$

where $IM(t)$ is the margin balance and $S(t)$ the (joint) survival probability. As above, the calculations can be simplified by discretizing the time interval and assuming independence between the margin balance and the funding spread.

Finally, let us look at the *Capital Valuation Adjustment—KVA* , which is supposed to reflect the cost of regulatory capital related to derivative transactions. Traditionally, there has been a market risk capital requirement, calculated for different product portfolios and market factors. The capital requirement is significant for proprietary positions, but can be neglected for back-to-back operations. Another component is the classical default CCR capital requirement defined as the RWA times 8%. The derivative exposures can be calculated according to Basel II rules by several methods: CEM—current exposure method, SM—standardized exposure method, and IMM—internal model method. The new Basel III component is the CVA charge described above. The total capital requirement $C(t)$ that needs to be calculated on a (derivative) portfolio level again represents a cost, in this case the cost of capital $CC(t)$ The cost of capital should be considered rather as the spread between the required capital return and the risk-free rate (since the capital per se can be invested into risk-free assets). Thus, the KVA mathematical definition is

$$KVA = E\left[\int_{0}^{T} e^{-rt} C(t) \cdot CC(t) \cdot S(t)dt\right].$$

As usual, the formula can be discretized and based on the expected capital requirements $EC(t)$ and other relevant expected future parameters.

While CVA and (more or less) FVA have become accounting standards, MVA and KVA are used rather for reporting and monitoring. There is an ongoing debate regarding the consequences of their accounting, possible side-effects, and overlaps. In any case, the debate around FVA mentioned above should be resolved first, before the institutions start to account for the other XVAs.

6 Stochastic Interest Rates and the Standard Market Model

In Chap. 4, we have presented the Black-Scholes option valuation model that has become a market standard. However, the model has several limiting assumptions including the one saying that the instantaneous interest rates are constant. But the interest rates are not constant at all in real financial markets. First, there is a term structure of interest rates, 1-year interest rates are usually greater than over-night interest rates, and 5-year interest rates are usually greater than 1-year interest rates. Evaluating a 1-year European stock option, which interest rate should be used? Recall that a European derivative value was obtained as the present value of the expected payoff. Hence, in the Black-Scholes formula, one could propose to use the 1-year interest rate instead of the presumably constant short rate. It turns out that this simple modification, leading to the so-called *Standard Market Model*, is correct, but in order to prove it we need to generalize significantly the risk-neutral valuation framework.

Moreover, we also want to value interest rate derivatives, and in that case the assumption of constant interest rates would be a complete nonsense. Again, it turns out that the expected payoff of a European interest rate option can be discounted by the interest rate corresponding to the option's maturity. The Standard Market Model, in addition, takes the simplifying assumption that the future interest rate defining the payoff has a lognormal distribution. We will argue, in Chap. 8, that the lognormal distribution does not reflect interest rate behavior optimally, yet the formula has become a useful market valuation tool.

© The Editor(s) (if applicable) and The Author(s), under exclusive license to Springer Nature Switzerland AG 2020

J. Witzany, *Derivatives*, Springer Texts in Business and Economics,
https://doi.org/10.1007/978-3-030-51751-9_6

6.1 Price of Risk, Numeraire, and the Equivalent Martingale Measure

We have shown in Sect. 4.2 that under the risk-neutral measure the discounted derivative value f_t/e^{rt} is a martingale, i.e., $f_t/e^{rt} = E[f_s/e^{rs}|\Omega_t]$ for any $t < s$. Because the instantaneous interest rate r was assumed to be constant, it follows that the derivative value at $t = 0$ can be expressed as the discounted expected payoff

$$f_0 = E\left[\frac{f_T}{e^{rT}}|0\right] = e^{-rT}E[f_T]. \tag{6.1}$$

Now, let us drop the assumption of constant instantaneous interest rate, admit that interest rates have a term structure, and, moreover, may, like other asset prices, behave stochastically. In Sect. 3.1, we have defined the time t market value of a riskless zero-coupon bond $P(t, T)$ paying one currency unit at maturity T. The factor is used to discount a fixed cash flow C_T, i.e., the present value $c_t = P(t, T)C_T$ must be the time t market value of the cash flow in an arbitrage-free market. Given a European type derivative paying f_T at time T, in an analogy to (6.1), we would like to express its market value as

$$f_0 = P(0, T)E_T[f_T]. \tag{6.2}$$

The time zero conditional expectation on the right-hand side of (6.2) cannot be taken under the real-world probability measure, because f_T is not a fixed cash flow and investors normally require an extra profit margin for the risk taken. However, (6.2) would be achieved if we found a probability measure that makes the fraction $f_t/P(t, T)$a martingale. Since $P(T, T) = 1$, indeed we would have

$$\frac{f_0}{P(0,T)} = E_T\left[\frac{f_T}{P(T,T)}\right] = E_T[f_T] \tag{6.3}$$

proving (6.2). The conditional expectation on the right-hand side of (6.2) and (6.3) has the subscript T in order to emphasize that the expectation is taken under a specific measure that makes $f_t/P(t,T)$ a martingale. The measure is called *forward risk-neutral with respect to* $P(t, T)$. We will show that there is such a probability measure.

Sometimes we need to measure values in new units. A positive price process $g(t)$ that is used as such a new "currency," is called a *numeraire*. In order to show the existence of a forward risk-neutral measure we will be more general; we will show that given a numeraire g there is a probability measure P_g so that $f(t)/g(t)$ is a martingale for any derivative security f (that has the same sources of uncertainty as g).

An important example of a numeraire, besides the zero-coupon bond, is the *money-market account* $g(t)$ defined as the balance on an account where we deposit one unit at time zero, i.e., $g(0) = 1$. In addition, it accrues the instantaneous (stochastic) interest rate $r(t)$ at time t. Hence,

$$dg = r(t)dt, \text{ and}$$

$$g(t) = \exp\left(\int_0^t r(s)ds\right), \tag{6.4}$$

given the history of interest rates $r(s)$ until time t. Note that $g(t)$ is unknown from the perspective of time 0 since we do not know the future of the instantaneous interest rates. The integral on the right-hand side of (6.4) represents a random variable, because its value depends on the path $\omega \in \Omega_t$ where Ω is the space (hyperfinite binomial tree), on which the price processes are modeled. Under a measure that is forward risk-neutral with respect to the money market account, a derivative value can be expressed as the expected discounted payoff

$$f_0 = \frac{f_0}{g_0} = E_g\left[\frac{f(T)}{g(T)}\right] = E_g\left[\exp\left(-\int_0^T r(s)ds\right) f_T\right]. \tag{6.5}$$

The measure is a generalization of the classical risk-neutral measure we used in (6.1) assuming constant interest rates. Further on, it will be also called *the traditional risk-neutral measure*. However, in this case, the discount factor unfortunately cannot be taken out of the expectation, since the interest rate $r(t)$ is stochastic. The expression (6.5) can be useful, for example, in a Monte Carlo simulation, where we simulate possible paths for $r(t)$, possible payoffs f_T, and calculate the empirical mean of the discounted simulated values.

6.1.1 Market Price of Risk

In the real world, investors require a higher than risk-free return on risky investments. Only the risk-free investments, like the money-market account, bear the risk-free rate. On the other hand, if two investment securities depend on the same source of risk, so that one can be hedged by the other, then the risk margins will be proportional to the sensitivities with respect to the common source of risk. Recall (Sect. 5.4) that the Capital Asset Pricing Model (CAPM) claims that a stock risk premium equals the beta times the general market risk premium, where the beta is the sensitivity with respect to the unique source of systematic risk.

Let us formalize the risk-return principle in the context of stochastic modeling. Assume that f_1, f_2 are two price processes of two (derivative) securities with the same source of risk. By this we mean that the derivatives depend on the same underlying asset and their price process can be described by the following stochastic differential equations with the same Wiener process increments dz:

$$\begin{aligned} df_1 &= \mu_1 f_1 dt + \sigma_1 f_1 dz, \\ df_2 &= \mu_2 f_2 dt + \sigma_2 f_2 dz. \end{aligned} \tag{6.6}$$

At this point, we make another important generalization allowing the drift and volatility to be stochastic as well, depending on the same source of uncertainty. Recall the one-step binomial tree, where the up and down scenarios are parameterized by the drift and volatility known at the beginning of the one-step period, and the only source of randomness is the increment dz. The drift and volatility remain constant during the one-step period, but may change for the next period. Of course, the drift and volatility stochastic processes must satisfy certain regularity conditions so that (6.6) makes sense not only for the elementary infinitesimal time step, but also for any infinitesimal dt (a multiple of the elementary time step).

Looking at the Eq. (6.6), it is clear that the two securities can be combined appropriately in order to eliminate dz over the period from t to $t + dt$. The argument is like the delta hedging approach when setting up the Black-Scholes partial differential equation. Let us invest into $\sigma_2 f_2$ of the security "f_1" and short $\sigma_1 f_1$ of the security "f_2." Hence, the value of the portfolio is

$$\Pi = (\sigma_2 f_2) f_1 - (\sigma_1 f_1) f_2,$$

where the weights $(\sigma_2 f_2)$ and $(\sigma_1 f_1)$ are fixed at time t, while f_1 and f_2 are the securities' values changing over time. It follows from (6.6) and by our choice of the portfolio weights that over the period from t to $t + dt$:

$$\begin{aligned} d\Pi &= (\sigma_2 f_2) df_1 - (\sigma_1 f_1) df_2 = \\ &= (\sigma_2 f_2)(\mu_1 f_1 dt + \sigma_1 f_1 dz) - (\sigma_1 f_1)(\mu_2 f_2 dt + \sigma_2 f_2 dz) = \\ &= (\sigma_2 \mu_1 - \sigma_1 \mu_2) f_1 f_2 dt. \end{aligned} \tag{6.7}$$

Since dz is canceled out, Π is a risk-less portfolio (over the period from t to $t + dt$) and its return must be r, i.e.,

$$d\Pi = r\Pi dt = r(\sigma_2 - \sigma_1) f_1 f_2 dt. \tag{6.8}$$

Putting the Eqs. (6.7) and (6.8) together we get

$$\begin{aligned} (\sigma_2 \mu_1 - \sigma_1 \mu_2) f_1 f_2 dt &= r(\sigma_2 - \sigma_1) f_1 f_2 dt, \\ \sigma_2 \mu_1 - \sigma_1 \mu_2 &= r(\sigma_2 - \sigma_1), \text{ and so} \\ \frac{\mu_1 - r}{\sigma_1} &= \frac{\mu_2 - r}{\sigma_2}. \end{aligned} \tag{6.9}$$

Therefore, we can define $\lambda = \frac{\mu - r}{\sigma}$ to be *the price of risk* separately from the security depending on the given source risk. The price of risk is an analogy of the Sharpe's ratio (5.37). Note that the expected return over the period from t to $t + dt$

can be decomposed into the risk-free rate and the price of risk times the volatility, in particular $\mu_1 = r + \lambda\sigma_1$ and $\mu_2 = r + \lambda\sigma_2$. Since r, μ, and σ are allowed to be stochastic (change over the hyperfinite binomial tree), we have to keep in mind that λ may be stochastic as well.

It is important to generalize the result for more sources of uncertainty. For example, evaluating a stock option and taking into account stochastic interest rates, we have at least two sources of uncertainty (the stock price and the interest rate) that play a role in the model. Following Hull (2018), let the sources of uncertainty be, in general, represented by certain underlying asset prices or basic market variables $\theta_1, \ldots, \theta_k$ following the processes

$$d\theta_i = m_i\theta_i dt + s_i\theta_i dz_i \text{ for } i = 1, \ldots, k \tag{6.10}$$

where dz_i are Wiener processes (on different hyperfinite binomial trees). For each process θ_i, there is a particular price of risk $\lambda_i = \frac{m_i - r}{s_i}$ where r is the risk-free interest rate. Consequently,

$$m_i = r + \lambda_i s_i.$$

Now, if f is the price of a derivative security depending on the k sources of uncertainty following the stochastic differential equation

$$df = \mu f dt + \sum_{i=1}^{k} \sigma_i f dz_i, \tag{6.11}$$

then we will show that

$$\mu = r + \sum_{i=1}^{k} \lambda_i \sigma_i. \tag{6.12}$$

Generalizing the one-dimensional argument, let $w_i = \frac{\sigma_i f}{s_i \theta_i}$ be the weights calculated at time t and set up a portfolio Π investing into one derivative security f and shorting w_i of the security θ_i. By construction, we eliminate all the sources of uncertainty dz_i, and so again Π is a riskless portfolio. Analogously to (6.9) we get

$$\left(\mu f - \sum_{i=1}^{k} w_i m_i \theta_i\right) dt = d\Pi = r\Pi dt = r\left(f - \sum_{i=1}^{k} w_i \theta_i\right) dt,$$

and rearranging the equation

$$\mu - r = \sum_{i=1}^{k} \frac{\sigma_i}{s_i}(m_i - r) = \sum_{i=1}^{k} \lambda_i \sigma_i$$

proving (6.12). Note that this argument holds only locally, over the period from t to $t + dt$. The riskless portfolio must be rebalanced for other periods since the interest rate, the drifts, volatilities, and prices of risk may change over time.

6.1.2 Equivalent Martingale Measure

In Sect. 4.2, we have argued that, for a time interval from t to $t + dt$, by changing appropriately the one-step binomial tree branching probability from the real-world probability p to the risk-neutral probability q, we assure not only that the drift of f is r, $df = rfdt + \sigma fdz$, but also that $dS = rSdt + \sigma Sdz$ for the underlying asset price process. Hence, while the real-world price of risk λ was positive, the price of risk has become zero in the risk-neutral world. This argument can be generalized to change λ to an arbitrary target price of risk λ' just by appropriately changing the branching probability.

Let θ be the only underlying variable corresponding to a single source of risk following the process $d\theta = m\theta dt + s\theta dz$. According to Witzany (2008), let us consider a one-step tree in the (hyperfinite) binomial tree of the process corresponding to the time interval from t to $t + dt$, where θ_t either go up to $\theta_t u$ with the probability p, or down to $\theta_t d$ with the probability $1 - p$ (Fig. 6.1).

The stochastic differential equation $d\theta = m\theta dt + s\theta dz$ means that

$$E_p[\theta_{t+dt}] = \theta_t(pu + (1-p)d) = \theta_t(1 + mdt) \text{ and} \tag{6.13}$$

$$\text{var}_p[\theta_{t+dt}] = \theta_t^2\left(pu^2 + (a-p)d^2\right) - \theta_t^2(1 + mdt)^2 = \theta_t^2 s^2 dt. \tag{6.14}$$

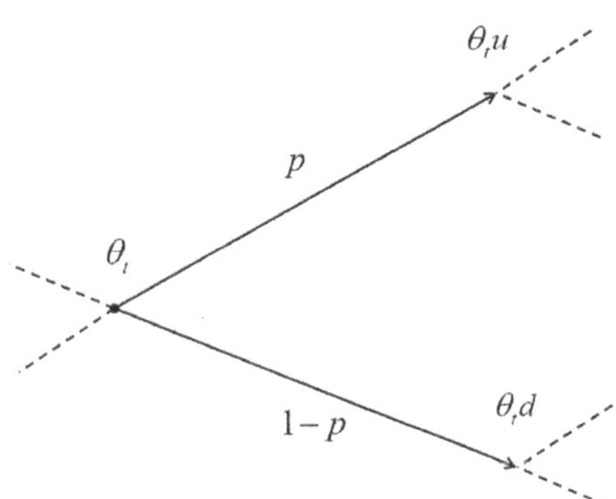

Fig. 6.1 One step of a binomial tree for a time interval from t to $t + dt$

It is enough if the equations hold up to an error of a lower order (infinitesimally smaller) than dt. Let, for simplicity, $u = 1 + s\sqrt{dt}$ and $d = 1 - s\sqrt{dt}$ be in the form[1] proposed by Cox, Ross, Rubinstein (1979) and $p = 0.5 + \gamma\sqrt{dt}$ where γ is a parameter. It is easy to see that the Eq. (6.13) is equivalent to $1 + 2\gamma s dt = 1 + m dt$, and so $\gamma = \frac{m}{2s}$. Since m is a finite number and s is a finite positive (non-infinitesimal), γ must be finite as well. It turns out, evaluating algebraically the variance, that by setting u and d as above, (6.14) will always hold (up to an infinitesimal error of a lower order) provided γ is finite:

$$\mathrm{var}_p[\theta_{t+dt}] = \theta_t^2(s^2 dt + 4\gamma^2 s^2 dt^2) \approx \theta_t^2 s^2 dt$$

as $4\gamma^2 s^2 dt^2$ is negligible (infinitely smaller) with respect to dt. This is a key result: by changing the parameter γ, i.e., the probability p, we adjust the mean of the stochastic differential equation, but the volatility s remains unchanged!

Now, we are ready to change the actual price of risk $\lambda = \frac{m-r}{s}$ to an arbitrary λ'. To achieve that, we just need to change $m = r + \lambda s$ to $m' = r + \lambda' s = m + (\lambda' - \lambda)s$, and so the probability p to the adjusted probability:

$$q = 0.5 + \frac{m'}{2s}\sqrt{dt} = 0.5 + \frac{m + (\lambda' - \lambda)s}{2s}\sqrt{dt} = p + \frac{\lambda' - \lambda}{2}\sqrt{dt}. \tag{6.15}$$

Note that by changing the probability we also change the mean of the original Wiener process dz by αdt where $\alpha = \lambda' - \lambda$. In order to define a Wiener process with respect to the changed probability measure, we have to adjust dz by the same amount, i.e., set

$$\widetilde{dz} = dz - \alpha dt.$$

Since r, m, and s may change over the hyperfinite binomial tree, the branching probabilities must be changed accordingly for all one-step subtrees. Specifically, if $\omega \in \Omega_n$, where n is the number of steps in the hyperfinite binomial tree, let $p(\omega \restriction k)$ denote the probability of the move from the predecessor to the node $(\omega \restriction k)$ (i.e., in the context of Fig. 6.1, it is either p if the move is up or $1 - p$ if the move is down). The original probability measure is then defined by

$$P(\omega) = \prod_{k=1}^{n} p(\omega \restriction k) \quad \text{for} \quad \omega \in \Omega_n, \tag{6.16}$$

while the changed measure is

[1] Generally, the coefficients u and d should have the form $1 + adt \pm s\sqrt{dt}$.

$$Q(\omega) = \prod_{k=1}^{n} q(\omega \restriction k), \tag{6.17}$$

where $q(\omega \restriction k)$ corresponds to the probabilities changed according to (6.15).

Now, we are ready to prove the existence of an equivalent martingale measure given a numeraire $g(t)$ with one source of uncertainty, i.e., following the process of the form

$$dg = \mu_g g dt + \sigma_g g dz. \tag{6.18}$$

Let P be the original probability measure (on the binomial tree where g lives) and λ be the corresponding price of risk, i.e., $\mu_g = r + \lambda\sigma_g$. According to the general result above we can change the price of risk λ to the new price of risk $\lambda' = \sigma_g$. It turns out that the resulting probability measure Q is the *forward risk-neutral measure with respect to the numeraire* g, i.e., the *equivalent martingale measure*. With respect to the measure Q, the drift of g is changed to $r + \sigma_g^2$ and the volatility remains unchanged, i.e., under Q it follows the stochastic differential equation[2]

$$dg = \left(r + \sigma_g^2\right) g dt + \sigma_g g dz. \tag{6.19}$$

Let f be another derivative security price process with the same source of uncertainty (on the same binomial tree) following, with respect to Q, the stochastic differential equation

$$df = (r + \sigma_g \sigma_f) f dt + \sigma_f f dz. \tag{6.20}$$

We want to prove that f/g is a martingale, i.e., that the drift of the process is zero. Let us apply the Ito's lemma to $\ln g$ and $\ln f$:

$$d(\ln g) = \left(r + \sigma_g^2 - \sigma_g^2/2\right) dt + \sigma_g dz = \left(r + \sigma_g^2/2\right) dt + \sigma_g dz,$$

$$d(\ln f) = \left(r + \sigma_g \sigma_f - \sigma_f^2/2\right) dt + \sigma_f dz.$$

Subtracting the two equations we get

[2]Note that dz in (6.19) viewed as a Wiener process on the hyperfinite binomial tree is not, in fact, the same as the dz in (6.18). Its mean has to be adjusted in order to compensate the changed probabilities and the change of drift.

$$d(\ln f/g) = d(\ln f - \ln g) = \left(\sigma_g\sigma_f - \sigma_f^2/2 - \sigma_g^2/2\right)dt + \left(\sigma_f - \sigma_g\right)dz =$$

$$= \frac{\left(\sigma_f - \sigma_g\right)^2}{2}dt + \left(\sigma_f - \sigma_g\right)dz.$$

The Ito's lemma can be used again to show finally that $f/g = \exp(\ln f/g)$ is a martingale:

$$d(f/g) = \left(\sigma_f - \sigma_g\right)(f/g)dz.$$

6.1.3 Girsanov's Theorem

The construction of the equivalent martingale probability measure Q defined by (6.17) from the original measure P given by (6.16) allows us to prove the Girsanov's well-known theorem. The theorem yields an analytical expression for the ratio between the new and the original elementary probability

$$\frac{dQ}{dP} = \frac{Q(\omega)}{P(\omega)} = \prod_{k=1}^{n} \frac{q(\omega\upharpoonright k)}{p(\omega\upharpoonright k)} = \prod_{k=1}^{n}\left(1 + \alpha(\omega\upharpoonright k)\omega_s(k)\sqrt{dt}\right). \tag{6.21}$$

Here $q(\omega\upharpoonright k) = p(\omega\upharpoonright k)\left(1 + \alpha(\omega\upharpoonright k)\omega_s(k)\sqrt{dt}\right)$, $\omega_s(k) = \pm 1$ depends on $\omega(k)$ going up or down, and the change of probability splitting parameter

$$\alpha(\omega\upharpoonright k) = \frac{\lambda' - \lambda}{2p(\omega\upharpoonright k)} \approx \lambda' - \lambda$$

also generally depends on the path. The Wiener process with respect to P has to be adjusted with respect to this price of risk increment, $d\tilde{z} = dz - \alpha dt$, to get a Wiener process with respect to Q.

In order to simplify the right-hand side of (6.21), let us take its logarithm and expand it using the second-order Taylor expansion of $\ln(1 + x) \approx x - x^2/2$ disregarding the third and higher order terms:

$$\ln\left[\prod_{k=1}^{n}\left(1 + \alpha(\omega\upharpoonright k)\omega_s(k)\sqrt{dt}\right)\right] = \sum_{k-1}^{n} \ln\left(1 + \alpha(\omega\upharpoonright k)\omega_s(k)\sqrt{dt}\right)$$
$$\approx \sum_{k=1}^{n}\left(\alpha(\omega\upharpoonright k)\omega_s(k)\sqrt{dt} - \alpha(\omega\upharpoonright k)^2 dt/2\right) \approx \int_0^T \alpha dz - \frac{1}{2}\int_0^T \alpha^2 dt. \tag{6.22}$$

We could disregard the terms of order $dt^{3/2}$ or higher because their summation in (6.22) would be still infinitesimal. The first integral on the right-hand side of (6.22) is a stochastic integral of α over the path ω, while the second is an ordinary integral, but still depends on the path. Consequently, we have proved the Girsanov result:

$$\frac{dQ}{dP} = \exp\left(\int_0^T \alpha dz - \frac{1}{2}\int_0^T \alpha^2 dt\right). \tag{6.23}$$

In particular, if α is a constant, then

$$\frac{dQ}{dP} = \exp\left(\alpha\int_0^T dz - \frac{1}{2}\alpha^2\int_0^T dt\right) = \exp(\alpha z(T) - \alpha^2 T/2). \tag{6.24}$$

6.1.4 Ito's Lemma and the Equivalent Martingale Measure for Several Factors

To generalize the equivalent martingale measure result to more sources of uncertainty (factors), we first need to state and prove two multivariate versions of the Ito's lemma.

Let $x_1, \ldots, x_m$ be m Ito processes following the stochastic differential equations

$$dx_i = a_i dt + b_i dz_i \text{ for } i = 1, .., m, \tag{6.25}$$

and let G be a differentiable function of m variables. Then $G = G(x_1, \ldots, x_m)$ is an Ito process satisfying the stochastic differential equation

$$dG = \left(\sum_{i=1}^m \frac{\partial G}{\partial x_i} a_i + \frac{\partial G}{\partial t} + \frac{1}{2}\sum_{i,j=1}^m \frac{\partial^2 G}{\partial x_i \partial x_j} b_i b_j \rho_{ij}\right) dt + \sum_{i=1}^m \frac{\partial G}{\partial x_i} b_i dz_i \tag{6.26}$$

where ρ_{ij} is the correlation between dz_i and dz_j. This m-dimensional form of the Ito's lemma in particular shows that if f is the price of a derivative security depending on several underlying factors then it follows an equation of the form (6.11).

To prove (6.26), we use the multivariate version of the Taylor's lemma for $G = G(x_1, \ldots, x_m)$ where $x_1, \ldots, x_m$ stand for real variables, not stochastic processes,

$$dG = \sum_{i=1}^m \frac{\partial G}{\partial x_i} dx_i + \frac{\partial G}{\partial t} dt + \frac{1}{2}\sum_{i,j=1}^m \frac{\partial^2 G}{\partial x_i \partial x_j} dx_i dx_j + \sum_{i=1}^m \frac{\partial^2 G}{\partial x_i \partial t} dx_i dt + \cdots. \tag{6.27}$$

To get (6.26) plug-in dx_i from (6.25) into (6.27), collect the terms of dt and dz_i, and analyze the terms of the form $dz_i dt$, and $dz_i dz_j$. First, the terms $dz_i dt$, and higher powers of dz_i and dt, can be neglected, because, on the level of an elementary time step δt, the value $\pm\delta t\sqrt{\delta t}$ is of higher order (infinitely smaller) with respect to δt. Regarding the term $dz_i dz_j$, we have to take into account a possible correlation between the two processes (on the two binomial trees). The value of the term $\delta z_i \delta z_j$ on the level of an elementary time step δt is $\pm\delta t$. The mean of the product of the two variables is

$$E\left[\delta z_i \delta z_j\right] = \rho_{ij}\sqrt{E[\delta z_i^2]E\left[\delta z_j^2\right]} = \rho_{ij}\delta t$$

by the definition of correlation between δz_i and δz_j. The variance of $\delta z_i \delta z_j$ is of the order of δt^2 and so infinitely smaller than δt. Now, if we consider an infinitesimal $dt = K\delta t$, although an infinite multiple of δt, then the mean of $dz_i dz_j$ still is $\rho_{ij} K\delta t = \rho_{ij} dt$ and the variance is of the order of $K\delta t^2$, and so the standard deviation (of the order $\sqrt{K}\delta t = dt/\sqrt{K}$) is infinitely smaller with respect to dt. Therefore $dz_i dz_j$ can be treated as the constant $\rho_{ij} dt$ explaining the relevant terms in (6.26).

Another simpler alternative of Ito's lemma is the following: Let $G = G(x)$ be a function of one variable and x be an Ito multivariate process following the equation $dx = adt + \sum_{i=1}^{m} b_i dz_i$. Then $G = G(x)$ is again an Ito process and we have

$$dG = \left(\frac{\partial G}{\partial x}a + \frac{\partial G}{\partial t} + \frac{1}{2}\frac{\partial^2 G}{\partial^2 x}\sum_{i,j=1}^{m} b_i b_j \rho_{ij}\right)dt + \frac{\partial G}{\partial x}\sum_{i=1}^{m} b_i dz_i. \tag{6.28}$$

In this case, we just use the univariate Taylor expansion and argue analogously as above. Now, we are ready to show the existence of an equivalent martingale measure with respect to a given numeraire g depending on several factors

$$dg = \mu_g g dt + \sum_{i=1}^{m} \sigma_{g,i} g dz_i. \tag{6.29}$$

Without loss of generality, we can assume that the factors dz_i are independent,[3] i.e., the correlations $\rho_{ij} = 0$. Then the price of risk can be changed independently for each dz_i to $\lambda_i = \sigma_{g,i}$. With respect to the resulting probability measure Q the stochastic differential equation for g can be written as

[3]Generally, the factors that are not collinear can always be orthogonalized, i.e., there are linear combinations dw_i of $dz_1, \ldots, dz_m$ so that their variance is dt and mutual correlations are 0. The linear combinations applied to the underlying securities can be interpreted as the prices of new assets defined by (dynamically) rebalanced portfolios. Therefore, dz_i can be expressed as linear combinations of $dw_1, \ldots, dw_m$, and substituting the combinations into (6.29) we have dg expressed in terms of uncorrelated sources of risk dw_i.

$$dg = \left(r + \sum_{i=1}^{m} \sigma_{g,i}^2\right) g dt + \sum_{i=1}^{m} \sigma_{g,i} g dz_i.$$

For any other derivative security f depending on the same sources of risk, we have

$$df = \left(r + \sum_{i=1}^{m} \sigma_{g,i}\sigma_{f,i}\right) f dt + \sum_{i=1}^{m} \sigma_{f,i} f dz_i.$$

Finally, we may apply, as usual, the log-transformation and proceed as in the univariate case

$$d(\ln(f/g)) = -\frac{1}{2}\sum_{i=1}^{m} (\sigma_{f,i} - \sigma_{g,i})^2 dt + \sum_{i=1}^{m} (\sigma_{f,i} - \sigma_{g,i}) dz_i.$$

Therefore, applying the backward exponential transformation and the Ito's lemma we get

$$d(f/g) = \sum_{i=1}^{m} (\sigma_{f,i} - \sigma_{g,i})(f/g) dz_i,$$

Proving, in general, that f/g is a martingale.

6.1.5 Change of Numeraire

To complete our theoretical foundations for various advanced applications in derivatives pricing, let us consider the situation when we start with a numeraire g and then, for some reason, decide to change it to a new numeraire h (with the same m sources of uncertainty). The forward risk-neutral measure Q with respect to g is given by the prices of risk $\lambda_i = \sigma_{g,i}$, while the forward risk-neutral measure Q' with respect to h is given by the prices of risk $\lambda_i' = \sigma_{h,i}$. Let f be the price of a market variable with the same sources of uncertainty. We know that the volatilities $\sigma_{f,i}$ of f remain unchanged when the measure is changed from Q to Q', but there is a change of drift from μ_f to $\mu'_f = \mu_f + \alpha_f$. The *Change of Numeraire Theorem* claims that the change of drift can be simply expressed as

$$\alpha_f = \rho \sigma_f \sigma_w, \tag{6.30}$$

where σ_f is the total volatility of f, σ_w the total volatility of the *numeraire ratio* $w = h/g$, and ρ the instantaneous correlation between df and dw. To interpret the notions of *total volatility* and correlation, and to show (6.30), we express the drift of f first with respect to the measure Q:

$$\mu_f = r + \sum_{i=1}^{m} \sigma_{g,i}\sigma_{f,i},$$

and then with respect to Q':

$$\mu'_f = r + \sum_{i=1}^{m} \sigma_{h,i}\sigma_{f,i}.$$

Consequently,

$$\alpha_f = \mu'_f - \mu_f = \sum_{i=1}^{m} (\sigma_{h,i} - \sigma_{g,i})\sigma_{f,i} = \sum_{i=1}^{m} \sigma_{w,i}\sigma_{f,i}. \tag{6.31}$$

It follows from the multivariate Ito's lemma applied to $\ln w = \ln h - \ln g$ that indeed $\sigma_{w,i} = \sigma_{h,i} - \sigma_{g,i}$ and so the equality on the right-hand side of (6.31) holds. It remains to show that

$$\sum_{i=1}^{m} \sigma_{w,i}\sigma_{f,i} = \rho\sigma_f\sigma_w = \frac{\mathrm{cov}(df, dw)}{fw\, dt}. \tag{6.32}$$

The processes f and w follow (with respect to Q) the equations

$$df = \mu_f f dt + \sum_{i=1}^{m} \sigma_{f,i} f dz_i \text{ and}$$

$$dw = \mu_w w dt + \sum_{i=1}^{m} \sigma_{w,i} w dz_i.$$

We assume that the factors dz_i are independent, and so

$$\mathrm{cov}(df, dw) = E\left[\left(\sum_{i=1}^{m} \sigma_{f,i} f dz_i\right)\left(\sum_{j=1}^{m} \sigma_{w,j} w dz_j\right)\right] = \left(\sum_{i=1}^{m} \sigma_{f,i}\sigma_{w,i}\right) fw dt$$

proving (6.32). Therefore, by the total volatility of f we mean the annualized standard deviation of df/f, i.e., $\sigma_f = \sqrt{\sum_{i=1}^{m} \sigma^2_{f,i}}$, and similarly for w. Note that the volatility and correlation definitions do not depend on the measure, since the volatilities of the factors remain unchanged. Hence, it is correct in applications to estimate the total volatilities and correlations of f and w from the observed historical returns (i.e., under the real-world probability measure).

6.2 Black's Standard Market Model

Now, finally, we are ready to generalize the Black-Scholes model to the context where interest rates are allowed to be stochastic. Let us consider a European call option with maturity T, strike price K, on an underlying asset with the price S following the general process

$$dS = \mu_S S dt + \sum_{i=1}^{m} \sigma_{S,i} S dz_i \tag{6.33}$$

where the drift μ_S, the interest rate r, and possibly even the factor volatilities $\sigma_{S,\,i}$ are allowed to be stochastic. The call option value f depends, according to the Ito's lemma, on the same sources of uncertainty, its value at time T is given by the payoff $f = \max(S_T - K, 0)$, and we want to find today's value f_0. The seemingly extremely complex valuation problem can be simplified by introducing the forward measure Q, which is risk-neutral with respect to the zero-coupon bond numeraire $g(t) = P(t, T)$. The ratio f/g becomes a martingale, in particular

$$\frac{f_0}{P(0,T)} = E_T\left[\frac{f_T}{P(T,T)}\right] = E_T[f_T], \text{ and so}$$

$$f_0 = P(0,T)E_T[\max(S_T - k, 0)],$$

where $E_T[\cdot]$ denotes the expectation with respect to the forward risk-neutral measure. Since $P(0, T)$ is known at time 0 (it is the zero-coupon bond market value), it remains only to evaluate the expected payoff $E_T[\max(S_T - K, 0)]$. At this point, the *Standard Market Model* makes an important simplifying assumption on S_T being lognormal (with respect to Q) with the mean $E_T[S_T]$ and log-variance $\sigma^2 T$, i.e.,

$$\ln S_T \sim N(\ln E[S_T] - \sigma^2 T/2, \sigma^2 T). \tag{6.34}$$

Although this assumption is approximately valid in most situations, we should keep in mind that it is inconsistent with the general model (6.33), where the stochastic drift and volatilities might cause S_T to have various other distributions. Nevertheless, given (6.34), exactly as in Sect. 4.2, we can integrate the call option payoff weighted by the lognormal density and obtain

$$E_T[\max(S_T - K, 0)] = E_T[S_T]N(d_1) - KN(d_2), \text{ where} \tag{6.35}$$

$$d_1 = \frac{\ln(E_T[S_T]/K) + \sigma^2 T/2}{\sigma\sqrt{T}} \quad \text{and}$$

$$d_2 = \frac{\ln (E_T[S_T]/K) - \sigma^2 T/2}{\sigma\sqrt{T}} = d_1 - \sigma\sqrt{T}.$$

It remains to express the expected value $E_T[S_T]$ in terms of the values known at time 0. But this is easy, for example, if S is the price of a non-income paying asset, then $S_t/P(t,T)$ is a martingale, and so

$$E_T[S_T] = E_T\left[\frac{S_T}{P(T,T)}\right] = \frac{S_0}{P(0,T)}.$$

If R is the maturity T interest rate in continuous compounding defined by the equation $P(0,T) = e^{-RT}$, that is, $R = -\frac{1}{T}\ln P(0,T)$, then the generalized standard market model formula can indeed be written as the Black-Scholes formula with the (constant) instantaneous interest rate r replaced by the maturity T interest rate R:

$$c_0 = S_0 N(d_1) - e^{-RT} K N(d_2), \text{where} \tag{6.36}$$

$$d_1 = \frac{\ln (S_0/K) + (R + \sigma^2/2)T}{\sigma\sqrt{T}} \quad \text{and}$$

$$d_2 = \frac{\ln (S_0/K) + (R - \sigma^2/2)T}{\sigma\sqrt{T}} = d_1 - \sigma\sqrt{T}.$$

For income paying assets and commodity contracts, it is more convenient to express the formula in the form proposed by Black (1976). Let F be the forward price of such an asset with maturity T. The time 0 forward price F_0 is the one that makes the value of the forward contract equal to zero. With respect to the numeraire $P(t,T)$—forward risk-neutral measure, it means that

$$0 = P(0,T)E[S_T - F_0] = P(0,T)(E[S_T] - F_0), \tag{6.37}$$

and so $E_T[S_T] = F_0$. Since $S_T = F_T$, we can work only with the forward price and assume that F_T is lognormal with the mean F_0 and volatility σ_F. Hence, we can replace $E_T[S_T]$ by F_0 and σ by σ_F in (6.35). The resulting formula is then called the *Black's formula* and the model is called the *Black's standard market model*:

$$c_0 = P(0,T)(F_0 N(d_1) - K N(d_2)), \text{where} \tag{6.38}$$

$$d_1 = \frac{\ln (F_0/K) + \sigma_F^2 T/2}{\sigma_F\sqrt{T}} \quad \text{and}$$

$$d_2 = \frac{\ln (F_0/K) - \sigma_F^2 T/2}{\sigma_F\sqrt{T}} = d_1 - \sigma_F\sqrt{T}.$$

For a put option with the same parameters, d_1 and d_2, we have

$$p_0 = P(0,T)(KN(-d_2) - F_0N(-d_1)). \tag{6.39}$$

If there is a futures or forward market, then the advantage of the formula is that it uses the directly observable market price F_0. The forward price volatility σ_F can also be estimated from historical futures prices (or obtained as an implied volatility from the Black formula and quoted premiums). In the case of interest rate derivatives, it is shown in Sect. 6.3 that the Black's standard market model is also applicable with F_0 replaced by an appropriate forward interest rate.

6.2.1 Options to Exchange one Asset for Another

An important application of the numeraire technique with the standard market model assumptions is for the valuation of options to exchange one asset for another. Such options are, for example, embedded in *convertible bonds*, where investors are given an option to exchange the bonds for the issuer's stocks at a conversion ratio.

Let us assume that an option holder has the right to exchange at time T an investment asset worth U for another investment asset worth V, i.e., to pay with the asset of value U and obtain, in exchange, an asset of value V. Note that we do not distinguish call and put, because there is no cash payment involved—the option could be viewed as a call option on V if U were the regular currency, and as a put on U if V were the currency. In either case, the payoff of the option is

$$f_T = \max\left(V_T - U_T, 0\right).$$

The option can be valued by introducing either U or V as our currency and applying the numeraire result. Choose $g_t = U_t$ to be the numeraire, then with respect to the U_t forward risk-neutral measure

$$f_0 = U_0 E_U\left[\frac{\max\left(V_T - U_T, 0\right)}{U_T}\right] = U_0 E_U\left[\max\left(\frac{V_T}{U_T} - 1, 0\right)\right]. \tag{6.40}$$

The expected value on the right-hand side will yield an analogy of the standard market model formula if the ratio $h_T = V_T/U_T$ is assumed to be lognormally distributed. This is a reasonable assumption, because $\ln h_T = \ln V_T - \ln U_T$ will be normally distributed, if $\ln V_T$ and $\ln U_T$ are jointly normally distributed. Moreover, if $\sigma_V^2 T$ is the variance of $\ln V_T$, $\sigma_U^2 T$ the variance of $\ln U_T$, and ρ their correlation, then $\sigma_V^2 T - 2\rho\sigma_V\sigma_U T + \sigma_U^2 T$ will be the variance of $\ln h_T$. Therefore

$$\sigma_h = \sqrt{\sigma_V^2 - 2\rho\sigma_V\sigma_U + \sigma_U^2} \tag{6.41}$$

is the volatility of h_T that can be used in the Black-Scholes formula. Regarding the expected value of h_T, note that $h = V/U$ is a martingale with respect to the forward risk-neutral measure, and so

$$E_U[h_T] = E_U\left[\frac{V_T}{U_T}\right] = \frac{V_0}{U_0}.$$

Combining (6.40) and (6.35), we finally get the following result:

$$f_0 = V_0 N(d_1) - U_0 N(d_2), \text{where} \tag{6.42}$$

$$d_1 = \frac{\ln\left(V_0/U_0\right) + \sigma_h^2 T/2}{\sigma_h \sqrt{T}} \quad \text{and}$$

$$d_2 = \frac{\ln\left(V_0/U_0\right) - \sigma_h^2 T/2}{\sigma_h \sqrt{T}} = d_1 - \sigma_h \sqrt{T}.$$

Example 6.1 Let us consider a convertible corporate 5-year bond which, after 3 years, gives the investors an option to exchange their bonds for the common stocks of the issuer at a conversion ratio of 1:40, i.e., one bond for forty stocks. The bond pays a relatively low 3% coupon, and without the conversion option its current value would be $U_0 = 950$USD. The current stock price is 22 USD, so we can set $V_0 = 40 \times 22 = 880$. A convertible bond investor gets the plain vanilla bond plus the conversion option. Therefore, to determine the fair price, we simply need to evaluate the option to convert the bonds into stocks. Let us use the standard market model outlined above. The estimated volatility of the stock is $\sigma_V = 15\%$, the volatility of the bond forward price is $\sigma_U = 5\%$, and the instantaneous correlation between the two assets is $\rho = 50\%$. The high correlation is because the bond and the stock have the same issuer. On the other hand, the bond price has limited growth potential, while the stock may grow without any limitation, and therefore the correlation is not perfect.[4] Plugging in the parameters to (6.41) and (6.42), we get $\sigma_h = 13.2\%$, and $c = 53$USD. Consequently, the fair price of the convertible is 1003 USD, a little bit above its par value, despite the relatively low coupon it offers.

6.3 Valuation of Caps and Floors, Bond Options, and Swaptions

In this section, we are ready to apply the technique of a numeraire and the standard market model to value the most common interest rate options. The difficulty of interest rate modeling is that, generally, one needs to model not just one price but the

[4]Note that this example shows a weak point of the standard market model: the correlation between the bond and stock price is asymmetric. If the stock price falls down, then the bond's price will go down as well, but slower, and the option will have no or negligible value. If the stock rises to high values, then the bond's price grows just a little bit, and the option becomes quite profitable. Hence, the ordinary linear correlation and the standard market model may in fact underestimate the value of the option embedded in convertible bonds.

dynamics of the full term-structure of interest rates. We will discuss the advantages and disadvantages of various interest rate models in Chap. 7. It seems that currently there is no "standard" stochastic model of interest rate dynamics. The standard market model for European interest rate options is based on a dramatic simplification: it looks only at one future interest rate and assumes that it has a lognormal distribution at time T. The advantage is that the Black's formula can then be used. On the other hand, the distributional assumption may be far from reality, especially for longer maturities. It is obvious that interest rates tend to return to a certain average value—there is a mean reversion property that is not reflected by the lognormal distribution; the rates should definitely be between 0% and 100%, and so the standard deviation of the rates should be bounded from above. On the other hand, the lognormal distribution has an unlimited standard deviation $\sigma\sqrt{T}$ for large values of T (if σ is kept constant). Despite this inconsistency, the standard market model is used in the derivative markets as a standard. Nowadays, it is more a standard means of communication than a standard tool for valuation. Black's model allows the easy calculation of the premium given volatility, and the implied volatility given a market premium. A bank may first apply an advanced interest rate model to determine the premium of an option, calculate the implied volatility using the standard market model, and then quote it or compare it to the volatility quotes of other market players.

6.3.1 Bond Options

Bond options have traditionally existed as embedded options in callable or puttable bonds. A callable bond allows the issuer to buy back the bond at a pre-specified price at one or more dates. If there is only one date when the bond can be called, then the embedded option is of European type. The bond price can be decomposed into the price of the bond without the option and the price of the call option on the bond sold by the investor to the issuer. Thus, to value the bond, besides valuing the no-option plain vanilla bond, we need to value the European call option. There are also many OTC and exchange-traded bond options. OTC options are usually of the European type, while exchange-traded bond options (based on the bond futures) are usually of the American type.

For European type bond options, the Black's formula (6.38) applies simply with F_0 equal to the time T bond forward price and σ_F equal to the bond price volatility. If a quoted bond forward price is not directly available, then it can be calculated as the implied (full) forward price:

$$F_0 = \frac{Q_0 - I}{P(0, T)} \tag{6.43}$$

where Q_0 is the full bond price at time 0 and I is the present value of the coupons to be paid over the life of the option. Note that the strike price K in (6.38) must also be given as the full cash price. If the strike was set up as a clean price, then it must be adjusted for the accrued interest.

Example 6.2 An AAA bank has issued a 5-year bond that can be called back after 2.5 years at the net price of 102%. Its annual coupon is 5%, and without the option the bond would initially be valued at 103% based on the yields of comparable AAA plain vanilla bonds. Let us estimate the fair price investors should pay for the bond. An investor who buys the callable bond is, in fact, buying the plain vanilla bond with the same coupon, and, at the same time, selling a call option on the bond, i.e., $Q_{\text{callable}} = Q_0 - c_0$ where $Q_0 = 103\%$. The call option value c_0 can be evaluated by Black's standard market model (6.38) with F_0 given by (6.43) and $K = 104.5\%$, since the accrued interest in 2.5 years will be 2.5%. Let us assume that the risk-free interest rates in continuous compounding for the 1, 2, and 2.5-year maturities are 2%, 2.5%, and 3%, respectively. The bond forward price volatility has been estimated at $\sigma_F = 5\%$. To calculate the forward price, we need to discount the first two coupons:

$$I = 5\%\left(e^{-0.02} + e^{-0.025\cdot 2}\right) = 9.66\%, \text{and so}$$

$$F_0 = (103\% - 9.66\%) \cdot e^{0.03\cdot 2.5} = 100.61\%.$$

Plugging in the inputs to (6.38), we get $c_0 = 3.93\%$. Consequently, the fair price an investor should initially pay for the callable bond is not 103%, but a significantly lower 99.07%. The price of the out-of-the-money call option is surprisingly high due to its long-time maturity.

The key input of the calculations in Examples 6.1 and 6.2 was the bond forward price volatility, which should not be confused with the yield volatility or with the bond spot price volatility. To understand the difference, recall that bond price Q changes can be approximated using duration and the market yield-to-maturity (YTM) changes:

$$\frac{\Delta Q}{\Delta_0} \approx -D\Delta y = -Dy_0\frac{\Delta y}{y_0}.$$

Hence, the volatility of Q depends on the duration, which decreases when the time of measurement approaches maturity, and on the YTM volatility. The YTM volatility, generally, also depends on the time to maturity, but if we consider only parallel shifts in the yield curve, then the volatility σ_y of y should be independent of the time to maturity, and so

$$\sigma_Q \approx Dy_0\sigma_y. \tag{6.44}$$

For example, the volatility of a bond that has 5 years to maturity would significantly differ from the volatility of the forward bond price, which has only 2 years to maturity. It would also be incorrect to base the volatility on a historical series of

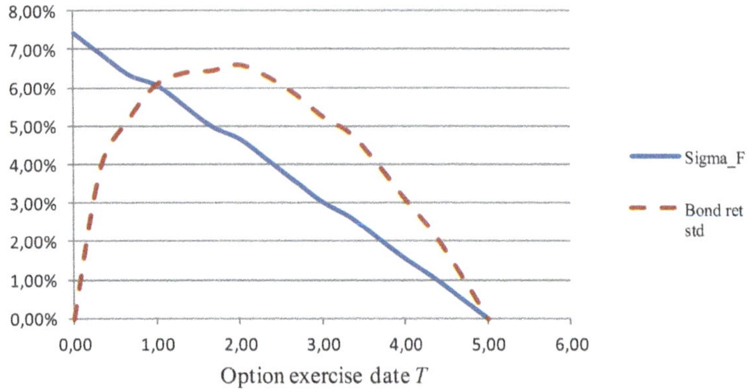

Fig. 6.2 Dependence of the forward bond price volatility and of the standard deviation of the forward bond log-price on the option maturity T

returns of the spot bond prices. One possibility could be to use a historical series of forward bond prices for the specified bond and forward maturity. However, the most practical approach is to estimate YTM volatility and apply (6.44).

Example 6.3 Let us re-estimate the volatility of the forward bond price used in Example 6.2. Given the actual price $Q_0 = 103\%$ of the plain vanilla bond, the yield to maturity is $y_0 = 4.07\%$, and the spot price duration is $D = 4.56$. The forward price duration in 3 years can be estimated as $D_F = 1.86$, provided the YTM remains unchanged.[5] The YTM volatility estimated from the historical data is 40% and so according to (6.44) the forward bond price volatility is

$$\sigma_F \cong 1.86 \times 4.07\% \times 40\% = 3.03\% \tag{6.45}$$

The estimated volatility is lower than the 5% volatility used in Example 6.2. The call option value with this volatility parameter comes out lower at $c_0 = 2.17\%$, and the estimated callable bond fair value is 100.83%.

In spite of the historical YTM volatility based estimate (6.45), the traders might use higher market volatility, but the example shows the sensitivity and importance of the volatility estimates. It is interesting to look at Fig. 6.2, where we see that the forward bond price volatility σ_F decreases according to (6.44), with the option exercise date T going from 0 to the bond's maturity 5, but the standard deviation $\sigma_F\sqrt{T}$ of the logarithm of the forward bond price initially goes up, and then down to zero as T approaches the bond's maturity. Consequently, the bond call options will have the greatest value if the exercise date lies somewhere in the middle between the options start date and the bond's maturity.

[5] More precisely, we should use the forward YTM. However, the difference in the forward bond price volatility estimate would be negligible.

6.3.2 Caps and Floors

Popular interest rate options traded on the OTC markets, and often embedded into floating-rate notes or ordinary commercial loans, are called *interest rate caps* and *floors*. An interest rate cap provides an upper bound K_u on floating rates $R_{M,i}$ periodically reset at times t_i, $i = 1, .., n$ with $t_{n+1} = T$ equal to the cap's maturity. The capped interest rate effectively paid at t_{i+1} is

$$\min(K_u, R_{M,i}) = R_{M,i} - \max(R_{M,i} - K_u, 0), \tag{6.46}$$

and so the cap provides an insurance against the floating rate rising to unexpected levels. In fact, the ordinary float rate is offset by the payoff of an interest rate call option. The times t_i are usually spaced evenly based on an interest rate tenor (e.g., 1 month or 3 months) with δ_i denoting the exact time factor between t_i and t_{i+1} depending on the day-count convention. Hence, if L is the principal of the capped loan then the payments are $L\delta_i$. $\min(K_u, R_{M,i})$. Similarly, the interest *rate floor* provides a lower bound K_l on periodically reset floating rates, and so the payments are $L\delta_i \max(K_i, R_{M,i})$. While an interest rate cap provides a protection to the debtor, and therefore has a cost, an interest rate floor provides a benefit to the creditor. Consequently, a floor will fully or partially offset the cost of a corresponding cap. The combination of a cap and a floor, called a *collar*, is usually constructed so that the initial cost equals to zero. For $K_l < K_u$, the loan payments protected by a collar can be expressed as

$$L\delta_i \times \max(K_l, \min(K_u, R_{M,i})).$$

Figure 6.3 shows an example of quarterly interest payments where the (highly volatile) floating rate is capped by 7% and floored by 3%.

Caps and floors are not only embedded in commercial loans, but also traded on the OTC derivative markets as interest rate options. To price a cap, we need to price the series of call options with the payoffs $L\delta_i \max(R_{M,i} - K_u, 0)$ paid at t_{i+1}. A single such option is called a *caplet*, while the option with the payoff $L\delta_i \max(K_l - R_{M,i}, 0)$ is, similarly, called a *floorlet*. Since there is no dependence between the exercise of the individual options, it is enough to price the caplets (and floorlets) separately. Note that this is not the same as for a series of option dates on a callable bond, where exercise at one of the dates excludes exercise at all the other dates.

In order to price the caplet paying $L\delta_i \max(R_{M,i} - K_u, 0)$ paid at t_{i+1}, applying the standard market model, we can use the measure which is forward risk neutral with respect to the zero-coupon bond $P(t, t_{i+1})$. Let F_i be the actual (time 0) forward interest rate for the period t_i to t_{i+1}. Similarly to (6.37), since it is the rate contracted today, in order for the forward contract payoff $R_{M,i} - F_i$ payable at time t_{i+1} to have today's value zero, we must have

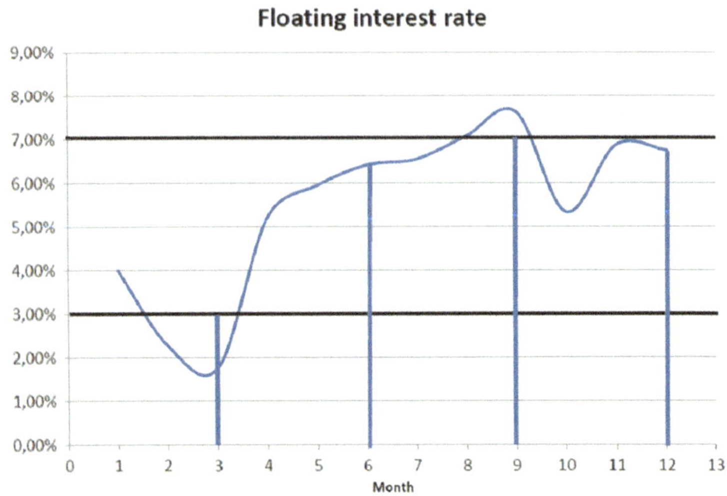

Fig. 6.3 Quarterly interest payments with a collar that guarantees the maximum rate at 7% and the minimum rate at 3%

$$0 = P(0, t_{i+1})E_{t_{i+1}}[R_{M,i} - F_i], \text{ and so}$$

$$F_i = E_{t_{i+1}}[R_{M,i}],$$

where the expectation is with respect to the $P(t, t_{i+1})$ forward risk-neutral measure. Alternatively, we can use the (arbitrage-free market) formula for the time t forward rate

$$F_i(t) = \frac{1}{\delta_i} \frac{P(t, t_i) - P(t, t_{i+1})}{P(t, t_{i+1})}.$$

Note that it is a martingale by the definition of the equivalent martingale measure. In particular, it means that

$$F_i(0) = E_{t_{i+1}}[F_i(t_i)] = E_{t_{i+1}}[R_{M,i}]$$

since $F_i(t_i) = R_{M,i}$. Consequently, the Black's Standard Market Model (6.38) can be applied with one modification: although the payment takes place at time t_{i+1}, the interest rate $R_{M,i}$ is set already set at t_i, and so the log-variance $\sigma_i^2 t_i$ of $R_{M,i}$ is conventionally scaled to the time t_i rather than t_{i+1}. Finally, the Black formula for the caplet is

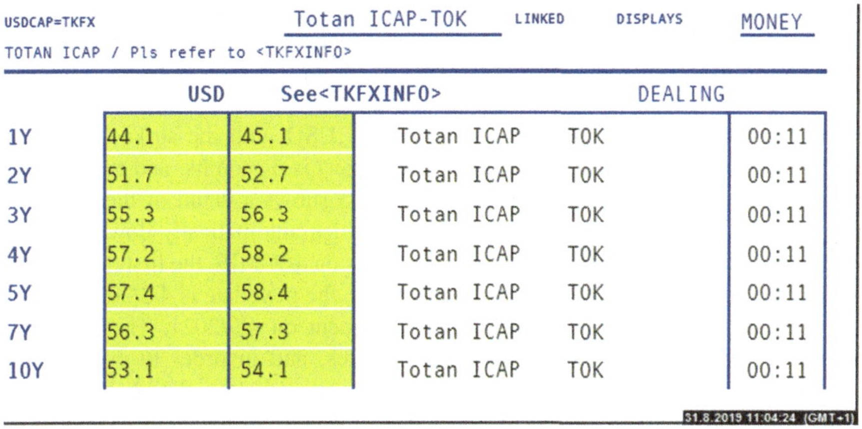

USDCAP=TKFX Totan ICAP-TOK LINKED DISPLAYS MONEY

TOTAN ICAP / Pls refer to <TKFXINFO>

	USD	See<TKFXINFO>			DEALING
1Y	44.1	45.1	Totan ICAP	TOK	00:11
2Y	51.7	52.7	Totan ICAP	TOK	00:11
3Y	55.3	56.3	Totan ICAP	TOK	00:11
4Y	57.2	58.2	Totan ICAP	TOK	00:11
5Y	57.4	58.4	Totan ICAP	TOK	00:11
7Y	56.3	57.3	Totan ICAP	TOK	00:11
10Y	53.1	54.1	Totan ICAP	TOK	00:11

31.8.2019 11:04:24 (GMT+1)

Fig. 6.4 USD at-the-money cap and floor volatilities (Source: Thomson Reuters/Tokyo Forex, 31.8.2019)

$$c_i = L\delta_i P(0, t_{i+1})(F_i N(d_1) - K_u N(d_2)), \text{ where} \tag{6.47}$$

$$d_1 = \frac{\ln(F_i/K_u) + \sigma_i^2 t_i/2}{\sigma_i\sqrt{t_i}} \text{ and}$$

$$d_2 = \frac{\ln(F_i/K_u) - \sigma_i^2 t_i/2}{\sigma_i\sqrt{t_i}} = d_1 - \sigma_i\sqrt{t_i}.$$

Similarly, the Black formula can be written down for a floorlet or we can use the put-call parity applicable to the caplet and floorlet with the same strike price K:

$$c_i - p_i = L\delta_i P(0, t_{i+1})(F_i - K). \tag{6.48}$$

The right-hand side of (6.48) is the market value of a forward paying K and receiving the float rate set for the period t_i to t_{i+1}. For caps and floors, the put-call parity can be generalized using a fix-paying interest rate swap value

$$cap(K) - floor(K) = irs(K). \tag{6.49}$$

The principal, maturity, tenor, and the strike (fix rate) of the cap, floor, and swap are the same. However, since the cap and the floor do not provide any protection on the first float interest rate (set at the start date), we also have to delete the first exchange of payments from the IRS valuation on the right-hand side of (6.49).

Figure 6.4 shows an example of volatility quotes for at-the-money caps. In practice, caplets are rarely traded separately, and so the same quoted cap volatility for maturity is used to value all the caplets through the Black's formula (6.47). The values are summed to obtain the correct value of the whole cap. Note that, in this case, the valuation of the individual caplets is not precise. "True" caplet volatilities may in fact vary for different maturities, but the cap volatility is always the same rate for all the caplets involved. Implied caplet volatility estimates could be obtained, for example, from the quoted cap volatilities by a technique similar to bootstrapping.

Example 6.4 Consider a 2Y cap on 10 million USD with the strike 2.2%. The standard tenor of caps with maturity 2Y (or longer) is 3 months, and to value the collar we need to value seven caplets. Figure 6.5 shows a detail of the valuation produced by the Thomson Reuters calculator, which automatically downloads the market volatilities and interest rates. The cap starts on 4/9/2019, the first reset date is 4/12/2019 (the payoff takes place 3 months later), the next date is 4/3/2013, and so on, with last reset on 4/6/2021 and final settlement on 4/9/2021. The calculator automatically calculates the forward interest rates, and in order to evaluate the individual caplets correctly, there are also bootstrapped caplet volatilities. The sum of the caplet premiums is around 14,271 EUR, corresponding to the flat cap volatility of 51.36% (see Fig. 6.4).

6.3.3 Normal and Shifted Lognormal Volatilities

The USD cap and floor quotes given in Fig. 6.4 are also called lognormal volatilities, since they indicate the variance of $\ln R_M$ in the context of the standard market (Black-Scholes) model. However, recently, for many currencies including EUR, JPY, or even CZK, the short-term interest rates have become negative, and so the Black-Scholes model is not applicable any more ($\ln R_M$ is always positive and it is not defined if the interest rate R_M is negative). There are two possible solutions accepted by the market as indicated by Fig. 6.6. The first one is to keep the Black-Scholes model or Black's formula, but to quote the lognormal volatility of interest rates shifted by a constant, so that they are not negative anymore. Another possibility is to apply a model which assumes that future interest rates are distributed normally, and so may attain negative value. This is the key assumption of the Bachelier (1900) model, which is, in fact, a predecessor of the Black-Scholes model. It is also consistent with the popular Vašíček interest rate model (see Sect. 7.1).

Specifically, regarding the *shifted lognormal* (or *displaced diffusion*) *model,* the stochastic forward $T \times T'$ interest rate F_T is replaced by a new variable $X_T = F_T - \Theta$, where Θ is a negative constant (e.g. -1%), which is assumed to be lognormally distributed with quoted volatility σ_{SL}, i.e., $\mathrm{var}[\ln X_T] = \sigma_{SL}^2 T$ and $E[X_T] = F_0 - \Theta$. Therefore, we can value the caplets and floorlets, using the Black's formula, as options on the shifted interest rate X_T with the shifted strike price $K - \Theta$

$$f_{SL}(T, T', K, F_0) = f_{BF}(T, T', K - \Theta, F_0 - \Theta, \sigma_{SL})$$

where f_{BF} is the Black's caplet or floorlet formula, and F_0 the actual $T \times T'$ forward rate for the caplet/floorlet interest rate period.

The disadvantage of the shifted lognormal model is that it is based on an arbitrary choice of the (shifting) constant Θ. There might be different volatility quotations for different levels of Θ, and in addition the lognormality assumption is inconsistent if

USD VANILLA CAP — UPDATED AT: 11:46:34 AM

USD | OTC | Swap Type: Cap and Floor: Vanilla Cap | Valuation Date: Aug 31, 2019

Main | Caplets & Floorlets | Volatilities | Zero Curves | CVA

Start Date	Cap Strike	Cap Vol (%)	Forward	Premium	Notional	Amortization	Delta	Gamma
Sep 4, 2019	2,20000	44,10	2,138	0,00	10 000 000,00	0,00	0,2743	337,6868
Dec 4, 2019	2,20000	44,10	1,915	-1 893,36	10 000 000,00	0,00	0,3074	81,5663
Mar 4, 2020	2,20000	44,10	1,600	-1 199,28	10 000 000,00	0,00	0,1962	55,0056
Jun 4, 2020	2,20000	44,31	1,418	-1 077,12	10 000 000,00	0,00	0,1725	46,6463
Sep 4, 2020	2,20000	50,43	1,299	-1 607,85	10 000 000,00	0,00	0,2162	44,4933
Dec 4, 2020	2,20000	53,32	1,267	-2 285,48	10 000 000,00	0,00	0,2666	43,3348
Mar 4, 2021	2,20000	56,69	1,176	-2 699,05	10 000 000,00	0,00	0,2908	41,8485
Jun 4, 2021	2,20000	59,51	1,145	-3 508,98	10 000 000,00	0,00	0,3327	40,1919
Sep 4, 2021					0,00	10 000 000,00		

Fig. 6.5 Valuation of individual caplets and floorlets of a 2Y cap on 10 million USD (Thomson Reuters cap and Floor calculator, 31.8.2019)

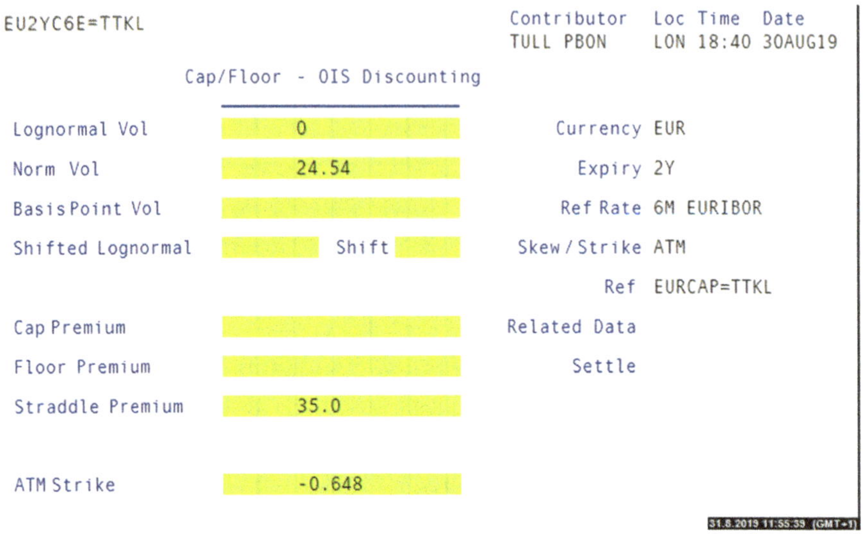

Fig. 6.6 EUR 2-year cap volatility quotation (Thomson Reuters, 31.8.2019)

taken for different values of Θ. The prevailing solution is indicated in Fig. 6.6, where the only volatility quote is denoted "Norm Vol." The assumption is that the forward interest rates are normally distributed, i.e., $F_T\tilde{N}\left(F_0, \sigma_N^2 T\right)$ where σ_N is the *normal volatility*. Note that σ_N represents the annualized standard deviation of absolute, not relative, changes in the interest rates. For example, the normal volatility quote "24.54" given in Fig. 6.6 should be interpreted in basis points, i.e., $\sigma_N = 0.2454\%$ With respect to the standard $P(t, T')$ forward risk-neutral measure the caplet (similarly floorlet) option value is

$$c(T, T', K) = P(0, T')E_{T'}[\max\,(F_T - K, 0)],$$

and we just need to evaluate the expectation on the right side which is, with the normality assumption, quite straightforward giving us the well-known Bachelier formula (Bachelier 1900):

$$c_{\text{Bach}}(T, T', K, \sigma_N) = P(0, T')\left[(F_0 - K)N(d) + \sigma\sqrt{T}N'(d)\right],$$
$$p_{\text{Bach}}(T, T', K, \sigma_N) = P(0, T')\left[(K - F_0)N(-d) + \sigma\sqrt{T}N'(d)\right],$$
$$d = \frac{F_0 - K}{\sigma_N\sqrt{T}},$$

and where $N'(d)$ denotes the standard normal density function.

Example 6.5 Figure 6.7 shows the valuation of three 6M caplets forming a 2-year 10 million EUR cap with the strike price $K = -03\%$. All the caplets are out-of-the-

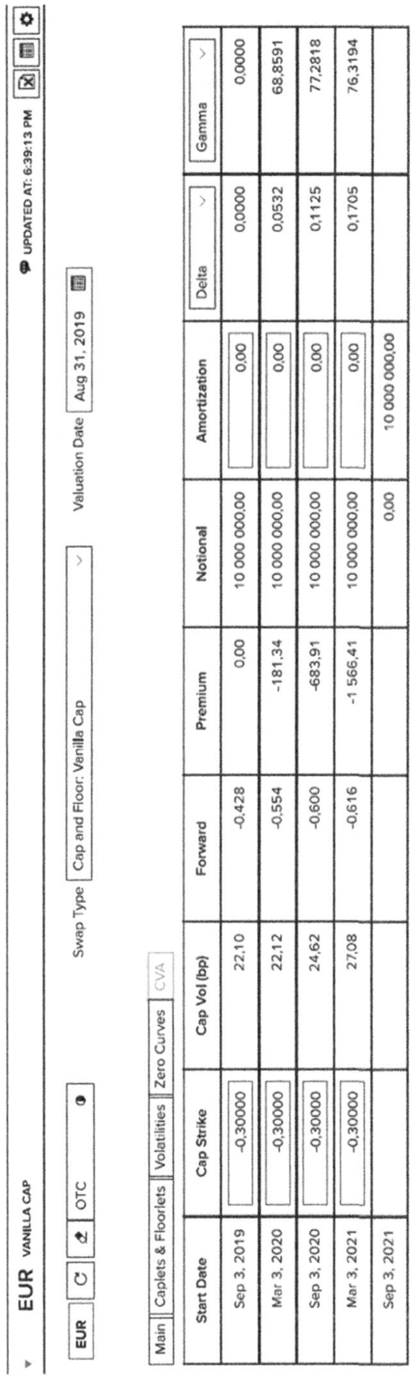

Start Date	Cap Strike	Cap Vol (bp)	Forward	Premium	Notional	Amortization	Delta	Gamma
Sep 3, 2019	-0,30000	22,10	-0,428	0,00	10 000 000,00	0,00	0,0000	0,0000
Mar 3, 2020	-0,30000	22,12	-0,554	-181,34	10 000 000,00	0,00	0,0532	68,8591
Sep 3, 2020	-0,30000	24,62	-0,600	-683,91	10 000 000,00	0,00	0,1125	77,2818
Mar 3, 2021	-0,30000	27,08	-0,616	-1 566,41	10 000 000,00	0,00	0,1705	76,3194
Sep 3, 2021					0,00	10 000 000,00		

Fig. 6.7 Valuation of individual caplets and floorlets of a 2Y cap on 10 million EUR (Thomson Reuters cap and Floor calculator, 31.8.2019)

money since the forward rates are even lower than the strike price. Notice that the caplet volatilities are, indeed, given in bps and the calculator automatically applies the Bachelier formula. The total cap value is 2342 EUR, corresponding to the flat volatility of 25.50 bps used by the calculator.

6.3.4 Swaptions

A European swap option, or *swaption*, gives to the holder the right (but not an obligation) to enter into an interest rate swap with specified parameters (notional, fix rate, maturity) at a certain time in the future (exercise date of the swaption). A company may use the fix-payer's swaption to limit the to-be-swapped fixed interest rate on a floating-rate loan agreement. If s_K denotes the swaption strike swap rate and s_T denotes the market swap rate at the exercise date T, then the company will exercise the swaption if $s_K < s_T$. Otherwise, it is better (or equivalent) to use the actual market swap rate $s_T \leq s_K$. The payoff rate is $\max(s_T - s_K, 0)$, but it is not paid at time T, or at a single future time, but effectively at all swap fix rate payment times $T_1, \ldots, T_N$. The swaption payoff at time T can be expressed as the cash flow present value

$$f_T = \sum_{i=1}^{N} P(T, T_i)\delta_i L \max(s_T - s_K, 0).$$

where L is the notional principal and δ_i is the time factor corresponding to the period from T_{i-1} to T_i (with $T_0 = T$). In this case, it is useful to introduce another numeraire, called *the annuity*,

$$A(t) = \sum_{i=1}^{N} \delta_i P(t, T_i). \tag{6.50}$$

Note that the $A(t)$ corresponds to the value of a portfolio of zero-coupon bonds, hence it is the price of a tradable asset. Since $f_T = A(T)L \max(s_T - s_K, 0)$, we have

$$f_0 = A(0) \times E_A\left[\frac{f_T}{A(T)}\right] = A(0) \times L \times E_A[\max(s_T - s_K, 0)]$$

where the expectation is taken with respect to the $A(t)$ forward risk-neutral measure. According to the standard market model principle, let us assume that s_T is lognormally distributed with the expected value $E_A[s_T]$ and log-variance $\sigma_F^2 T$. The forward swap rate s_t of a swap starting at T and paying the fixed coupons s_t at $t_1, \ldots, T_N$ is the rate that makes the time t present value of the fixed and float leg of the swap with a unit principal equal to zero, i.e.,

$$0 = P(t, T_0) - P(t, T_N) - \sum_{i=1}^{N} P(t, T_i)\delta_i s_t = P(t, T_0) - P(t, T_N) - A(t)s_t.$$

Therefore,

$$s_t = \frac{P(t, T_0) - P(t, T_N)}{A(t)}. \tag{6.51}$$

For $t = T$ the Eq. (6.51) gives the time T spot market swap rate. Moreover, it follows from (6.51) and the fact that $A(t)$ is a numeraire that the time zero forward swap rate

$$s_0 = \frac{P(0, T) - P(0, T_N)}{A(0)} = E_A\left[\frac{P(T, T_0) - P(T, T_N)}{A(t)}\right] = E_A[s_T].$$

Finally, we can write down the Black formula for a swaption, where the holder has the option to pay the fixed swap rate s_K:

$$f_0 = A(0)(s_0 N(d_1) - s_K N(d_2)), \text{where} \tag{6.52}$$

$$d_1 = \frac{\ln(s_0/s_k) + \sigma_F^2 T/2}{\sigma_F\sqrt{T}} \quad \text{and}$$

$$d_2 = \frac{\ln(s_0/s_K) - \sigma_F^2 T/2}{\sigma_F\sqrt{T}} = d_1 - \sigma_F\sqrt{T}.$$

Example 6.6 Let us calculate the value of an out-of-the-money swaption with 1Y exercise on a 4Y swap with 10 million USD principal, if the 1Y × 5Y forward swap rate is $s_0 = 1.195\%$ and strike $s_K = 1.5\%$. Let us assume that the actual annuity value is $A(0) = 3.8$. We will use the volatility quotes given in Fig. 6.8.[6] The table has two dimensions, exercise times and tenors, thus the volatility corresponding to our swaption is $\sigma_F = 60.9\%$. According to (6.52) and using the given parameters we obtain $f_0 = 72659$ EUR from the perspective of the fix-rate payer. Regarding the annuity factor (6.50), we need to construct the swap zero-coupon curve as outlined in Sect. 3.1. Some applications, like the Thomson Reuters calculator, provide the discount factor values automatically.

Remark A European swaption is theoretically equivalent to a bond option. According to our analysis in Sect. 3.1, a fix paying swap position market value is the same as the value of a short position in the corresponding fixed coupon bond and a long position in the floating rate note. Since the floating rate note value always

[6]To get a more precise valuation we should use "volatility smile" quotes (see Chap. 8) since the swaption is out-of-the-money (OTM) and the volatility quotes in Fig. 6.8 are for ATM options.

USDSWPTNS=TTKL USD ATM Swaption LINKED DISPLAYS TTKLIN

USD 1Y		USD 2Y		USD 3Y		USD 4Y	
TULL PBON	LON	TULL PBON	LON	TULL PBON	LON	TULL PBON	LON
1M EX	46.03	1M EX	57.18	1M EX	63.91	1M EX	66.51
3M EX	52.13	3M EX	60.53	3M EX	65.65	3M EX	66.88
6M EX	55.47	6M EX	61.20	6M EX	62.54	6M EX	67.79
1Y EX	60.46	1Y EX	62.27	1Y EX	61.62	1Y EX	60.90
2Y EX	62.01	2Y EX	61.78	2Y EX	60.79	2Y EX	59.32
3Y EX	61.88	3Y EX	60.47	3Y EX	57.96	3Y EX	55.98
4Y EX	60.29	4Y EX	56.71	4Y EX	55.04	4Y EX	53.58
5Y EX	54.95	5Y EX	53.40	5Y EX	52.14	5Y EX	51.02
7Y EX	49.94	7Y EX	48.30	7Y EX	47.04	7Y EX	46.28
10Y EX	44.12	10Y EX	43.92	10Y EX	44.12	10Y EX	43.84
15Y EX	40.31	15Y EX	40.26	15Y EX	41.09	15Y EX	42.19
20Y EX	43.38	20Y EX	44.23	20Y EX	44.18	20Y EX	44.03

Fig. 6.8 At-the-money USD swaption volatilities (Source: Thomson Reuters, 2.9.2019)

equals to 100% (at the issue date, and if, theoretically, there is no credit risk), the swaption contract is equivalent to a put option on the fixed coupon bond with the strike price equal to its principal. Thus, one could value the swaption using the bond option standard market model formula. However, note that the two approaches are not internally consistent: the bond option valuation formula uses the assumption that the bond price $Q_T = s_T A(T) + P(T, T_N)$ is lognormally distributed with respect to the $P(t, T)$ forward risk-neutral measure, while the swaption valuation formula is based on the assumption of s_T being lognormal with respect to the $A(t)$ forward risk-neutral measure. A similar inconsistency can be observed in the case of caps (and floors) since a caplet (or floorlet) is theoretically equivalent to an option on a corresponding zero-coupon bond. In practice, the swaption (cap/floor) formulas are preferred, because there is always some credit and liquidity risk and the behavior of the swap (interest) rates is not the same as the behavior of the bond prices and their yields.

6.4 Quantos and Convexity Adjustments

Plain vanilla interest rate swaps and FRAs can be valued as discounted cash flows with unknown float rates replaced by the corresponding forward rates. The Black's formula also replaces the expected value of a future market rate with the forward rate. It turns out that this principle is not always applicable and one has to be careful when using it, especially if a rate is paid at the wrong time (e.g., *Libor in arrears*), or in the wrong currency (*quantos derivatives*). In those cases, a convexity adjustment first proposed by Brotherton-Ratcliffe and Iben (1993) should be applied (see also Witzany 2009 for a review of alternative approaches).

6.4.1 Interest Rate Convexity Adjustments

For example, let us consider a modified $T_1 \times T_2$ forward rate agreement (similar to STIR futures) paying a fixed rate R_K not at T_2, but at T_1, i.e., the settlement of $L\delta$ $(R_M - R_K)$ takes place at time T_1 without the effect of discounting (as above L is the notional principal and δ is the time factor). The standard FRA would pay $\frac{L\delta(R_M - R_K)}{1+\delta R_M}$ at time T_1, or $L\delta(R_M - R_K)$ at time T_2. The modified contract apparently simplifies the settlement procedure, but its correct valuation surprisingly gets more complex. Let R_F be the actual $T_1 \times T_2$ forward rate, then an outstanding standard FRA contract can be simply valued (see also Sect. 3.2) as follows:

$$f_0 = L\delta P(0, T_2)(R_F - R_K), \tag{6.53}$$

since $R_F = E_{T_2}[R_M]$ under the $P(t, T_2)$ forward risk-neutral measure. The modified contract pays the payoff $L\delta(R_M - R_K)$ prematurely, as early as at time T_1, and the equivalent time T_2 payoff is

$$\text{payoff}_{T_2} = L\delta(R_M - R_K)(1 + \delta R_M).$$

The modified FRA can be expressed using the $P(t, T_2)$ forward risk-neutral measure as

$$\widetilde{f}_0 = P(0, T_2)E_{T_2}\left[\text{payoff}_{T_2}\right] = L\delta P(0, T_2)E_{T_2}[(R_M - R_K)(1 + \delta R_M)] \tag{6.54}$$

The expected value on the right-hand side of (6.54) cannot, unfortunately, be solved by replacing R_M with $R_F = E_{T_2}[R_M]$ since the expression inside of the expectation operator is not a linear function of R_M. It is a quadratic, i.e., a convex function of R_M.

Generally, according to Jensen's inequality, if X is a random variable and G a convex function, then $G(E[X]) \leq [G(X)]$, and the inequality is strict, if X is non-trivial and G strictly convex (see Fig. 6.9). In the case of (6.54) the function $G(X) = (X - R_K)(1 + \delta X)$ is quadratic in $X = R_M$, moreover $E[X] = R_F$, and thus

$$(R_F - R_K)(1 + \delta R_F) = G(E_{T_2}[R_M]) < E_{T_2}[G(R_M)] = E_{T_2}[(R_M - R_K)(1 + \delta R_M)].$$

In this case, the expected value in (6.54) can be evaluated in the context of the standard market model, assuming that R_M is lognormally distributed with the mean R_F and variance

$$R_F^2 \sigma_F^2 T_t = \text{var}[R_M] = E_{T_2}\left[R_M^2\right] - R_F^2.$$

Since the expected value of the quadratic term on the right-hand side of (6.54) can be expressed in terms of the variance, $E_{T_2}\left[R_M^2\right] = R_F^2 + R_F^2 \sigma_F^2 T_1$, we obtain

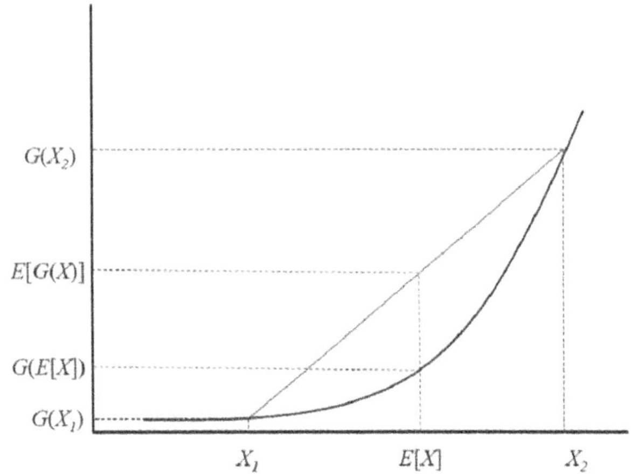

Fig. 6.9 Relationship between the expected value of a random variable X transformed by a convex function G and the transformed expected value of X

$$\begin{aligned}\tilde{f}_0 &= L\delta P(0,T_2)E_{T_2}[(R_M - R_K)(1+\delta R_M] = \\ &= L\delta P(0,T_2)((R_F - R_K)(1+\delta R_F + R_F^2\sigma_F^2 T_1\delta).\end{aligned} \tag{6.55}$$

Hence, R_M could be replaced by R_F, but there is an additional interest rate *convexity adjustment* $R_F^2\sigma_F^2 T_1\delta \approx R_F^2\sigma_F^2 T_1(T_2 - T_1)$. This adjustment becomes more significant if $T_1(T_2 - T_1)$ is large.

In general, the convexity adjustment gives only an approximation based on the second-order Taylor expansion. Let $\theta_T = G(y_T)$ be a market variable (e.g., an asset value) that is expressed as a function of a rate y_T, or vice versa let the rate $y_T = G^{-1}(\theta_T)$ be expressed as a function of θ_T. Assume that the expected value $E_T[\theta_T] = \theta_0$ with respect to the $P(t, T)$ forward risk-neutral measure is known (and equal to the forward value). A typical example is a bond value depending on the yield to maturity (see Fig. 6.10):

$$Q_T = G(y_T) = \sum_{t_i} \frac{CE_i}{(1+y_t)^{t_i}}. \tag{6.56}$$

In order to value a forward contract on y_T, one needs to evaluate its expected value. It seems that the expected value could be replaced by the forward rate defined as $y_0 = G^{-1}(\theta_0)$, but this would be incorrect, since we know that if G is strictly convex and decreasing then

$$G(E_T[y_T]) < E_T[G(y_T)] = \theta_0, \text{therefore (applying } G^{-1})$$

$$E_T[y_T] > G^{-1}(\theta_0) = y_0. \tag{6.57}$$

If G were concave, then the inequality would be opposite. Since the function G is usually convex, like the one in (6.56), we need to add a *convexity adjustment* to y_0 in

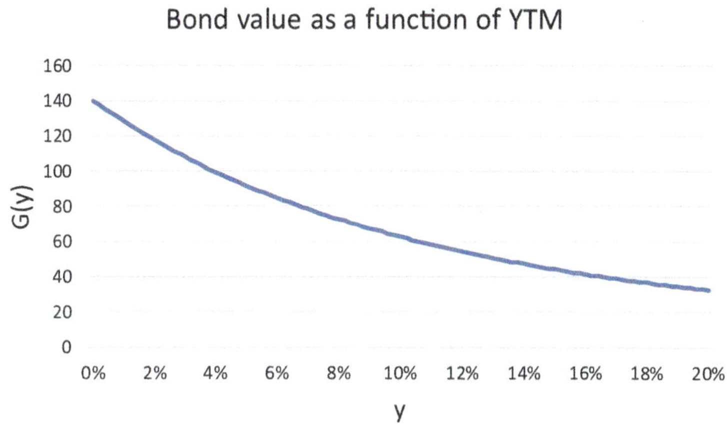

Fig. 6.10 Example of a 10-year bond with 4% annual coupon value as a function of the yield-to-maturity

order to obtain a better approximation of $E_T[y_T]$. We know that $E_T[G(y_T)] = \theta_0$, and the idea is to expand $G(y_T)$ at y_0 using Taylor's expansion:

$$G(y_T) = G(y_0 + (y_T - y_0)) \approx G(y_0) + G'(y_0)(y_T - y_0) + \frac{G''(y_0)}{2}(y_T - y_0)^2.$$

Applying the expectation operator to the left-hand side and to the right-hand side of the second-order Taylor's expansion yields

$$\theta_0 = E_T[G(y_T)] = G(y_0) + G'(y_0)(E_T[y_T] - y_0) + \frac{G''(y_0)}{2} E_T\left[(y_T - y_0)^2\right].$$

This equation can be easily solved for $E_T[y_T]$, since $G(y_0) = \theta_0$ and $E_T[(y_T - y_0)^2]$ can be approximated by the variance of y_T in the form $y_0^2 \sigma_y^2 T$, where σ_y is the volatility of the rate y_T:

$$E_T[y_T] \approx y_0 - \frac{1}{2} y_0^2 \sigma_y^2 T \frac{G''(y_0)}{G'(y_0)}. \tag{6.58}$$

In particular, if G is the function given by (6.56) valuing a bond from its yield-to-maturity y, then the first derivative $G'(y_0) = -D_0 Q_0$, where D_0 is the forward (modified) duration and Q_0 is the forward bond price, and, similarly, the second-order derivative $G''(y_0)\, C_0 Q_0$, where C_0 is the forward bond convexity, and so

$$E_T[y_T] \approx y_0 + \frac{1}{2} y_0^2 \sigma_y^2 T \frac{C_0}{D_0}. \tag{6.59}$$

Example 6.7 Consider an instrument that pays in 1 year the market yield-to-maturity of a specified Czech government 10-year bond against a fixed rate K multiplied by the notional principal of 100 million CZK. Calculate the rate K that makes the present value of the contract equal to zero. The forward yield of the bond is $y_0 = 3.3\%$, the forward (modified) duration is 7.9, and the convexity is given as 76. The yield volatility (similar to swaption volatility, see Fig. 6.8) is $\sigma_y = 35\%$ and so applying (6.59) the equilibrium rate is

$$K = E_1[y_1] \approx y_0 + \frac{1}{2} y_0^2 \sigma_y^2 \frac{C_0}{D_0} = 3.3\% + (3.3\% \times 35\%)^2 \frac{76}{7.9} = 3.43\%.$$

The convexity adjustment itself is around 13 bps, i.e., quite a significant amount of 130,000 CZK due to the high notional principal.

The convexity adjustment result (6.58) is particularly important for *constant maturity swaps (CMS)*, where a fixed rate is exchanged against constant maturity (N-year) swap rates reset at the payment times. For example, a 3-year constant maturity swap with the tenor $N = 10$ will exchange a fixed rate K against the effective 10-years IRS rate quoted in 1, 2, and 3 years. The example above demonstrates that the convexity adjustment is not negligible, and in fact, it increases with the maturity T due to (6.59).

In case of a $T_1 \times T_2$ short rate R_M paid at T_1 (and not T_2), the popular convexity adjustment is based on the transformation of the rate to the zero-coupon bond value

$$G(R_M) = \frac{1}{1 + \delta R_M}.$$

Differentiating G at R_F and applying (6.58) we get

$$E_{T_1}[R_M] \approx R_F + \frac{R_F^2 \sigma_F^2 \delta T}{1 + \delta R_F}. \tag{6.60}$$

Although we have chosen a different approach the result can be verified to be equivalent to (6.55). However, in (6.55), we have used a precise expression for the variance of R_M based on the lognormality assumption. Slightly inconsistent forms of the convexity adjustments can sometimes be derived by applying different approaches (see Pelsser 2003 or Witzany 2009).

The convexity adjustment (6.60) is also important for the valuation of *Libor-in-arrears* swaps where the interest rate is reset at the end (i.e., in arrears) not at the beginning of the regular interest rate periods.

Example 6.8 Let us estimate the convexity adjustment of a 6M Libor paid in arrears in 5 years, if the forward rate is $R_F = 2.5\%$ and the volatility is $\sigma_F = 40\%$:

$$E_{T_1}[R_M] - R_F \approx \frac{R_F^2 \sigma_F^2 \delta T}{1 + \delta R_F} = \frac{(2.5\% \times 40\%)^2 \times 0.5 \times 5}{1 + 0.5 \times 2.5\%} = 2.5 \text{ bps}.$$

The seemingly negligible adjustment of 2.5 bps will be certainly important for contracts with large notional principals.

6.4.2 Quantos

A *quanto derivative* (or just a *quanto*) is a derivative instrument that pays in one currency a payoff based on an underlying asset that is traded and quoted in another currency. A well-known example is the Dollar Nikkei 225 index futures contract traded on the CME, where the settlement payoff is defined as \$5 × (I_T − K), the variable I_T denotes the Nikkei stock index value at the futures maturity (quoted in JPY), and K is the futures price. To value the contract and find the equilibrium futures price, we may use the $P_{USD}(t, T)$ forward risk-neutral measure to get

$$f_0 = P_{USD}(0, T)(5 \times E_{USD,T}[I_T] - K),$$

but the problem is that we cannot correctly replace the expected value $E_{USD,\,T}[I_T]$ with the forward index value, since the underlying Nikkei index stocks are traded in Yen and not in Dollars. Nevertheless, we know that the expected value $E_{YEN,\,T}[I_T]$ with respect to the $P_{YEN}(t, T)$ forward risk-neutral measure equals to the index forward value F_0 denominated in JPY. What we need to estimate is the difference between the two expected values $E_{YEN,\,T}[I_T]$ and $E_{USD,\,T}[I_T]$ under the two different measures.

In order to give a general answer to the question stated above, let us consider two currencies X and Y, and a price value I_T of a security traded in Y. To value a forward contract on I_T paid in X, we need to find the expected value $E_X[I_T]$ with respect to the $P_X(t, T)$ forward risk-neutral measure. In currency Y, the expected value equals to the forward price $F_0 = E_Y[I_T]$, and to get the expected value in the currency X we need to change the measure. In this case we can use the change of measure technique from Sect. 6.1.

Let us measure all prices in the "home" currency X. Therefore, let $g(t) = P_Y(t, T)/S(t)$ be the first numeraire and $h(t) = P_X(t, T)$ the second numeraire, where $S(t)$ is the spot exchange rate of X in terms of Y. Since $g(t) = P_Y(t, T)/S(t)$ is the currency X market price of the Y zero-coupon bond, it is a correct numeraire. Consequently, with respect to the forward risk-neutral measure, we can use the martingale property and argue that the currency Y forward value F_0 solves the equation:

$$\frac{0}{g(0)} = E_g\left[\frac{I_T/S_T - F_0/S_T}{g(T)}\right] = E_g[I_t - F_0], \text{therefore}$$

$$F_0 = F_g[I_T].$$

In order to estimate the difference between $E_h[I_T] = E_X[I_T]$ and $E_g[I_T] = F_0$ we need to look at the volatility σ_w of the numeraire ratio,

$$w(t) = \frac{h(t)}{g(t)} = S(t)\frac{P_X(t,T)}{P_y(t,T)} = F(t,T). \tag{6.61}$$

In particular, we need its instantaneous correlation ρ with I_t and at the volatility σ_1 of I_t. These parameters can be easily estimated from historical data since according to (6.61) the numeraire ratio $w(t)$ simply equals to the X/Y forward exchange rate $F(t, T)$ with the maturity T and quoted at the time t. Following (6.30) the change of drift is $\alpha = \rho\sigma_F\sigma_I$. Consequently,[7]

$$E_h[I_T] \approx E_g[I_T]e^{\alpha T} = F_0 e^{\rho\sigma_F\sigma_1 T} \tag{6.62}$$

Example 6.9 The CME lists Nikkei 225 index futures settled in Yens and in Dollars. On Sep 2, 2012, the Dec 2019 Yen contract closed at 20,475, while the Dollar contract closing price was 20,525. Let us check that the observed difference corresponds to the theoretical result (6.62). If the volatility of the index estimated from historical data was around 20%, of the USD/YEN forward rate 12%, and the instantaneous correlation of returns 40%, then

$$E_{USD}[I_T] \approx F_0 e^{\rho\sigma_F\sigma_I T} = 20475e^{0.4\times 0.12\times 0.2\times 0.25} \approx 20524.2$$

indeed, comes out close to the quoted Dollar contract value.

6.4.3 Timing Adjustments

The change of numeraire technique can also be applied to find an appropriate valuation adjustment if a market quantity V_T payable at time T is paid later, at a "wrong" time $\widetilde{T}$. Again, in order to value such a forward contract, we need to estimate the difference between $E_{\widetilde{T}}[V_T]$, the expected value with respect to the $P(t,\widetilde{T})$ forward risk-neutral measure, and $E_T[V_T] = F_0$ equal to the forward rate. Let $W(t) = P(t,\widetilde{T})/P(t,T)$ be the numeraire ratio, σ_W its volatility, ρ_{WV} the

[7]Consider a price processes θ_1 following $d(\ln\theta_1) = \mu dt + \sigma dz$ and θ_2 following the process $d(\ln\theta_2) = (\mu + \alpha)dt + \sigma dz$ with $\theta_1(0) = \theta_2(0)$. Then $d(\ln(\theta_1/\theta_2)) = \alpha dt$, $E[\ln(\theta_1(T)/\theta_2(T)) = \alpha T]$, and so $E[\theta_1(T)] \approx E[\theta_2(T)]e^{\alpha T}$ assuming that the drift and volatility parameters are constant. Since this is not always exactly true, we should in any case use the "approximately equal" sign $\approx$. The expected values and their difference remain unchanged even if θ_1 and θ_2 have different the sources of uncertainty.

instantaneous correlation with V, and σ_V the volatility of V. Then the drift caused by the change of numeraire is $\alpha = \rho\sigma_W\sigma_V$ and

$$E_{\widetilde{T}}[V_T] \approx E_T[V_T]e^{\alpha T} = F_0 e^{\rho_{WV}\sigma_W\sigma_V T}. \tag{6.63}$$

If $R = R(t, T, \widetilde{T})$ is the time t forward rate $T \times \widetilde{T}$ in continuous compounding, then the numeraire ratio is $W(t) = e^{-R(\widetilde{T}-T)}$. It can be shown by the Ito's lemma that $\sigma_W = R(\widetilde{T} - T)\sigma_R$ and $\rho_{WV} = -\rho_{RV}$, where σ_R is the volatility of R and ρ_{RV} the instantaneous correlation between R and V. Hence, to estimate (6.63), it is enough to find the forward interest rate volatility and its correlation with the asset returns. The time adjustment (6.63) can be applied to ordinary asset prices, but also to get an alternative form of the interest rate convexity adjustment (6.60).

Example 6.10 Consider a derivative that provides a payoff defined as the EUR/CZK spot exchange rate S_2 observed in 2 years multiplied by CZK 100 million notional, but paid in 5 years. In order to value properly the strange contract, the 2-year forward exchange rate $F_0 = 25$ needs to be adjusted according to (6.63). The 2Y × 5Y forward rate currently is 2.5% and its volatility can be estimated at 40%. The exchange rate volatility is around 15% and the instantaneous correlation between the interest rates and the exchange rate is −20% (an increase in the interest rate statistically causes an immediate appreciation of CZK), consequently,

$$E_5[S_2] \approx F_0 e^{-\rho_{RS}R(\widetilde{T}-t)\sigma_R\sigma_S T} = 25e^{0.2\times(0.025\times3\times0.4)\times0.15\times2} \approx 25.045.$$

The adjustment is only a few basis points, but in terms of the notional amount the impact is more than 4.5 million CZK. In other words, a trader unaware of the time adjustment effect could easily lose the amount against a trader who understands and knows how to calculate it.

Interest Rate Models 7

The Standard Market Model developed and applied in the previous chapter assumes that interest rates or bond prices are lognormally distributed. The model does not describe the stochastic dynamics of interest rates over time, and so it cannot be applied to value American-style options, callable bonds, or other more complex interest rate derivatives. In this chapter, we are going to introduce the most important interest rate models, which can be classified into two categories: *short-rate* and *term-structure models* (see, e.g., Brigo and Mercurio 2006 or Málek 2005). The short-rate models focus on the instantaneous interest rate stochastic dynamics. The rest of the term-structure is derived from the short rate at a point in time, and from the model parameters. Term-structure models, on the other hand, specify equations for (forward) interest rates in all maturities, and these equations are tied by certain consistency (non-arbitrage) conditions. In both cases, the models are developed and applied under a risk-neutral measure, but can be calibrated from the real-world data.

7.1 Short-Rate Models

The short-rate models are specified by stochastic differential equations for the instantaneous interest rate $r(t)$ in the form

$$dr = m(r,t)dt + s(r,t)dz, \tag{7.1}$$

where m and s are functions of the short rate r and time t. There are no equations for the dynamics of interest rates with longer maturities. However, if the model (7.1) is set up with respect to the money market account

© The Editor(s) (if applicable) and The Author(s), under exclusive license to Springer Nature Switzerland AG 2020
J. Witzany, *Derivatives*, Springer Texts in Business and Economics,
https://doi.org/10.1007/978-3-030-51751-9_7

$$g(t) = exp\left(\int_0^t r(\tau)d\tau\right) \tag{7.2}$$

risk-neutral measure (or if the measure is changed and $m(r,t)$ is adjusted accordingly), then we can proceed as follows: Let $P(t,T)$ denote, as usual, the value of a zero-coupon bond paying one currency unit at time T. Then $P(t,T)/g(t)$ is a martingale, i.e.,

$$\frac{P(t,T)}{g(t)} = E_g\left[\frac{P(T,T)}{g(T)}\right] = E_g\left[\exp\left(-\int_0^T r(\tau)d\tau\right)\right], \quad \text{and thus}$$

$$P(t,T) = E_g\left[\exp\left(-\int_t^T r(\tau)d\tau\right)\right] = E_g\left[e^{-\bar{r}(T-t)}\right], \quad \text{where} \tag{7.3}$$

$$\bar{r} = \frac{1}{T-t}\int_t^T t(\tau)d\tau$$

is the average instantaneous rate from t to T. Given a zero-coupon bond price $P(t,T)$, the continuously compounded interest rate $R(t,T)$ at time t and maturing at T (i.e., with time to maturity $T-t$) can, by definition, be calculated as

$$R(t,T) = -\frac{1}{T-t}\ln P(t,T). \tag{7.4}$$

The Eqs. (7.3) and (7.4) show that once the process (7.1) is given, the dynamics of the term-structure is fully determined. Notice that since the process (7.1) is Markov, i.e., the probability distribution of future scenarios depends only on the value of r at time t, the expected value (7.3) depends only on $r(t)$, and therefore $P(t,T)$ and $R(t,T)$ are indeed only functions of $r(t)$, t, and of the model parameters.

The short-rate models could be further classified into *equilibrium* and *non-arbitrage models*; or from another perspective, as *one-factor* or *multi-factor models*. Equilibrium models start with a relatively simple specification of (7.1) and then derive the term-structure dynamics according to the (equilibrium) conditions (7.3) and (7.4). The problem of the models is that the initial model term-structure does not have to match exactly the observed interest rates. If simple interest rate instruments like bonds or swaps were valued by the model, then there could be an arbitrage opportunity. This issue is solved by non-arbitrage models where the specification becomes more complex, but the initial model term-structure exactly fits the market interest rates and the arbitrage opportunities are eliminated. Both types of models exist in a one-factor form (7.1) with one source of uncertainty, or in a multi-factor form with two or more sources of uncertainty. The disadvantage of one-factor models is that over all maturities the rates can basically move in only one direction, while in practice we may observe opposite movements in different maturities (e.g., twisting of the curve when short-term interest rates go up while long-term interest rates go down). Multi-factor models can capture better the complex dynamics of interest rates, but not surprisingly they become more difficult to work with.

7.1.1 The Rendelman–Bartter (Dothan) Model

It is natural to propose (Dothan 1978; Rendelman and Bartter 1980) the geometric Brownian motion model that works well for stocks and foreign currencies, i.e.,

$$dr = \mu dt + \sigma r dz, \tag{7.5}$$

where $r = r(t)$ is the instantaneous interest rate and μ, σ are constants. The distribution of the short rate r at a future point of time T under the Rendelman–Bartter model is lognormal in line with the standard market model. However, in practice, interest rates tend to return to a certain long-term level and rarely exceed 50 or 100%. The *mean-reversion property* has a classical macroeconomic explanation: if interest rates are too high, then the economy tends to slow down and a low demand for loans causes the rates to go down. Similarly, if interest rates are very low, then there is a higher demand for loans, the economy expands, and the rates tend to go up. The model (7.5) implies that the variance $\sigma^2 T$ of $\ln r(T)$ is unlimited (if σ were constant). In the case of the standard market model, the volatility can be calibrated to fit the distribution of interest rates at a particular point of time T. For the Rendelman–Bartter model, the volatility σ must fit all maturities, and so, if it is calibrated well for short-time maturities, the model implied variance of r will be too large for longer maturities, and if it is calibrated for longer maturities, the variance will be too low on short maturities. Consequently, the model is not appropriate for capturing the dynamics of interest rates over a longer time horizon or for pricing interest rate instruments with longer maturity. The model has another surprising contradictory consequence: an explosion of the bank money market account, meaning that the expected value of the money market account (7.2) is infinite for any $T > 0$, i.e., $E[g(T)] = +\infty$. The argument is based on the fact that the growth of the account over a short time interval from time t to $t + \Delta t$ is a double-exponential function of a normal variable.[1] Note that this phenomenon does not arise in the standard market model, where we deal with a particular future (non-instantaneous) interest rate $R(T, T_1)$, with a bond value, or other ordinary instrument valued at a future point of

[1]The growth of the money market account from a time t to $t + \Delta t$ can be bounded from below by an expression of the form $\exp(\exp(y)\Delta t)$, where $y = N(m, s^2)$ is obtained by averaging the generalized Wiener process $\ln r(s)$. The expected value of the growth lower bound is calculated as an integral, where the values are weighted by the normal density of y which is exponential in a negative quadratic function of y. Since $\exp(y)$ overgrows any polynomial of y, the expected value of the money market account must be infinite.

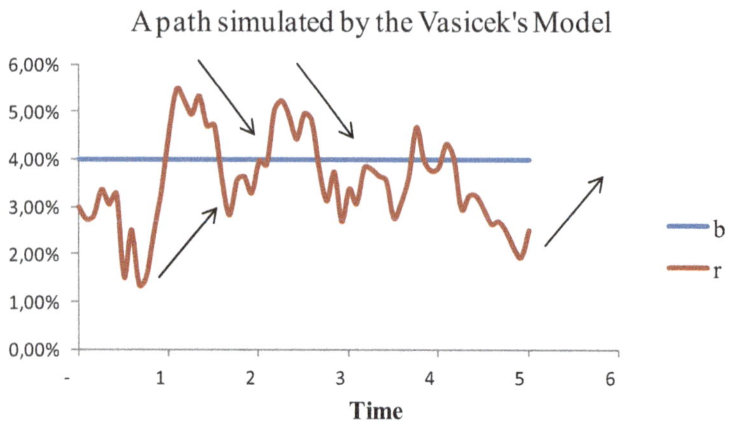

Fig. 7.1 Mean reversion of a simulated Vasicek process path ($a = 10\%$, $b = 4\%$, $\sigma = 2\%$)

time T. The bank account explosion problem, in fact, does not arise in the numerical implementations of the Rendelman–Bartter model using binomial trees, or Monte Carlo simulations, provided the time step is not too small.

Dothan (1978) assumes that there is no drift in (7.5) under the classical (money market account forward) risk-neutral measure (hence, there would be a positive drift under the real-world measure) and finds an analytical formula for $P(t, T)$. The analytical tractability of the model is only partial, since the formula involves the integration of complex hyperbolic functions; moreover, there is no analytical formula for zero-coupon bond options.

7.1.2 Vasicek's Model

Most of the disadvantages of the Rendelman–Bartter model are addressed in the Vasicek (1977) model. The interest rates have a mean-reversion property under the model, there is no bank account explosion, and, moreover, the model is analytically quite tractable. The main disadvantage of the model is that the interest rates are normally distributed and can attain negative values with a nonzero probability. On the other hand, this might be an advantage at times of negative interest rates. Vasicek's model assumes that the instantaneous interest rate evolves according to an *Ornstein–Uhlenbeck process* with constant coefficients a, b, and σ:

$$dr = a(b - r)dt + \sigma dz. \tag{7.6}$$

The coefficient b can be interpreted as a long-term interest rate level and a as the speed of reversion (see Fig. 7.1). The diffusion coefficient σ is not the same as the volatility of the geometrical Brownian motion model—it reflects the annualized standard deviation of the absolute, not relative changes of the modeled rate. If the short rate follows the process (7.6) under the real-world measure, then the same type of process is followed under the risk-neutral measure[2] and vice versa. Further on, we will work with the traditional risk-neutral measure.

The stochastic differential equation (7.6) can be relatively easily solved for r. Let us apply the Ito's lemma to the process $G = e^{at}r$. Then we get the equation:

$$dG = abe^{at}dt + \sigma e^{at}dz, \tag{7.7}$$

where G turns out to be canceled out on the right-hand side. Now, the coefficients of dt and dz on the right-hand side of (7.7) are deterministic, and consequently we can solve the equation for G by integration:

$$G(t) - G(0) = \int_0^t abe^{as}ds + \int_0^t \sigma e^{as}dz = b(e^{at} - 1) + \sigma x(t), \quad \text{where} \tag{7.8}$$

$$x(t) = \int_0^t e^{as}dz \tag{7.9}$$

is a stochastic integral that can be interpreted as an infinite sum of independent random normal increments $e^{as}dz$. Therefore, $x(t)$ is normally distributed with mean zero and variance equal to

$$\mathrm{var}[x(t)] = \int_0^t e^{2as}ds = \frac{1}{2a}\left(e^{2at} - 1\right).$$

Finally, the instantaneous short rate

$$\begin{aligned} r(t) &= e^{-at}G(t) = e^{-at}(G(0) + (e^{at} - 1) + \sigma x(t)) \\ &= e^{-at}r(0) + b(1 - e^{-at}) + \sigma e^{-at}x(t) \end{aligned} \tag{7.10}$$

is indeed normally distributed with the mean $E[r(t)] = e^{-at}r(0) + b(1 - e^{-at})$ and variance

$$\mathrm{var}[r(t)] = \sigma^2 e^{-2at}\mathrm{var}[x(t)] = \frac{\sigma^2}{2a}\left(1 - e^{-2at}\right).$$

The Vasicek model belongs to the class of *affine term-structure* models, where the continuously compounded interest rate $R(t, T)$ can be expressed as a linear function of the short rate $r(t)$, i.e., in the form

---

[2]Changing the measure from the real-world one to the risk-neutral one, the coefficient of dt is reduced by $-\lambda\sigma$. Provided the price of risk λ is constant the parameter b is changed to $b' = b - \lambda\sigma/a$.

$$R(t,T) = \alpha(t,T) + \beta(t,T)r(t) \tag{7.11}$$

with coefficients being deterministic functions of t and T. Generally, if (7.1) is a short rate risk-neutral dynamics, such that $m(r, t) = \lambda(t)r + \eta(t)$ and $s^2(r, t) = \gamma(t)r + \delta(t)$ are affine functions of r, then the model has an affine term-structure. To prove the claim, it is enough to show that the zero-coupon bond can be written in the form

$$P(t,T) = A(t,T)e^{-B(t,T)r(t)}. \tag{7.12}$$

Indeed, then $\alpha(t, T) =- (\ln A(t, T))/(T - t)$ and $\beta(t, T) = B(t, T)/(t - t)$. According to Ito's lemma the zero-coupon bond (7.12) must satisfy the stochastic differential equation

$$dP = \left(A'e^{-Br} - AB're^{-Br} - ABe^{-Br}m + \frac{1}{2}AB^2s^2e^{-Br}\right)dt - ABe^{-Br}sdz.$$

Since we work under the risk-neutral measure, the drift of P must be equal to r, and consequently, by the classical argument, we obtain the differential equation

$$A'e^{-Br} - AB'e^{-Br} - ABe^{-Br}m + \frac{1}{2}AB^2s^2e^{-Br} = rAe^{-Br},$$

i.e., dividing by Ae^{-Br} and collecting the terms of r

$$\left(\frac{A'}{A} - B\eta + \frac{1}{2}B^2\delta\right) + \left(-B' - B\lambda + \frac{1}{2}B^2\gamma - 1\right)r = 0. \tag{7.13}$$

Because Eq. (7.13) must hold for all possible values of r, it is necessary and sufficient to solve the following two ordinary differential equations for A and B:

$$\begin{aligned} -B' - B\lambda + \frac{1}{2}B^2\gamma - 1 = 0, \\ (\ln A)' - B\eta + \frac{1}{2}B^2\delta = 0. \end{aligned} \tag{7.14}$$

The first (Riccati) differential equation can, in general, be solved numerically. In specific cases it might be solved analytically, for example, if the coefficients λ and γ are constant. Integrating the second equation then easily gives us the function A. The solutions must also satisfy the initial conditions $A(T, T) = 1$ and $B(T, T) = 0$ because $P(T, T) = 1$.

In the case of the Vasicek's process, $\lambda = -a$ and $\gamma = 0$, therefore we simply need to solve the linear first-order differential equation with constant coefficients $-B' + ab = 1$. The general solution is $B = (1 + ce^{at})/a$, and solving the initial condition for c we obtain

$$B(t,T)=\frac{1-e^{-a(T-t)}}{a}. \tag{7.15}$$

Finally, by integrating the second differential equation in (7.14) and solving the initial condition $A(T,T)=1$ we get

$$A(t,T)=\exp\left(\frac{(B(t,T)-T+t)(a^2b-\sigma^2/2)}{a^2}-\frac{\sigma^2 B(t,T)^2}{4a}\right). \tag{7.16}$$

The analytical expression for $P(t,T)$ can be used to calibrate the Vasicek's model. For example, based on a historical short-rate series, we can estimate the diffusion coefficient σ, and then calibrate the coefficients a and b in order to fit best the current term-structure of interest rates. The term-structure of Vasicek's model can be increasing, decreasing, or slightly humped, but, generally, we can never exactly fit the actual term-structure of interest rates. In order to find optimal a and b, certain metrics need to be defined. For example, given market zero-coupon rates $R^M(0,T_i)$ for $i=1,\ldots,n$ and the diffusion coefficient σ (estimated from the historical data), we need to find a and b minimizing the sum of squared errors between the model implied rates $R^{Vas}(0,T_i;a,b,\sigma)$ and the observed rates:

$$SSE(a,b)=\sum_i\left(R^M(0,T_i)-R^{Vas}(0,T_i;a,b,\sigma)\right)^2.$$

The Vasicek model also yields analytical formulas (Jamshidian 1989) for European type options on zero-coupon bonds, fixed coupon bonds, caps, and floors. Let us consider a European call option exercised at T with the strike price K on a zero-coupon bond maturing at T^*. According to (7.12) the payoff

$$\text{payoff}(T)=(P(T,T^*)-K)^+=f(r(T))$$

is a function of the short rate at time T. The option value can be expressed as

$$c_0=E\left[\exp\left(-\int_0^T r(s)ds\right)f(r(T))\right]=E\left[e^{-\bar{r}(T)T}f(r(T))\right] \tag{7.17}$$

where the expectation is with respect to the (money market account) risk-neutral measure, and where

$$\bar{r}(T)=\frac{1}{T}\int_T^1 r(s)ds \tag{7.18}$$

is the average interest rate from time 0 to T. The discount factor $e^{-\bar{r}(T)T}$ unfortunately cannot be taken out of the expectation, and to calculate the expected value we need to know the joint distribution of the random variables $r(T)$ and $\overline{r}(T)$. The rate $r(T)$ is normally distributed with mean and variance given by (7.10). Since each $r(s)$ in (7.18) is, according to (7.10), a combination of a mean depending on s and on a weighted (hyperfinite) sum of $dz(\tau)$ for $\tau \in [0, s)$ it follows that the average rate $\overline{r}(T)$ is also a combination of a mean and a weighted sum of $dz(\tau)$ for $\tau \in [0, T)$:

$$\begin{aligned}\overline{r}(T) &= \frac{1}{T}\int_0^T \left(e^{-as}r(0) + b(1-e^{-as}) + \sigma e^{-as}\int_0^s e^{a\tau}dz(\tau)\right) = \\ &= \frac{1}{T}\int_0^T (e^{-as}r(0) + b(1-e^{-as}))ds - \frac{1}{T}\int_0^T \left(\sigma e^{a\tau}\int_0^T e^{-as}ds\right)dz(\tau)\end{aligned} \tag{7.19}$$

Therefore, $\overline{r}(T)$ is also normally distributed and, moreover, $r(T)$ and $\overline{r}(T)$ have a joint normal distribution. Both integrals on the right-hand side of (7.19) can be easily expressed analytically; similarly, we can express the variance and covariance of $\overline{r}(T)$ and $r(T)$ using the facts that $\text{cov}(dz(\tau_1), dz(\tau_2)) = 0$ if $\tau_1 \neq \tau_2$ and $\text{var}(dz(\tau)) = dt$. Finally, the expected value can be analytically expressed as a double integral of the discounted payoff weighted by the bivariate normal density. The integral can, with a bit of luck and good analytical skills, be simplified to a nicer form. The Jamshidian's (1989) result is the following formula similar to the one of Black and Scholes:

$$c_0 = LP(0, T^*)N(h) - KP(0, T)N(h - \sigma_P) \tag{7.20}$$

where L is the bond principal, K is the strike price,

$$h = \frac{1}{\sigma_P}\ln\frac{LP(0, T^*)}{P(0, T)K} + \frac{\sigma_P}{2},$$

and the standard deviation of the logarithm of the bond price at time T

$$\sigma_P = \frac{\sigma}{a}\left(1 - e^{-a(T^*-T)}\right)\sqrt{\frac{1 - e^{-2aT}}{2a}}.$$

Likewise, a put option on the zero-coupon bond with the same parameters is valued by

$$p_0 = KP((0, T)N(-h + \sigma_P) - LP(0, T^*)N(-h).$$

The zero-coupon bond valuation formula can also be used to value caplets and floorlets: Let us consider a caplet on the interest rate $R_M(T, T^*)$ expressed in the money market compounding, exercised at time T^*, and with the fixed exercise rate R_K. The payoff of the caplet on a principal L and discounted to the time T is

$$\frac{L\delta(R_M - R_K)^+}{1 + R_M\delta} = \left(L - \frac{L(1 + R_K\delta)}{1 + R_M\delta}\right) = (L - L(1 + R_K\delta)P(T, T^*))^+,$$

where δ is the time factor from T to T^*. Therefore, the caplet is, in terms of the valuation, equivalent to a European put option on the zero-coupon bond $P(T, T^*)$ multiplied by the face value $L(1 + R_K\delta)$ and with the strike price L. Similarly, a floorlet can be valued through the call on a zero-coupon bond valuation formula.

Vasicek's single-factor model can also be used to value a European call (or put) on a fixed coupon bond using Jamshidian's (1989) trick. The value of the bond Q at time T is a (cash-flow $C_1, \ldots, C_n$) weighted sum of zero-coupon bonds $P(T, T_1), \ldots, P(T, T_n)$, each depending monotonically on the short rate $r(T)$. Therefore, considering a time T European call on $Q = Q(r(T))$ with an exercise price K, there is a rate r^*, so that $Q(r^*) = K$, and the call option will be exercised if and only if $r(T) < r^*$. Moreover, the payoff on the fixed-coupon bond $(Q(r(T)) - K)^+$ equals to a weighted sum of payoffs on the zero-coupon bond call options $C_i(P(T, T_i) - K_i)^+$ where the strike prices $K_1, \ldots, K_n$ are calculated as the zero-coupon bond values determined by r^* at time T. Therefore, the fixed-coupon bond call option value equals to the sum of zero-coupon bond call option values calculated by (7.20) with appropriate parameters as outlined above.

7.1.3 Cox–Ingersoll–Ross (CIR) Model

The Cox, Ingersoll, Ross (1985) model, based on a general equilibrium asset pricing model, introduces a "square-root" term into the diffusion coefficient of the Vasicek's model:

$$dr = a(b - r)dt + \sigma\sqrt{r}dz. \tag{7.21}$$

Contrary to the Vasicek's model, the instantaneous interest rate is always positive, provided $2ab > \sigma^2$ (also known as Feller's condition, Feller 1968). Because of this property and relatively good analytical tractability, the model has been a benchmark for many years, replacing the Vasicek's model. However, its analytical solutions are not as "nice" as in the case of the Vasicek's model. The distribution of $r(t)$ is non-central chi-squared (see Cox et al. 1985 or Brigo and Mercurio 2006 for a full description of the distribution and its parameters), making the model analysis more difficult. Nevertheless, it still has an affine term-structure, since the square of the diffusion coefficient equals $\sigma^2 r$. The differential equations (7.14) can be solved, yielding relatively feasible solutions:

$$B(t, T) = \frac{2\left(e^{\gamma(T-t)} - 1\right)}{(\gamma + a)\left(e^{\gamma(t-t)} - 1\right) + 2\gamma},$$

$$A(t,T)=\left(\frac{2\gamma e^{(a+\gamma)(T-t)/2}}{(\gamma+a)(e^{\gamma(T-t)}-1)+2\gamma}\right)^{2ab/\sigma^2}\quad\text{with}\quad\gamma=\sqrt{a^2+2\sigma^2}.$$

The affine form of the zero-coupon bond prices (7.12) can be used to calibrate the model parameters similarly to the case of the Vasicek's model. The valuation formula for zero-coupon bond options is then given in an analytic form involving the complex noncentral chi-squared distribution density function (Cox et al. 1985).

To summarize, the CIR model is still analytically tractable, and it is more consistent with empirical interest rate behavior (no negative values) compared to the Vasicek model, but the trade-off is a higher complexity of its analytical solutions and of its distributional properties (involving the noncentral chi-squared distribution).

7.1.4 Ho–Lee Model

The Vasicek's and CIR models are equilibrium models, where the actual term-structure is an output depending on a few parameters, rather than an input of the model. The parameters can be calibrated to fit approximately the term-structure, but the approximation might be unsatisfactory for the pricing of interest rate derivatives. For example, if the model implied that the forward interest rates were lower than the market forward interest rates, then an at-the-money caplet would be valued as if it were in-the-money, and a relatively small difference between the model-implied rates and the market forward rates would cause a relatively large valuation error. The Vasicek or CIR model can be calibrated to fit interest rates in two specific maturities T_I and T_2, so that options on the rate $R(T_1,T_2)$ can be valued consistently (there are two parameters a and b, with σ reflecting the historical or expected volatility). However, it is generally impossible to calibrate the two parameters to fit precisely the interest rates in more than two maturities in order to price a series of interest rate options, e.g., a cap, a fix-coupon bond option, a swaption, etc.

The Ho–Lee model is the simplest *non-arbitrage model* that is able to fit exactly today's term-structure of interest rates. It has the form

$$dr=\theta(t)dt+\sigma dz\ , \tag{7.22}$$

where σ is a constant, but $\theta(t)$ is a deterministic function of time chosen to fit precisely the today's term-structure. The model can be relatively easily solved analytically. Integrating the right-hand side of (7.22), we see that $r(t)=r(0)+\int_0^t\theta(s)ds+\sigma z(t)$ is normally distributed. It turns out that options can be valued by the same formulas (7.20) as for the Vasicek model with $\sigma_P=\sigma(T^*-T)\sqrt{T}$.

It is also an affine model with the differential equations (7.14) taking the form

$$-B' = 1,$$
$$(\ln A') - B\theta + \frac{1}{2}B^2\sigma^2 = 0.$$

Since $B(T,T) = 0$ and $A(T,T) = 1$, it follows that $B(t,T) = T - t$,

$$\ln A(t,T) = -\int_t^T (T-s)\theta(s)ds + \frac{1}{6}\sigma^2(T-t)^3, \quad \text{and} \tag{7.23}$$

$$P(t,T) = A(t,T)e^{-(T-t)r(t)}. \tag{7.24}$$

If we are given today's term-structure by the zero-coupon bond prices $P(0,T)$, for $T > 0$, and the instantaneous standard deviation of the short rate σ, then the goal is to calibrate the function $\theta(t)$. The factor $A(0,T)$ can be expressed from (7.24), and so (7.23) for $t = 0$ it can be written in the form

$$\ln P(0,T) + T \cdot r(0) = -\int_0^T (T-s)\theta(s)ds + \frac{1}{6}\sigma^2 T^3.$$

Differentiating the equation twice with respect to T we obtain

$$\frac{\partial^2}{\partial T^2}\ln P(0,T) = -\theta(T) + T\sigma^2.$$

Note that

$$f(t,T_1,T_2) = \frac{\ln P(t,T_1) - \ln P(t,T_2)}{T_2 - T_1} \tag{7.25}$$

defines the $T_1 \times T_2$ forward rate in continuous compounding, and so the time T instantaneous forward rate is

$$F(t,T) = \lim_{T_2 \to T} f(t,T,T_2) = -\frac{\partial}{\partial T}\ln P(t,T). \tag{7.26}$$

Hence, the Ho–Lee model is exactly calibrated with respect to today's term-structure of interest rates if

$$\theta(t) = F'(0,t) + \sigma^2 t,$$

where $F'(0,t)$ is the derivative of the today's instantaneous forward curve with respect to t.

The *STIR futures convexity adjustment* with respect to forward interest rates mentioned in Sect. 3.2 can be derived from the Ho–Lee model. Let us consider a STIR futures rate $F(t,T_1,T_2)$ and the forward rate $f(t,T_1,T_2)$ in continuous compounding given by (7.25). The futures contract is settled daily (i.e., almost continuously), and so the futures rate is a martingale with respect to the traditional risk-neutral measure. This is not the case of the forward rate. Firstly, according to Ito's lemma, the zero-coupon bond price (7.24) follows (with respect to the traditional risk-neutral measure) the equation:

$$dp(t,T) = r(t)P(t,T)dt - (T-t)\sigma P(t,T)dz,$$

and thus the forward interest rate (7.25) follows:

$$df = \sigma^2 \frac{(T_2-t)^2 - (T_1-t)}{2(T_2-T_1)} dt + \sigma dz. \tag{7.27}$$

Therefore, the expected change of $f(t, T_1, T_2)$ between time 0 and time T_1 equals the integral of the dt term on the right-hand side of (7.27)

$$\begin{aligned} E[f(T_1,T_1,T_2)] - f(0,T_1,T_2) &= \int_0^{T_1} \sigma^2 \frac{(T_2-t)^2 - (T-t)^2}{2(T_2-T_1)} dt = \\ &= \frac{\sigma^2}{6(T_2-T_1)} \left[(T_1-t)^3 - (T_2-t)^3\right]_0^{T_1} \\ &= \frac{\sigma^2}{2} T_1 T_2 \end{aligned} \tag{7.28}$$

If we assume that the forward contract settlement is based on interest rates in continuous compounding and that the futures rate equals the forward rate at the time of settlement, then

$$F(T_1,T_1,T_2) = f(T_1,T_1,T_2), \quad \text{and so} \quad E[F(T_1,T_1,T_2)] = E[f(T_1,T_1,T_2)],$$

where the expectation is taken from the perspective of time zero. Since the futures rate is a martingale, i.e., $F(0, T_1, T_2) = E[F(T_1, T_1, T_2)]$, we obtain from (7.28) that the futures rates are higher than the forward rates, and the difference is

$$F(0,T_1,T_2) - f(0,T_1,T_2) = \frac{\sigma^2}{2} T_1 T_2. \tag{7.29}$$

The argument was based on rates in continuous compounding. The market futures and forward rates are expressed in simple money market compounding and they are slightly larger than the rates in continuous compounding by approximately the same difference. Consequently, the convexity adjustment (7.29) also remains valid if the rates are expressed in the simple compounding. Alternatively, we could start with the forward rate in simple compounding

$$f_{MM}(t,T,{}_1T_2) = \frac{P(t,T_1) - P(t,T_2)}{(T_2-T_1)P(t,T_2)},$$

which is a martingale under the $P(t, T_2)$ forward risk-neutral measure. When the numeraire is changed to the money market account, i.e., the measure is changed to the traditional risk-neutral measure (where the futures rate is a martingale), then there will be a drift that can be expressed using the change of numeraire theorem (see Sect. 6.1).

7.1.5 Hull–White Model and Other Non-arbitrage Models

The Ho–Lee model can be criticized from the same viewpoint as the Rendelman–Bartter model, since the variance of the short rate $\sigma^2 t$ is potentially unlimited. Moreover, the modeled short rate may attain negative values, since the distribution is normal. This deficiency is partially solved by the Hull–White (1990a) model, which generalizes the Vasicek model with the mean-reversion property and allows it to fit exactly the initial term-structure of interest rates. The long-term rate which is constant in the Vasicek model is allowed to be a function of time in the Hull–White model:

$$dr = (\theta(t) - ar)dt + \sigma dz. \tag{7.30}$$

The model can be solved using similar techniques to those used by the Vasicek model and the Ho–Lee model. Options on zero-coupon bonds are valued according to exactly the same formula (7.20) as for the Vasicek model. It is an affine model with the following analytic formulas valuing zero-coupon bonds:

$$P(t, T) = A(t, T)e^{-B(t,T)r(t)}$$

where

$$B(t, T) = \frac{1 - e^{-a(T-t)}}{a}$$

and

$$\ln A(t, T) = \ln \frac{P(0, T)}{P(0, t)} + B(t, T)F(0, t) - \frac{1}{4a^3}\sigma^2\left(e^{-aT} - e^{-at}\right)^2\left(e^{2at} - 1\right).$$

The function $\theta(t)$ can, similarly as in the Ho–Lee model, be calculated from the forward rate term-structure:

$$\theta(t) = F'(0, t) + aF(0, t) + \frac{\sigma^2}{2a}\left(1 - e^{-2at}\right).$$

We can show analogously to the case of the Ho–Lee model, that in the context of the Hull–White model, the convexity adjustment between futures and forward rates is

$$F(0,T_1,T_2)-f(0,T_1,T_2)=\frac{B(T_1,T_2)}{T_2-T_1}\times\left(B(T_1,T_2)\left(1-e^{-2aT_1}\right)+2aB(0,T_1)^2\right)\times\frac{\sigma^2}{4a}. \quad (7.31)$$

The *CIR model* can be generalized in a similar way, allowing the long-term rate to depend on time:

$$dr=(\theta(t)-ar)dt+\sigma\sqrt{r}dz.$$

The disadvantage of this model is that there is no analytical solution for $\theta(t)$ in terms of the initial term-structure and there is no guarantee that a numerical approximation of $\theta(t)$ will keep the short rate positive (Brigo and Mercurio 2006). This is one of the reasons why the Hull–White model has been more popular than the generalized CIR. However, there is an analytically tractable version of the non-arbitrage CIR proposed by Jamshidian (1995), where he assumes that the CIR volatility also depends on time, $\sigma(t)=\sqrt{\theta(t)/\delta}$, where $\delta>1/2$ is a constant.

Another well-known model that tries to deal with the negative rates is the *Black-Karasinski* (1991) *model*. It generalizes the *Exponential-Vasicek model* where $y=\ln r$ follows the mean-reverting stochastic differential equation with constant coefficients. The Black–Karasinski model allows the coefficients to depend on time:

$$dy=(\theta(t)-a(t)y)dt+\sigma(t)dz.$$

This model has become quite popular among practitioners and financial engineers due to its ability to fit the swaption volatility surface relatively well. However, like the Exponential–Vasicek model or the Rendelman–Bartter model, it is not analytically tractable, and, in it, the money-market account explodes.

7.1.6 Two-Factor Models

In a one-factor affine model, the whole term-structure depends on the short rate $r(t)$ due to the equation $R(t,T)=\alpha(t,T)+\beta(t,T)r(t)$. Consequently, from the perspective of time zero, the correlation between the two rates $R(t,T_1)$ and $R(t,T_2)$ with different maturities is always 100%. This contradicts empirical reality, where interest rates with different maturities are positively, but less than perfectly correlated. This issue becomes especially important if we need to price a derivative, whose time t payoff depends on the two rates $R(t,T_1)$ and $R(t,T_2)$, and so the valuation depends on the joint distribution between the two variables. This issue could be partially solved by generalizing the one-factor models to two-factor affine models

$$R(t,T) = \alpha(t,T) + \beta_1(t,T)x_1(t) + \beta_2(t,T)x_2(t) \tag{7.32}$$

where $x_1(t)$ and $x_2(t)$ follow one-factor models with different sources of uncertainty, which can possibly be correlated. The first factor may represent the short rate and the second one the long rate. The correlation between $R(t, T_1)$ and $R(t, T_2)$ given by (7.32) is then, generally, less than one, and the correlation between the factors gives the model an additional flexibility. For example, the *Vasicek's two-factor model* can be set up as follows:

$$\begin{aligned} r &= x_1 + x_2, \\ dx_1 &= k_1(\theta_1 - x_1)dt + \sigma_1 dz_1, \\ dx_2 &= k_2(\theta_2 - x_2)dt + \sigma_2 dz_2, \end{aligned}$$

with instantaneous correlation $dz_1 dz_2 = \rho dt$. The model can be extended with a deterministic shift

$$r = x_1 + x_2 + \varphi(t)$$

in order to fit exactly the initial zero-coupon term-structure. Due to the reasons explained above, essentially all one-factor models have been generalized to two or more factor versions with an increasing level of complexity.

One may also ask whether two factors are enough. The answer is provided by a number of studies [see Brigo and Mercurio (2006) for a review] using the principal component analysis and showing that the first two components (typically the parallel shift and the change of slope) explain around 90% of the variation in the yield curve, while the first three components (the first two plus the change of curvature) explain 95–99% of the variation. Consequently, the two or three-factor models should capture sufficiently well the dynamics of the whole yield curve.

7.2 Term-Structure Models

The short-rate models presented in the previous sections focus on the short-rate dynamics. The initial term-structure and its dynamics are a consequence of the model specification. The Ho–Lee and Hull–White non-arbitrage models allow the exact fitting of the initial term-structure of interest rates, but it turns out that there is only a limited flexibility in setting the interest rate volatility term-structure. Term-structure models try to capture the dynamics of a set of interest rates with different maturities by giving more freedom in the volatility structure specification. The general Heath–Jarrow–Morton (Heath et al. 1992) framework models instantaneous forward interest rates for all maturities. The Libor Market Model (LMM) focuses on forward interest rates with a defined periodicity, while the Swap Market Model (SMM) deals with forward swap rates. The LMM is compatible with the standard market model used to value caps and floors, while the SMM is compatible with the standard market model valuing swaptions. Therefore, they are called "market models."

7.2.1 Heath–Jarrow–Morton Model

The HJM one-factor model assumes that instantaneous forward interest rates $F(t, T)$ evolve under the traditional risk-neutral measure according to the following family of diffusion processes:

$$dF(t,T) = m(t,T)dt + s(t,T)dz(t), \text{for } t \le T. \tag{7.33}$$

The current term-structure is an input of the model setting

$$F(0,T) = F_M(0,T)$$

where $F_M(0, T)$ is the market instantaneous forward rate. The instantaneous drift $m(t, T)$ and the diffusion coefficient $s(t, T)$ depend on time and are allowed to be generally stochastic as well. In a multi-factor model $\mathbf{s}(t, T)$ and $d\mathbf{z}(t)$ would be N-dimensional vectors.

The main result of HJM is that in order to define a consistent non-arbitrage model, a modeler needs to specify only the volatility term-structure, i.e., the function $s(t, T)$ that is then used to calculate the drift $m(t, T)$. To illustrate the result, let us consider the Ho–Lee model with constant coefficients $dr = \theta dt + \sigma dz$. Then according to (7.24)

$$P(t,T) = \exp\left[\frac{\sigma^2}{6}(T-t)^3 - \frac{\theta}{2}(T-t)^2 - (T-t)r\right].$$

Recall that $F(t,T) = -\frac{\partial}{\partial T}\ln P(t,T)$, i.e.,

$$F(t,T) = -\frac{\sigma^2}{2}(T-t)^2 + \theta(T-t) + r,$$

and thus, according to the Ito's lemma

$$dF(t,T) = \sigma^2(T-t)dt + \sigma dz.$$

Hence, the drift of F depends only on the diffusion coefficient σ and time t. It cannot be specified independently of σ and there is no dependence on the short rate drift coefficient θ. This result is not a coincidence due to simplicity of the model—it is a general fact that can be proved in the HJM framework.

The general argument starts with the zero-coupon bond dynamics under the traditional risk-neutral measure:

$$dP(t,T) = r(t)P(t,T)dt + v(t,T)P(t,T)dz.$$

The volatility $v(t,T)$ depending on time t and maturity T is allowed to be, in general, stochastic. Since the bond price volatility declines to zero at maturity, as $P(T,T) = 1$, it must satisfy the condition $v(T,T) = 0$ for all T. The instantaneous forward rate $F(t,T)$ is a limit of the $T_1 \times T_2$ forward rates in continuous compounding:

$$f(t,T_1,T_2) = \frac{\ln P(t,T_1) - \ln P(t,T_2)}{T_2 - T_1}. \tag{7.34}$$

Ito's lemma can be applied in order to write down the stochastic differential equation for $\ln P(t,T_1)$, for $\ln P(t,T_2)$, and then combine the two equations for the forward rate (7.34)

$$df(t,T_1,T_2) = \frac{v(t,T_2)^2 - v(t,T_1)^2}{2(T_2 - T_1)}dt + \frac{v(t,T_1) - v(t,T_2)}{T_2 - T_1}dz. \tag{7.35}$$

The equation holds for every T_1 and T_2, hence if $T_1 = T$ and $T_2 = T + dT$ where dT is infinitesimal, then the coefficient of dz in (7.35) equals (up to an infinitesimal error) to the minus derivative $v_T(t,T)$ of $v(t,T)$ with respect to T, provided the derivative exists. Note that the derivative depends on the path ω, if $v(t,T)$ is stochastic. Similarly, the coefficient of dt approximates the derivative of $v(t,T)^2/2$ which is defined, if $v_T(t,T)$ exists,

$$\frac{1}{2}\frac{\partial v(t,T)^2}{\partial T} = v(t,T)v_T(t,T).$$

Therefore, the instantaneous forward rate follows the process

$$dF(t,T) = v(t,T)v_T(t,T)dt - v_T(t,T)dz. \tag{7.36}$$

Comparing the Eqs. (7.33) and (7.36), provided we are given the diffusion coefficient function, we obtain

$$s(t,T) = v_T(t,T), \text{ and } v(t,T) = -\int_t^T s(t,\tau)d\tau \tag{7.37}$$

as $s(t,t) = 0$. Therefore, the instantaneous drift of $F(t,T)$ is given by

$$m(t,T) = v(t,T)v_T(t,T)dt = s(t,T)\int_t^T s(t,\tau)d\tau. \tag{7.38}$$

This is the main HJM result. The advantage of the HJM approach is that the volatility term-structure can be separately specified for each instantaneous forward rate maturity T. This is not possible in the case of the short-rate models. It can be shown that the HJM model with deterministic and separable diffusion coefficients in the form $s(t,T) = \xi(t)\psi(T)$ is equivalent to the Hull–White model (7.30) with all parameters possibly depending on time (Brigo and Mercurio 2006). Nevertheless, the HJM framework becomes more general in a multivariate setting and with stochastic $s(t,T)$.

In practice, implementation of the HJM model is difficult. We have to use a discrete time scale and calibrate the model with respect to forward rates interpolated from the market interest rates given a limited number of maturities. Generally, there are no analytical formulas pricing zero-coupon bonds and options, and so all calculations must be performed through computer time-intensive simulations.

Moreover, the short-rate process in the general HJM framework is non-Markovian, which means that the binomial trees become non-recombining. To understand the unpleasant property of the HJM framework, note that $r(t) = F(t,t)$ and

$$F(t,t) = F(0,t) + \int_0^t dF(\tau,t).$$

Replacing dF by (7.36) we obtain

$$r(t) = F(0,t) + \int_0^t v(\tau,t)v_t(\tau,t)d\tau - \int_0^t v_t(\tau,t)dz(\tau). \tag{7.39}$$

The third term depends on the path of z from time 0 to time t; in addition, the second term also becomes path dependent if $v(\tau,t)$ is stochastic. Consequently, a binomial tree for $r(t)$ can hardly be recombining. To explain the issue further, let us differentiate (7.39) and use the assumption that $v(t,t) = 0$, then

$$dr(t) = F_t(0,t)dt + \left[\int_0^t \left(v(\tau,t)v_t(\tau,t) + v_t(\tau,t)^2\right)d\tau\right]dt - \left[\int_0^t v_{tt}(\tau,t)dz(\tau)\right]dt + v_t(t,t)dz(t)$$

The first and the last terms are Markovian, but the second and the third generally depend on the path, especially if $v(\tau,t)$ is stochastic. This means that the increment of $r(t)$ from t to $t + dt$ could depend on the path on which the process arrived at $r(t)$ contradicting the Markov property.

7.2.2 Libor Market Model

Brace, Gatarek, Musiela (1997) proposed a discrete time alternative to the HJM model, where the forward rates used in the market play the role of key modeled variables. In addition, the stochastic equation for each forward rate is set up in a way consistent with the standard market model. Specifically, following the notation of Hull (2018), let $0 = t, t_1, t_2, \ldots$ be a sequence of the reset times of an IRS contract, FRN, or a cap (floor). The periodicity may, for example, be 3 or 6 months. The key rates, needed to value the caps, floors, and other instruments, are the forward rates between times t_k and t_{k+1} in the simple money market compounding from the perspective of time t, i.e.,

$$F_k(t) = \frac{1}{\delta_k} \frac{P(t, t_k) - P(t, t_{k+1})}{P(t, t_{k+1})}, \tag{7.40}$$

where δ_k is the appropriate day fraction for the compounding period t_k to t_{k+1}. It follows from (7.40) that $F_k(t)$ is a martingale with respect to the $P(t, t_{k+1})$ forward risk-neutral measure and evolves according to the equation

$$dF_k(t) = \zeta_k(t) F_k(t) dz \tag{7.41}$$

where $\zeta_k(t)$ is the volatility of $F_k(t)$, generally depending on time t. The distribution of $F_k(t)$ is lognormal in line with the standard market model, provided volatility is a deterministic function of time. The Eqs. (7.41) are stated with respect to different probability measures corresponding to different periods t_k to t_{k+1}. To formulate an alternative of HJM, the numeraires $P(t, t_{k+1})$ need to be changed to a common numeraire. Let h be the *rolling CD account* given by the periods $0 = t_0, t_1, t_2, \ldots$. It is a practical alternative to the money-market account; it starts with a unit deposit (investment) of $h(0) = 1$ into the zero-coupon bonds maturing at t_1, i.e., for $t \leq t_1$ its value is $h(t) = P(t, t_1)P(0, t_1)^{-1}$, then at time t_1 the proceeds are invested into zero-coupon bonds maturing at t_2, and so on. Therefore, for $t_k \leq t \leq t_{k+1}$ the rolling CD account

$$h(t) = P(t, t_{k+1}) h(t_k) P(t_k, t_{k+1})^{-1}$$

is proportional to the zero-coupon bond $P(t, t_{k+1})$. Consequently if $m(t)$ denotes the index of the next reset time, i.e., the least m such that $t \leq t_m$, then at time t the rolling CD measure is risk-neutral with respect to $P(t, t_{m(t)})$. In order to transform (7.41) to the rolling CD risk-neutral measure framework, we need to change the numeraire $P(t, t_{k+1})$ to $P(t, t_{m(t)})$. According to Sect. 6.1, the corresponding change of drift is $\alpha = \rho\sigma_w\sigma_f$, where $w = P(t, t_{m(t)})/P(t, t_{k+1})$ is the numeraire ratio, $f = F_k(t)$, and ρ is the instantaneous correlation between w and f. If $v_k(t)$ denotes the volatility of $P(t, t_k)$

then it is easy to show[3] that the volatility of w is $v_{m(t)} - v_{k+1}$. Moreover, the instantaneous correlation is 1 since the changes in w and f are driven by one source of uncertainty dz, and so

$$dF_k(t) = \zeta_k(t)\left(v_{m(t)}(t) - v_{k+1}(t)\right)F_k(t)dt + \zeta_k(t)F_k(t)dz \tag{7.42}$$

with respect to the rolling CD risk-neutral measure. The zero-coupon bond volatilities in (7.42) cannot be specified independently of the forward rate volatilities. According to (7.40) there is a link that needs to be taken into account. It follows from (7.40) that

$$\ln P(t, t_k) - \ln P(t, t_{k+1}) = \ln(1 + \delta_k F_k(t)),$$

and so according to Ito's lemma

$$v_k(t) - v_{k+1}(t) = \frac{\delta_k \zeta_k(t) F_k(t)}{1 + \delta_k F_k(t)}.$$

By induction, for $m < k + 1$

$$v_m(t) - v_{k+1}(t) = \sum_{i=m}^{k} (v_i - v_{i+1}) = \sum_{i=m}^{k} \frac{\delta_i \zeta_i(t) F_i(t)}{1 + \delta_i F_i(t)}.$$

Therefore, the drift in (7.42) is completely determined by the forward rate volatilities $\zeta_i(t)$ and the rates $F_i(t)$:

$$dF_k(t) = \sum_{i=m}^{k} \frac{\delta_i \zeta_i(t) \zeta_k(t) F_i(t)}{1 + \delta_i F_i(t)} F_k(t)dt + \zeta_k(t)F_k(t)dz. \tag{7.43}$$

It follows from (7.43) that the multidimensional process $\langle F_1, F_2, \ldots \rangle$ is Markovian since the drift and volatility depend only on the information available at time t. Unfortunately, the bond and option prices do not have any analytical solution under LMM, but its numerical implementation is easier than in the case of the HJM model. Similarly to HJM, the Libor Market Model can be easily extended to a multivariate setting.

[3]If $dP(t, t_m) = (\ldots)dt + v_m P(t, t_m)dz$ then according to Ito's lemma $d \ln P(t, t_m) = (\ldots)dt + v_m dz$. Hence $d \ln (P(t, t_m)/P(t, t_{k+1})) = (\ldots)dt + (v_m - v_{k+1})dz$, and so the volatility of $P(t, t_m)/P(t, t_{k+1})$ is $v_m - v_{k+1}$.

7.2.3 Implementation of the LMM

If there is a market for caps and floors, then it is natural to calibrate the LMM model to the market-implied caplet and floorlet volatilities (see, e.g., Fig. 6.4). The volatilities do depend on time, but are quoted only for several maturities and can be assumed to be piecewise constant. Similarly, it is reasonable to assume that the instantaneous volatilities $\zeta_k(t) = \Lambda_{k\ -\ m(t)}$ are piecewise constant between $t_{m(t)\ -\ 1}$ and $t_{m(t)}$ and depend only on $k - m(t)$. Hence, if σ_1 is the quoted caplet volatility for the period $t_1 \times t_2$ then we set $\Lambda_0 = \sigma_1$. Next, if σ_2 is the quoted caplet volatility for the period $t_2 \times t_3$ then we equate the variances

$$\sigma_2^2 t_2 = \Lambda_0^2 t_1 + \Lambda_1^2 (t_2 - t_1),$$

solve for Λ_1, and so on in a bootstrapping-like procedure. If there are no quoted caplet volatilities, then $\Lambda_{i\ -\ 1}$ can be estimated from historical data looking at the series of quoted (implied) forward rates $F_i(t)$ for a fixed period $t_i \times t_{i+1}$ in the past with t going from t_0 to t_1.

The advantage of the piecewise constant volatility assumption is that the interest rates can be sampled by a Monte Carlo simulation going over the time grid $t = t_0, t_1, \ldots$. For $t = 0$ the rates $F_k(0)$ are set equal to the current market forward rates. Next, let us assume that we have already simulated $F_k(t_j)$ for $k = j, \ldots$ and we want to sample $F_k(t_{j+1})$ for $k = j+1, \ldots$. Note that $F_{j\ +\ 1}(t_j)$ does not need to be sampled, since the rate has been reset at time t_j and is not active anymore. Applying Ito's lemma to (7.43) we obtain

$$d \ln F_k(t) = \left(\sum_{i=m(t)}^{k} \frac{\delta_i F_i(t) \Lambda_{i-m(t)} \Lambda_{k-m(t)}}{1 + \delta_i F_i(t)} - \frac{\Lambda_{k-m(t)}^2}{2} \right) dt + \Lambda_{k-m(t)} dz. \tag{7.44}$$

For $t \in (t_j, t_{j+1})$ the diffusion coefficient in (7.44) is constant. If we assume, as an approximation, that $F_k(t) = F_k(t_j)$, then the drift term is constant as well, and we can sample the next forward rate from a lognormal distribution

$$F_k(t_{j+1}) = F_k(t_j) e^{\eta}, \quad \text{where}$$

$$\eta = \left(\sum_{i=j+1}^{k} \frac{\delta_i F_i(t_j) \Lambda_{i-m(t)} \Lambda lk - m(t)}{1 + \delta_i F_i(t_j)} - \frac{\Lambda_{k-m(t)}^2}{2} \right) \delta_j + \Lambda_{k-j-1} \varepsilon_j \sqrt{\delta_j}$$

and $\varepsilon_j \sim N(0, 1)$ is sampled independently for each $j = 0, 1, \ldots$.

7.2.4 Principal Component Analysis

A multifactor extension of the LMM can be implemented using the method of *Principal Components Analysis* (PCA). We are going to give an overview of this important method applicable in various parts of financial engineering.

Generally, if $\mathbf{X}$ is an $N \times m$ matrix with N rows of observations (e.g., historical returns of m market factors that are assumed to have zero means), then the idea is to define a linear transformation $\mathbf{Y} = \mathbf{XW}$ of $\mathbf{X}$ so that $\mathbf{W}$ is an $m \times m$ matrix and the transformed returns $\mathbf{Y}$ become uncorrelated. The columns of the transformed matrix then correspond to the returns of linear combinations of the m market factors given by the columns of $\mathbf{W}$, and the goal is to make them orthogoval, i.e., $\mathbf{Y'Y} = \mathbf{W'}(\mathbf{X'X})\mathbf{W}$ should be a diagonal matrix. This is achieved if $\mathbf{W}$ is an orthogonal matrix of column eigenvectors obtained from the covariance matrix $\mathbf{X'X}/N$ The covariance matrix is always positive semidefinite. If the factors are not collinear, then it is positive definite and we can assume that $\mathbf{W}$ is orthonormal. The covariance matrix of $\mathbf{XW}$ equals to the eigenvalue diagonal matrix $\mathbf{D}$. The eigenvalues $\lambda_1, \ldots, \lambda_m$ are positive and can be ordered from the largest to the smallest one. That is, if $\mathbf{x}$ denotes a $1 \times m$ random vector of the market factors, then the elements of the transformed $1 \times m$ random vector $\mathbf{y} = \mathbf{xW}$ can be assumed to be uncorrelated (based on the transformed historical data) with variances given by $\mathbf{D}$. The elements of the q-th eigenvector $w_{k,q}$ $(k = 1, \ldots, m)$ are known as *factor loadings*, while the elements y_q $(q = 1, \ldots, m)$ of $\mathbf{y}$ are called the *factor scores*. The variances of the factor scores, i.e., $\lambda_1, \ldots, \lambda_m$, add up to the total variance of the data (the sum of the variances of x_q, $q = 1, \ldots, m$).[4] In order to reduce the dimension of the data, we choose a $p < m$ so that the total variance given by the first p factors $\sum_{q-1}^{p} \lambda_q$ is sufficiently close (e.g., at least 95%) to the full variance $\sum_{q=1}^{m} \lambda_q$. Then, since $\mathbf{x} = \mathbf{yW'}$, we can use the approximation $x_k = \sum_{q=1}^{p} y_q w_{k,q}$. The factor loadings $w_{k,q}$ should be rescaled to $\alpha_{k,q} = c \cdot w_{k,q}$ so that the observed variance of data equals to the model variance after a reduction in the number of PCA factors:

$$\operatorname{var}[x_k] = \operatorname{var}\left[\sum_{q=1}^{p} \alpha_{k,q} y_q\right] = \sum_{q=1}^{p} \alpha_{k,q}^2 \lambda_q, \text{ i.e.}$$

$$\alpha_{k,q}^2 = \frac{\operatorname{var}[x_k]}{\sum_{q=1}^{p} w_{k,q}^2 \lambda_q} w_{k,q}^2.$$

[4]Note that $\mathbf{x} = \mathbf{yW'}$ and $\operatorname{var}[\mathbf{x}] = E[\mathbf{xx'}] = E[\mathbf{yW'Wy'}] = E[\mathbf{yy'}] = \operatorname{var}[\mathbf{y}]$, since $\mathbf{WW'} = \mathbf{I}$.

If x_k represents the return of a market factor F_k over a period of length Δt, then $\zeta_{k,q} = \alpha_{k,q}\sqrt{\lambda_q}/\sqrt{\Delta t}$ would be the q-th component of the volatility of F_k. Finally, the multi-factor model for F_k can be specified as

$$\frac{dF_k}{F_k} = (\ldots)dt + \sum_{q=1}^{p} \zeta_{k,q} dz_q$$

with p independent Wiener processes dz_q.

7.2.5 Swap Market Model

The LMM is appropriate for pricing caps, floors, and other related derivatives. In the case of swaptions, we should, rather, calibrate the model to quoted swaption volatilities, which is an additional complication. It appears that it would be more natural to model directly the forward swap rates underlying the swaptions in the context of the standard market model. In the case of emerging markets with no cap or swaption volatilities, IRS swaps are the most liquid instruments, and, again, we can argue that it would be more natural to calibrate a model working directly with the swap rates, rather than with forward "Libor" rates which must be recalculated from the quoted swap rates (although, in fact, the model uses the forward rather than the spot swap rates). Such a model, called the *Swap Market Model* (SMM), was proposed by Jamshidian (1997). Let us again fix the times $t_0 = 0, t_1, t_2, \ldots, t_n$ and let $S_i(t)$ be the forward swap rate between t_i and t_n as seen at time t. Let us assume, for the simplicity of the formulas, that the periodicity of swap payments is 1 year and $\delta_i = 1$.Then

$$S_i(t) = \frac{P(t, t_i) - P(t, t_n)}{A_i(t)}$$

is a martingale with respect to the annuity $A_i(t) = \sum_{j=i+1}^{n} P(t, t_j)$ forward risk-neutral measure. Therefore, $dS_i(t) = \sigma_i(t)S_i dz$ with respect to the $A_i(t)$ measure. In order to unify the measures, we choose the $P(t, t_n)$ forward risk-neutral measure. The change of numeraire result leads, after tedious, but straightforward algebraic calculations (see Jamshidian 1997), to the equation

$$\frac{dS_i}{S_i} = -\sum_{j=i+1}^{n-1} \frac{s_{i,j}\sigma_i\sigma_j S_j}{(1 + S_j)s_i} dt + \sigma_i dz, \quad \text{where} \tag{7.45}$$

$$s_{ij} = \sum_{k=j}^{n-1} \prod_{l=i+1}^{k} (1 + S_l) \quad \text{for} \quad 1 \le i \le j \le n-1 \quad \text{and} \quad s_i = s_{i,i}.$$

To estimate the swap rate volatility from historical forward swap rates or swaption quoted volatilities we again assume that $\sigma_i(t) = \Lambda_{i\ -\ m(t)}$ depends only on $i - m(t)$. Note that under the SMM probability measure, we have to take into account that the numeraire zero-coupon bond matures at t_n, hence to calculate the market value of a payoff depending on the swap rate S_i paid at t_i we need to evaluate (by Monte Carlo simulations)

$$P(0, t_n) \cdot E\left[\frac{\text{payoff}(S_i)}{P(t_i, t_n)}\right].$$

The model can, similarly to LMM, be implemented in the multi-factor form using the principal component analysis.

We have already mentioned the fact that the standard market model assumption of the lognormal swap rates is not generally compatible with the lognormality of forward interest rates. There is a similar issue when comparing SMM and LMM. However, as discussed in Brigo and Mercurio (2006), the models should yield similar results, if properly calibrated.

7.2.6 A Comparison of the Models: A Case Study

We have presented several short-rate and a few term-structure models. There are many other interest models that can be found in literature [see Brigo and Mercurio (2006) for an extensive review]. A financial engineer who needs to value a certain type of interest rate derivatives faces the puzzling question: which model should be chosen? Currently, there is no clear consensus among researchers and practitioners on how to answer that question. The LMM is often considered the best candidate for a universally acceptable interest rate model. In any case, a modeler must keep in mind that the models might lead to quite different distributional properties of the future interest rates, and to different valuation results.

For example, Černý (2011) compares Vasicek, Ho–Lee, Hull–White, and LMM theoretically and empirically in a case study valuing a complex real-life interest rate swap. The swap was entered into between a large Czech city and a large Czech bank in 2006 in order to offset another exotic swap between the city and an international bank. The notional amount of the swap was around 5.4 billion CZK and the maturity 2013. According to the confirmation, the city paid 12-month CZK Pribor capped at 6% and floored with increasing strikes 3–4% specified for individual accrual periods. The bank paid annually 5.55% − Spread, where the Spread = $\text{IRS}_{10} - \text{IRS}_2$ with the 10- and 2-year CZK swap rates being fixed just before the payment dates. Disregarding the question why a public entity should enter into such a contract, we see that the valuation of the swap involves the task of valuing a collar, applying a

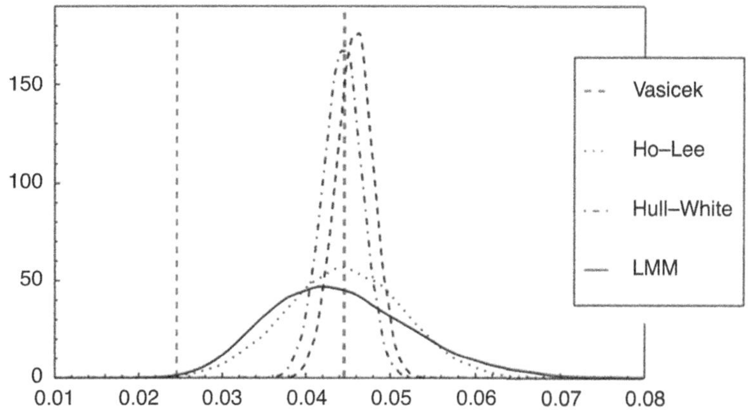

Fig. 7.2 Empirical distribution of simulated 12M Pribor (19.3.2008 from the perspective of 19.6.2006, source: Cerny 2011)

convexity adjustment, and, in addition, analyzing the issue of correlations between different rates, because the payoffs also depend on the differences between future rates with different maturities. The derivative was valued by Černý (2011) from the perspective of the trade date (19.6.2006), and since there were no public quotes of cap or swaption volatilities, the models had to be calibrated from available historical data. Without going into much detail of the calibration and valuation exercise, we may look, for example, at Fig. 7.2 with the empirical distribution of the future 12-month Pribor (as of 19.3.2008) simulated by different models. Note that the distributions of the mean reverting models (Vasicek and Hull–White) are not as wide as the distributions of the models without the mean-reversion (Ho–Lee and LMM). The discrepancy could be solved for one, but not for all the maturities at the same time.

The market value of the exotic swap is evaluated for each simulated scenario of interest rate development. The calculated cash flows must be discounted consistently with the numeraire used by the models. The mean of the simulated present values gives an estimation of the market value. Inspecting the distributional properties of the modeled interest rates in the four models, it is not surprising that the distributions of their present value, their means, standard deviations, and other characteristics might significantly differ (see Fig. 7.3 and Table 7.1).

The negative value of the swap (around -100 million CZK from the perspective of the city) indicates that the city made a bad deal. Unfortunately, this is a common situation, in which a financial institution might misuse the know-how asymmetry: while the financial institution has quantitative analysts and models to value such a complex derivative, the nonfinancial counterparty does not know exactly what the value is and what the risks of the transaction are. A public sector entity should first analyze its needs and use derivatives only for hedging in a very conservative (risk-averse) approach. The parameters of the transactions should be verified beforehand

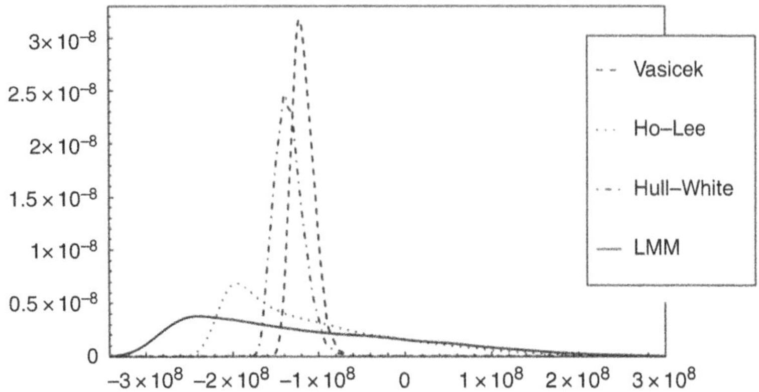

Fig. 7.3 Empirical distribution of the present value (source: Cerny 2011)

Table 7.1 Swap present value distribution mean and standard deviation

Model	Mean PV (million CZK)	Std. Dev. (million CZK)
Vasicek	−118.5	13.1
Hull–White	−131.8	18.6
Ho–Lee	−108.6	99.5
LMM	−98.3	124.1

by an independent party, especially if the public entity does not have the internal analytical skills for the task.

7.2.7 Calibration and Validation of Interest Rate Model

The example above illustrates the problem of interest model selection, calibration, and the importance of a validation process. In Sect. 7.1 we have mentioned that Vasicek's model could be calibrated by minimizing the sum of square differences between the actual market interest rates and the model implied interest rates for a set of selected maturities. Generally, by calibration of a model with a set of parameters Θ we mean an optimization procedure where the goal is to minimize the differences between observed prices P_i, and the corresponding model implied prices $P_i^M(\Theta)$, $i = 1, \ldots, n$, on a set of calibration instruments. The difference may involve a general loss function and varying weights can be assigned to different instruments:

$$TL(\Theta) = \sum_{i=1}^{n} w_i L\left(P_i, P_i^M(\Theta)\right).$$

The loss function may be the standard squared difference, $L(x, y) = (x - y)^2$, the absolute difference, $L(x, y) = |x - y|$, or another function. The choice of calibration instruments depends on the model, its use, and the availability of market quotes. For example, if we need to price an exotic interest rate swap, like the one above, and if there are plain vanilla cap and/or swaption quotations, then a selection of these would be typically used as the benchmark calibration set. Note that the standard cap or swaption quotes are in terms of the standard market model volatility, and so the prices must be calculated applying the standard market model on a set of defined contracts. In case of non-arbitrage short rate models (Ho–Lee or Hull–White) or the term-structure models (HJM, LMM, . . .), the initial term-structure is automatically fit by construction of the model, but in case of non-arbitrage models the calibration set must include some instruments characterizing the term-structure, e.g., bond prices or directly the quoted interest rates. The varying weights might be necessary to scale appropriately the prices of instruments with different notional amounts, or in order to combine price differences with interest rate differences, etc. If there are no quoted interest rate volatility instruments then the volatility parameters need to be estimated based on historical data in a way consistent with the structure like in case of the Vasicek's model.

It is obvious that the process of selection and validation of an interest rate model involves many expert decisions. Since many trading and risk management decisions depend on the quality of such pricing and risk management models, there is a need for thorough testing and validation. By validation on the level of the development team we mean a comparison of the model with respect to other benchmark models, testing of its stability with respect to various calibration inputs, an inspection of its distributional characteristics with respect to observable (actual or historical) data, etc. In terms of organization, it is now required by the Basel regulation (BCBS 2019) that all internal models used by banks are validated independently of the model development process. The independent validation may be performed internally or externally and includes, besides the quantitative validation, a qualitative assessment of the methodology and its implementation in terms of correctness, consistency, and appropriateness with respect to the priced instruments. The models must be validated prior to their initial implementation, and then periodically, particularly where there have been structural changes in the market or in the portfolio of instruments for which the model is used.

8 Exotic Options, Volatility Smile, and Alternative Stochastic Models

This chapter starts with an overview of the zoology of exotic options, i.e., with options that are more complex than plain vanilla ones. Some exotic options can be valued by a modification of the Black-Scholes formula, while for some there are more complicated formulas, developed in the context of the geometric Brownian motion, and the others can be valued only numerically using Monte Carlo simulations, binomial tree techniques, or partial differential equations. For most of the exotic derivatives, it turns out that the geometric Brownian motion model calibrated to value correctly the plain vanilla options might give quite imprecise results. The empirical phenomenon called *the volatility smile* (or *surface*) demonstrates that the market does not, in fact, believe in lognormal returns and the volatility constant over time. This fact has led to the development of various alternative stochastic models that try to capture better the behavior of market prices, especially the jumps and stochastic volatilities of the underlying asset returns. We will discuss some of the best-known models in the last section.

8.1 Exotic Options

Plain vanilla European and American options are the most popular OTC and exchange-traded options. Other types of options, more complex, less frequently traded, and more difficult to value are called *exotic*. In fact, only the European options that can be valued by the analytical Black-Scholes formula are truly *plain vanilla*. The American options can, in general, be valued only numerically and have some of the features of exotic options. Due to their popularity (most exchange-traded options are of American type) they are, nevertheless, also classified as plain vanilla. Exotic options can be developed for different reasons. They can match specific hedging or market speculation needs. However, due to their difficult valuation, they can also bring more profit to a financial institution utilizing know-how asymmetry. Unfortunately, the client may not realize the magnitude of the profit margin the

© The Editor(s) (if applicable) and The Author(s), under exclusive license to Springer Nature Switzerland AG 2020

J. Witzany, *Derivatives*, Springer Texts in Business and Economics,
https://doi.org/10.1007/978-3-030-51751-9_8

institution is effectively charging when it sells the complex product. In any case, exotic options have become a market reality, and so they should be understood by all counterparties who want to trade them.

We will consider a very simple classification of exotic options into two broad groups: *path-dependent options* and *miscellaneous exotic options* including *compound options*, *binary options*, or *non-standard American options* (see Wilmott 2006 for a more detailed classification).

8.1.1 Path-Dependent Options

Path-dependent options are options, whose payoffs depend not only on the underlying asset value at maturity, but also on the path taken by the asset price. These options can be divided into two subclasses: *weakly path-dependent* and *strongly path-dependent*. Weakly path-dependent options are options that can be triggered, or cancelled due to the path taken by the underlying asset, yet their value at any point of time depends only on the underlying asset price, and no additional variables need to be introduced. The most important type of weakly path-dependent options are *barrier options*, which are among the most popular exotic options. They can be *knocked-out* or *knocked-in* if the underlying asset price hits a predefined barrier. Their value depends only on the underlying asset price plus the information whether the knock-in or knock-out event has already happened. Note that American options are also weakly path-dependent. The decision to exercise an American option depends on the path of the underlying asset. The payoff, when the option has been exercised, and the market value at a point in time, when the option has not yet been exercised, depend, nevertheless, only on the current price of the underlying asset. *Strongly path-dependent options* are options whose value depends on an additional variable, not only on the underlying asset price. The most popular ones are *look-back* and *Asian options*.

8.1.2 Barrier Options

Barrier options can be classified as *knock-in* or *knock-out* call or put options. Besides the strike price K, there is also a specified barrier B. The relationship between the initial asset price and the barrier indicates from which direction the barrier needs to be hit. For example, an *up-and-out call option* pays $(S_T - K)^+$ if $S_t < B$ for all $t \in [0, T]$ and it pays nothing otherwise (see Fig. 8.1). It is implicitly assumed that $K < B$, otherwise the option contract does not make too much sense. The option can be valued analytically in the Black-Scholes framework, i.e., assuming that the spot price follows the geometric Brownian motion and that the risk-free interest rate is constant. With respect to the traditional risk-neutral measure the time zero option value can be expressed as

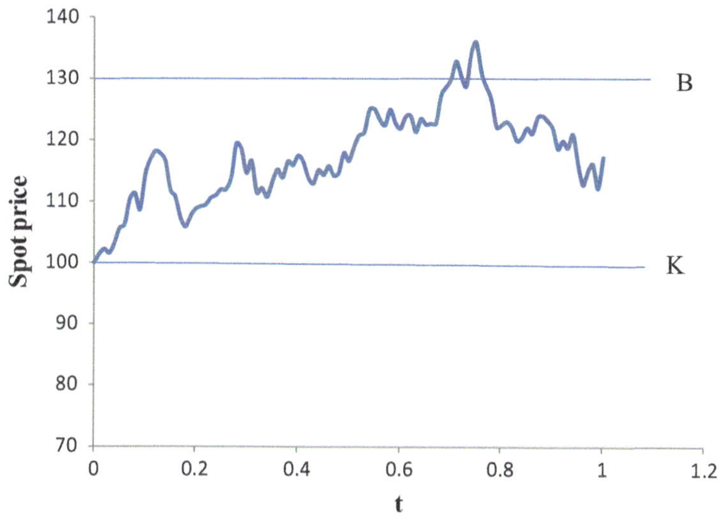

Fig. 8.1 Example of an up-and-out call option that has been knocked-out

$$c_0 = e^{-rT} E\left[(S_T - K)^+ I(M_T < B)\right], \tag{8.1}$$

where $M_t = \max\{S_\tau; 0 \le \tau \le t\}$, and the indicator function $I(M_T < B) = 1$ if $M_T < B$ and $I(M_T < B) = 0$ otherwise.

The expectation (8.1) can be evaluated in the context of a hyperfinite binomial tree (or Monte Carlo simulation) as an average of the payoffs $(S - K)$, for $S \in (K, B)$, weighted by the probabilities

$$\Pr[S_T \in [S, S + dS) \ \&\ M_T < B] = \Pr[\tilde{z}(T) \in [w, w + dw) \ \&\ \tilde{m}(T) < b]. \tag{8.2}$$

The variables on the right-hand side of (8.2) corresponding to the left-hand side through the monotonous one-to-one transformation $G(x) = e^{\sigma x}$, i.e.,

$$S_T = G(\tilde{z}(T)) = S_0 e^{\widetilde{\sigma z}(T)},$$

$w = G^{-1}(S)$, $\tilde{m}(T) = G^{-1}(M_T)$, and $b = G^{-1}(B)$. The probability on the right-hand side of (8.2) can be expressed using the *Reflection Principle.*

Since $\tilde{z}(t) = G^{-1}(S) = \alpha t + z(t)$ follows the generalized Wiener process with the drift $\alpha = (r - \frac{\sigma^2}{2})/\sigma$, we first need to change the risk-neutral measure Q to the measure $\tilde{Q}$ so that $\tilde{z}$ becomes a martingale. Every path of $\tilde{z}(t)$ that starts at zero, hits the barrier b, ends up at time T at a value w below the barrier, and can be reflected at the first intersection with the barrier, so that it ends up at the value $b + (b - w) = 2b - w$ above the barrier (see Fig. 8.2). On the other hand, any path that starts at zero and ends up above the barrier at time T can be reflected to the path that crosses the barrier, but ends up below the barrier. Consequently,

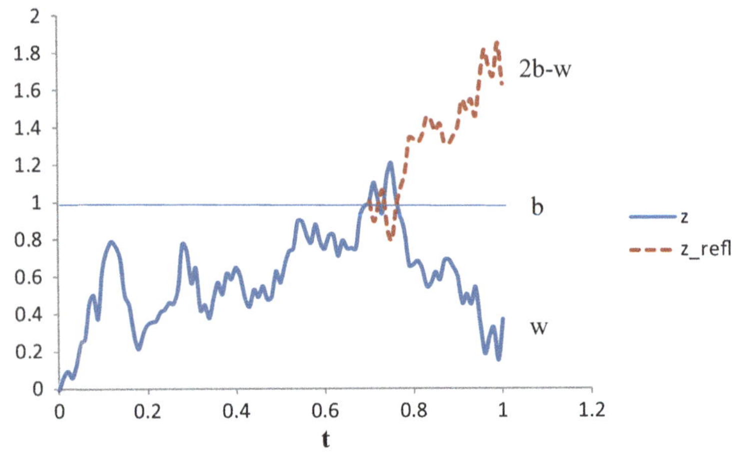

Fig. 8.2 Illustration of the reflection principle for a path of the Brownian motion

$$\begin{aligned}&\tilde{P}\mathrm{r}[\tilde{z}(T) \in [w, w + dw) \ \& \ \tilde{m}(T) \geq b] = \\ &= \tilde{P}\mathrm{r}[\tilde{z}(T) \in [2b - w, 2b - w + dw]] = \varphi(2b - w; 0, T)dw.\end{aligned} \tag{8.3}$$

Differentiating the right-hand side of (8.3) with respect to b, we obtain the joint density function of $\tilde{z}(T)$ and $\tilde{m}(T)$

$$\begin{aligned}&\tilde{P}\mathrm{r}[\tilde{z}(T) \in [w, w + dw) \ \& \ \tilde{m}(T) \in [b, b + db)] = \\ &= -\frac{d}{db}\varphi(2b - w; 0, T)dwdb = \frac{2(2b - w)}{T\sqrt{2\pi T}}\exp\left(-\frac{(2b - w)^2}{2T}\right)dwdb.\end{aligned}$$

According to the Girsanov's Theorem (6.24)

$$\frac{d\tilde{Q}}{dQ} = \exp\left(-\alpha z(T) - \frac{1}{2}\alpha^2 T\right) = \exp\left(-\alpha\tilde{z}(T) + \frac{1}{2}\alpha^2 T\right). \tag{8.4}$$

Hence, the density function with respect to the original measure is given by

$$\begin{aligned}&\mathrm{Pr}[\tilde{z}(T) \in [w, w + dw) \ \& \ \tilde{m}(T) \in [b, b + db)] = \\ &= \frac{2(2b - w)}{T\sqrt{2\pi T}}\exp\left(-\frac{(2b - w)^2}{2T} + \alpha w - \frac{1}{2}\alpha^2 T\right)dwdb.\end{aligned} \tag{8.5}$$

Finally, the probability that $\tilde{z}$ ends up at w and does not hit the barrier b is

$$\Pr[\tilde{z}(T) \in [w, w+dw) \,\&\, \tilde{m}(T) < b] =$$
$$\left[\int_w^b \frac{2(2x-w)}{T\sqrt{2\pi T}} \exp\left(-\frac{(2x-w)^2}{2T} + \alpha w - \frac{1}{2}\alpha^2 T \right) dx \right] dw = \tag{8.6}$$
$$= (\varphi(w;0,T) - \varphi(2b-w;0,T)) \exp\left(\alpha w - \frac{1}{2}\alpha^2 T\right) dw.$$

Note that since $\frac{d\tilde{Q}}{dQ}$ does not depend on b we could avoid the differentiation and integration with respect to b and simply multiply by $\frac{d\tilde{Q}}{dQ}$ the probability complementary to (8.3)

$$\tilde{\Pr}[\tilde{z}(T) \in [w, w+dw) \,\&\, \tilde{m}(T) < b] = (\varphi(w;0,T) - \varphi(2b-w;0,T))dw.$$

The joint density (8.5) will be useful, for example, in the case of the look-back options.

Hence, the up-and-out barrier option expected payoff can be written as

$$E[(S_T - K)^+ (M_T < B)] = \int_K^B (S-K) \cdot \Pr[z(T) \in [w, w+dw) \,\&\, m(T) < b] =$$
$$= \int_{G^{-1}(K)}^{b} (G(w) - K)(\varphi(w;0,T) - \varphi(2b-w;0,T)) \exp(\alpha w - \frac{1}{2}\alpha^2 T) dw.$$

Although the integration is not particularly nice, it is, in principle, the same as in the case of the Black-Scholes formula derivation. The integral can be split into four parts where we need to integrate the exponentials of various quadratic functions of w. The result, after multiplication by the discount factor from (8.1), can be written as follows (Shreve 2004):

$$c_{uo}(S_0, 0) = S_0(N(\delta_+(T, S_0/K) - N(\delta_+(T, S_0/B))-$$
$$-e^{-rT} K(N(\delta_-(T, S_0/K) - N(\delta_-(T, S_0/B))-$$
$$-B(S_0/B)^{-2r/\sigma^2} (N(\delta_+(T, B^2/(KS_0)) - N(\delta_+(T, S_0/B)) +$$
$$+e^{-rT} K(S_0/B)^{-2r/\sigma^2+1} (N(\delta_-(T, B^2/(KS_0)) - N(\delta_-(T, S_0/B)),$$

where $\delta_\pm(\tau, s) = \frac{1}{\sigma\sqrt{\tau}}\left[\ln s + \left(r \pm \frac{\sigma^2}{2}\right)\tau\right]$. To value $c_{uo}(S_t, t)$, the time T is, as usual, replaced by $T - t$ and S_0 by S_t, provided the barrier has not been hit by the time t. Analogous formulas can be obtained for other types of barrier options (see Hull 2018). Indeed, the barrier options are only weakly path dependent, their value is a function of S_t and t, plus the information, whether the barrier has been hit or not.

Barrier options also satisfy the Black-Scholes partial differential equation

$$df = \left(\frac{\partial f}{\partial S}\mu S + \frac{\partial f}{\partial t} + \frac{1}{2}\frac{\partial^2 f}{\partial S^2}\sigma^2 S^2\right)dt + \frac{\partial f}{\partial S}\sigma S dz,$$

but with different boundary conditions. For example, in the case of the up-and-out call, the conditions are $f(S,T) = (S - T)^+$, $f(0,t) = 0$ and $f(B,t) = 0$ with f defined only on the rectangle $[0,B] \times [0,T]$. The valuation can also be obtained by solving the PDE with boundary conditions empirically, e.g., using the method of finite differences (see, e.g., Hull and White 1990b).

8.1.3 Look-Back Options

Options, whose payoff is based on the maximum or minimum of the underlying price attained from the start date until maturity, are called *look-back options*. The payoff from a *floating* (strike price) *look-back call* is defined as $S_T - S_{\min}$, where $S_{\min}$ is the minimum price achieved during the life of the option. Similarly, the payoff from a *floating look-back put* is $S_{\max} - S_T$ where $S_{\max}$ is the maximum price achieved during the life of the option. In a fixed look-back option, the strike price is specified, but the final asset price is replaced by the maximum or minimum price. A *fixed* (strike price) *look-back call* pays $(S_{\max} - K)^+$ while a *fixed look-back put* pays $(K - S_{\min})^+$.

Note that in all cases, at time t, the value of a look-back option depends not only on the actual asset value S_t but also on the maximum (or minimum price) $S_{\max}(t)$ (or $S_{\min}(t)$) attained from the start date until the time t.

Let us discuss the problem of valuing a floating strike look-back put option. Other look-back options can be valued analogously, or by using a modified put-call parity relationship. With respect to the traditional risk-neutral measure and a constant interest rate r, the floating strike look-back put value is

$$p_{\text{fl}}(t) = e^{-r(T-t)}E[S_{\max} - S_T|t]. \quad (8.7)$$

Here, we must perform the calculation from the perspective of time t since the option value depends on two variables, S_t and $S_{\max}(t)$, while at time zero there is only one variable $S_0 = S_{\max}(0)$. To evaluate the expected value, we need to know the joint distribution function of $S_{\max}(T)$ and S_T. Similarly to the barrier options, this, in principle, reduces to the problem of finding the joint density of $\tilde{z}(T) = \frac{1}{\sigma}\ln S_T/S_0$ and $\tilde{m}(t) = \max\{\tilde{z}(s); 0 \le s \le t\}$ which is given by (8.5). As $\tilde{m}(t)$ is non-decreasing,

$$S_{\max}(T) = S_0 e^{\sigma\tilde{m}(T)} = S_{\max}(t)e^{\sigma\left(\tilde{m}(T) - \tilde{m}(t)\right)},$$

and $e^{-rt}S_t$ is a martingale with respect to the risk-neutral measure, (8.7) reduces to

$$p_{\text{fl}}(t) = e^{-r(T-t)}E[S_{\max} - S_T|t] = e^{-rT}S_{\max}E\left[e^{\sigma(\widetilde{m}(T)-\widetilde{m}(t))}\right] - S_t. \tag{8.8}$$

Note that the increment of the Wiener process maximum from t to T can be expressed as

$$\widetilde{m}(T) - \widetilde{m}(t) = [\max\{\widetilde{z}(s) - \widetilde{z}(t); t \leq s \leq T\} - (\widetilde{m}(t) - \widetilde{z}(t))]^+$$

and $\max\{\widetilde{z}(s) - \widetilde{z}(t); t \leq s \leq T\}$ has obviously the same unconditional distribution as

$$\max\{\widetilde{z}(s) - \widetilde{z}(0); 0 \leq s \leq T - t\} = \widetilde{m}(T - t).$$

Therefore, all we need to know is the density of $\widetilde{m}(\tau)$ for $\tau = T - t$, which can be obtained by integrating the joint density (8.5) with respect to the variable w. This and the remaining part of the integration is again a tedious exercise, but the resulting analytic formula appears "reasonably simple" (Shreve 2004):

$$\begin{aligned} p_{\text{fl}}(t) = & \left(1 + \frac{\sigma^2}{2r}\right)S_tN\left(\delta_+\left(T - t, \frac{S_t}{S_{\max}(t)}\right)\right) + \\ & + e^{-r(T-t)}S_{\max}(t)N\left(-\delta_-\left(T - t, \frac{S_t}{S_{\max}(t)}\right)\right) - \\ & - \frac{\sigma^2}{2r}e^{-r(T-t)}\left(\frac{S_{\max}(t)}{S_t}\right)^{2r/\sigma^2}S_tN\left(-\delta_-\left(T - t, \frac{S_{\max}(t)}{S_t}\right)\right) - S_t. \end{aligned}$$

8.1.4 Asian Options

Options, where the payoff depends on the average price of the underlying asset during their life, are called *Asian options*. The average price is theoretically defined in the continuous setting

$$S_{\text{ave}}(T) = \frac{1}{T}\int_0^T S(s)ds,$$

but in practice, it is calculated in a discrete setup (typically based on the daily closing prices) as

$$\widetilde{S}_{\text{ave}} = \frac{1}{m}\sum_{j=1}^{m} S(t_j).$$

The payoff of a *fixed strike Asian call* is $(S_{\text{ave}} - K)^+$, while the payoff of a *fixed strike put* is $(K - S_{\text{ave}})^+$. The options are strongly path-dependent since their value at time t depends not only on S_t but also on the average price $S_{ave}(t)$ realized from the start date until time t.

The advantage of the Asian options compared to the European plain vanilla options is that the final price that determines the payoff is more difficult to manipulate. The options might also be more appropriate for hedging purposes. For example, if a treasurer wants to hedge foreign currency expenses evenly spread over a time period then an Asian FX option depending on the average exchange rate over the period can match the risk better than a European option that depends only on the FX rate at maturity. The disadvantage of the Asian options is that they are much more difficult to value. In fact, the price of Asian options is not known in a closed form. One possible approach to their valuation is to set up and solve numerically an appropriate partial differential equation (see Večeř 2001 or Shreve 2004). Asian options can be valued using a generalized Black standard market model formula, if S_{ave} is assumed to be lognormal (see Turnbull and Wakeman 1991 or Hull 2018). For example, if $\widehat{S}_{ave}$ were defined as a geometric (not arithmetic) average of S_t following the geometric Brownian motion, then it would, indeed, be lognormal. Since the arithmetic average approximately equals the geometric average $S_{ave} \approx \widehat{S}_{ave}$, lognormality is a reasonable assumption.

8.1.5 Miscellaneous Exotic Options

There is a large variety of exotic options; some of them look exotic, but are relatively simple to value, while the others are difficult or even impossible to value analytically.

For example, *binary options* are options that pay an amount, or nothing. The amount can be defined as a fixed cash amount Q or as the asset price S_T (or equivalently the asset delivered at time T). Hence, we can distinguish *cash-or-nothing* put/call, and *asset-or-nothing* put/call options. The valuation formulas are easily obtained in the context of the Black-Scholes model. For example, assuming, for simplicity, that no income is paid by the asset, the cash-or-nothing call paying Q (if $S_T > K$) can be valued by.

$c_{bin,\ cash} = e^{-rT}E[Q \cdot I(S_T > K)] = Qe^{-rT}\Pr[S_T > K] = Qe^{-rT}N(d_2)$, where as usual

$$d_2 = \frac{\ln(S_0/K) + (r - \sigma^2/2)T}{\sigma\sqrt{T}}.$$

Similarly, $p_{bin,\ cash} = Qe^{-rT}N(-d_2)$. On the other hand, the asset-or-nothing call paying S_T if $S_T > K$ is valued by

$$c_{bin,asset} = e^{-rT}E[S_T \cdot I(S_T > K)] = S_0N(d_1)$$

and the asset-or-nothing put by $p_{bin,\ asset} = S_0N(-d_1)$.

Note that a European call payoff $(S_T - K) \cdot I(S_T > K)$ can be decomposed into a difference between the binary asset-or-nothing call payoff and the binary cash-or-nothing call payoff paying K, and so $c = c_{\text{bin, asset}} - c_{\text{bin, cash}}$. Similarly, a European put is the difference between the two corresponding binary puts. Therefore, the exotically looking binary options are, in a sense, more elementary than the plain vanilla options.

Warrants are, seemingly, just plain vanilla European (or American) options to buy stocks issued by a company. The complication in the pricing of the options comes from the fact that warrants are written by the company itself, which has a dilution effect on the stock price. Warrants are often issued with preferred stocks, giving current stockholders a prior right to subscribe to newly issued stocks. There might also be a secondary market with warrants, which facilitates the successful issue of additional company stocks.

To see how the valuation works more formally, let N denote the number of stocks outstanding prior to the exercise date of the warrants, M the number of (European style) warrants, E_t the value of equity at time t, and K the exercise prices. If all the warrants are exercised at maturity T, then the company issues M new stocks and its equity increases to $E_T + MK$. The warrant holders will exercise the option if, and only if, the strike price is lower than the market stock price including the dilution effect, i.e.,

$$K < \frac{E_T + MK}{M + N},$$

where $\frac{E_T+MK}{M+N}$ is the stock price when all the warrants are exercised. Therefore, the payoff function is

$$w_T = \max\left(\frac{E_T + MK}{N + M} - K, 0\right) = \max\left(\frac{E_T - NK}{N + M}, 0\right) = \\ = \frac{N}{N + M} \max\left(\frac{E_T}{N} - K, 0\right)$$

Therefore, the warrant can be valued as a call option on E_T/N, which is E_0/N today. But it would be a mistake to replace E_0/N by S_0, since the equity value today comprises of the market value of outstanding stocks and the market value of outstanding warrants, i.e.,

$$E_0 = NS_0 + Mw_0.$$

Therefore, the warrant value today must satisfy the following equation:

$$w_0 = \frac{N}{N + M} c\left(S_0 + \frac{M}{N} w_0, T, K, r, \sigma\right)$$

where $c(S, T, K, r, \sigma)$ is the plain vanilla call option valuation formula. The problem of the equation is that w_0 also appears on the right-hand side, and so it must be found numerically. A standard approach is to use, for example, the call option value $\widehat{w}_0$ without the dilution effect, as a first approximation, and then iteratively calculate

$$\widehat{w}_{i+1} = \frac{N}{N+M} c\left(S_0 + \frac{M}{N}\widehat{w}_i, T, K, r, \sigma\right).$$

It can be shown that the sequence $\widehat{w}_0, \widehat{w}_1, \widehat{w}_2, ..$ will relatively quickly converge to the "fixed point" w_0.

The same trick can be used to value American warrants using binomial trees. In this case, given an approximation $\widehat{w}_0$, the tree must be built for an initial underlying asset price $(NS_0 + M\widehat{w}_i)/(M+N)$ and the strike price needs to be adjusted to $NK/(N+M)$. Again, the binomial tree valuation needs to be repeated until the result of the valuation sufficiently coincides with the initial estimate, $\widehat{w}_i \approx \widehat{w}_{i+1}$.

Another type of exotic options that look complex, but that is relatively simple to value, are *chooser options*. The holder of a maturity T_2 chooser option can choose, at a time $T_1 < T_2$, whether the option will be a call or put. If the underlying options are of European type maturing at T_2 (and again the asset for simplicity pays no income) then the put-call parity can be used to decompose the chooser option value at the time T_1 into two parts

$$\begin{aligned} f(T_1) &= \max(c,p) = \max\left(c, c + e^{-r(T_2-T_1)}K - S_1\right) \\ &= c + \max\left(0, e^{-r(T_2-T_1)}K - S_1\right). \end{aligned}$$

Therefore, the chooser option can be valued as the sum of the European call with strike K and maturity T_2, and of the European put option with the strike $e^{-r(T_2-T_1)}K$ and the maturity T_1, i.e.,

$$f(t) = c(t; K, T_2) + p\left(t; e^{-r(T_2-T_1)}K, T_1\right) \quad \text{for} \quad t \le T_1.$$

Forward options are options that will start at some time in the future with the strike price yet to be determined by a mechanism. For example, a forward-start at-the-money European call will start at T_1 with the strike S_1 and will pay $(S_2 - S_1)^+$ at time T_2. An important property of the geometric Brownian motion is that the unconditional distribution of $S(T)/S(t)$ depends only on $T - t$. Consequently, the time T_1 value of an at-the-money option that lasts from T_1 to T_2 and pays $S_1(S_2/S_1 - 1)^+$ is equal to c_0S_1/S_0, where c_0 is the time zero value of an at-the-money option with maturity $T_2 - T_1$ and that pays $S_0(S(T_2 - T_1)/S_0 - 1)^+$. Since c_0 and S_0 are known at time zero, the value of the forward start call option is

$$c_{\text{forw}} = e^{-rT_1}E\left[c_0\frac{S_1}{S_0}\right] = e^{-rT_1}\frac{c_0}{S_0}E[S_1] = e^{-rT_1}\frac{c_0}{S_0}S_0e^{rT_1} = c_0.$$

Therefore, the surprising conclusion is that the forward start at-the-money option value equals to the plain vanilla at-the-money option value with the maturity equal to $T_2 - T_1$, provided the asset pays no income. If the income were paid continuously with a rate q then we would have $c_{\text{forw}} = e^{-qT_1}c_0$.

A related type of option is called *cliquet options* (alternatively *racket* or *strike reset options*). The idea is that there is a sequence of dates $T_0, T_1, \ldots, T_n$ and of forward start options from T_{i-1} to T_i. If all the options are set as at-the-money, then the total cliquet option value can be obtained as a sum of the forward options' values applying the result above. However, if the definition of the strike prices is more complicated, or if there are some additional conditions, e.g., an upper limit on the total payoff, then the cliquet option valuation becomes difficult, generally requiring a Monte Carlo simulation.

Compound options are second-generation derivatives, i.e., derivatives derived from certain underlying plain vanilla derivatives. For example, a European call on a call will have two exercise dates $T_1 < T_2$ and two strike prices K_1 and K_2. On the first exercise date T_1 the compound option holder will have the right to pay K_1 in order to buy a European call option on the underlying asset with the strike price K_2 and the maturity T_2. Compound options can be valued analytically if we assume that the underlying asset follows the geometric Brownian motion (Geske 1979). Note that the call-on-call time T_2 payoff depends on the prices $S(T_1)$ and $S(T_2)$. The first price $S(T_1)$ determines whether the compound option is exercised or not, while the second price $S(T_2)$ determines the payoff if the compound option is exercised. Hence, the valuation involves the integration of a function of the two prices multiplied by their joint density distribution. It is easy to see that the two variables are jointly lognormally distributed with correlation $\rho = \sqrt{T_1/T_2}$ and so it is not surprising that the analytic formula involves the bivariate cumulative standard normal distribution function $M(x, y; \rho)$ with that correlation. Specifically,

$$\begin{aligned} c_{\text{comp}} &= S_0 M\left(a_1, b_1; \sqrt{T_1/T_2}\right) - K_2 e^{-rT_2} M\left(a_2, b_2; \sqrt{T_1/T_2}\right) \\ &\quad - K_1 e^{-rT_1} N\left(a^2\right), \quad \text{where} \\ a_1 &= \frac{\ln\left(S_0/S^*\right) + \left(r - \sigma^2/2\right)T_1}{\sigma\sqrt{T_1}}, \quad a_2 = a_1 - \sigma\sqrt{T_1}, \\ b_1 &= \frac{\ln\left(S_0/K_2\right) + \left(r - \sigma^2/2\right)T_2}{\sigma\sqrt{T_2}}, \quad b_2 = b_1 - \sigma\sqrt{T_2}, \end{aligned}$$

and S^* is the price that makes the call option value at time T_1 equal to K_1. This is a nice result. However, we have seen that in practice option values depend not only on the underlying asset price, but also significantly on the volatility, which changes over time. In that respect, the compound option is, rather, an option on the implied volatility at T_1. For example, a call-on-call becomes profitable if the volatility goes up, even though the underlying price stays at the expected level. This consideration leads to a need for models with stochastic volatility, which will be discussed in Sect. 8.3.

In Chap. 7, we have already explained the valuation of various *quanto derivatives*, *options to exchange one asset for another*, and of selected *exotic interest rate derivatives*. Last but not least, we should mention *non-standard American options*. Even standard American options are in a sense exotic, because there is no general precise analytic valuation formula for them. A non-standard American option exercise may be restricted only to a certain time interval, e.g., to an initial part of the life of the option. For example, a *Bermudan option* is an American-type derivative, where early exercise is restricted only to certain dates. Moreover, the strike price may change during the life of the option. We have seen that bond embedded options often have some of these features. Similarly to the standard American options, the non-standard American options usually have to be valued numerically using a binomial tree.

8.2 Volatility Smile

The Black-Scholes model, which values options on an underlying asset, assumes that there is a constant volatility applicable to all future asset price movements and governed by the geometric Brownian motion. Our estimation of the volatility might change over time, but it should be used today as a constant applied to value European put and call options with all possible strike prices and maturities. But Fig. 8.3 demonstrates that the market reality is not consistent with that theoretical assumption. It shows quoted volatilities of FX options (three-month EUR/CZK and GBP/USD) with different deltas corresponding to different strike prices. The volatilities should be understood as input parameters of the Black-Scholes formula giving the correct market premiums. In terms of the call options' prices, the left-hand

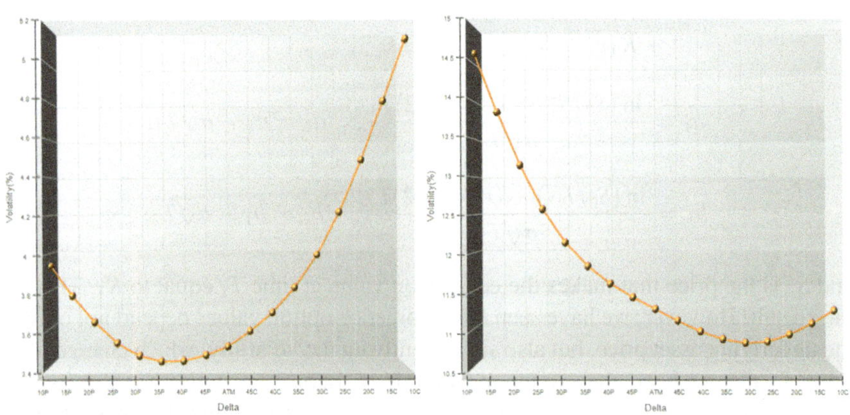

Fig. 8.3 3M EUR/CZK and volatility smile and 3M GBP/USD volatility skew (Source: Thomson Reuters, 9. 9. 2019)

side of the charts can be interpreted as in-the-money call volatilities, while the right-hand side corresponds to out-of-the-money call volatilities. The chart, however, indicates that the left-hand-side volatilities are related to out-of-the-money puts. This can be explained by the put-call parity according to which

$$c - p = e^{-r_2 T} S_0 - e^{-r_1 T} K. \tag{8.9}$$

The put-call parity is based on a simple arbitrage argument independently of any option valuation model, and so the Eq. (8.9) must hold for the market option prices c_{mkt} and p_{mkt}, as well as for the Black-Scholes model prices $c_{\mathrm{BS}}(\sigma)$ and $p_{\mathrm{BS}}(\sigma)$ given any volatility σ. Therefore,

$$c_{\mathrm{mkt}} - p_{\mathrm{mkt}} = c_{\mathrm{BS}}(\sigma) - p_{\mathrm{BS}}(\sigma). \tag{8.10}$$

If σ_{mkt} is the market volatility of a call option with strike K and maturity T, i.e., the volatility satisfying $c_{\mathrm{mkt}} = p_{\mathrm{BS}}(\sigma_{\mathrm{mkt}})$, then according to (8.10) we necessarily have the result that $p_{\mathrm{mkt}} = p_{\mathrm{BS}}(\sigma_{\mathrm{mkt}})$. Hence, given a strike price and maturity, the volatility implied from the call option price must be the same as the volatility implied from the corresponding put option price (with the same strike price). Since the deltas of call and put options are complementary, the in-the-money call volatilities can be interpreted as out-of-the-money put volatilities, and vice versa.

Inspecting the EUR/CZK volatility chart Fig. 8.3 (left-hand side), we see that out-of-the-money put and call volatilities are higher than at-the-money volatility. This phenomenon, called the *volatility smile*, is quite common for FX options. It means that out-of-the-money options are more expensive than they would be if we used the Black-Scholes model and at-the-money volatility. This empirical fact can be explained as an expectation of extreme movements in the exchange rate in one direction or another, or at least as a higher level of uncertainty in out-of-the-money option pricing relative to at-the-money options. For some currency pairs, like GBP/USD on the right-hand side chart in Fig. 8.3, the volatility smile becomes a *volatility skew,* with volatilities going up only in one direction. In the case of the GBP/USD currency pair, it has a clear explanation: the market is (currently) afraid of a sudden GBP depreciation due to Brexit uncertainty and the ongoing political turmoil, while a sudden GBP appreciation does not look feasible. The volatility skew is quite normal in the case of equity options—the equity market prices more often jump down than jump up. The phenomenon can also be explained by the leverage effect—if a company's equity decreases, there is an increase in leverage, and the company becomes riskier. In certain situations, the volatility "smile" looks rather like *a frown*, for example, if there is an expected important decision, like a merger or acquisition, whose outcome significantly impacts the prices in a positive or negative direction.

The volatility smile can also be explained by a probability distribution of the future prices that differs from the theoretical lognormal distribution (Breeden and Litzenberger 1978). Recall that the price of a European option with the strike price

K and maturity T, with respect to the traditional risk-neutral probability measure, is given by

$$c(K) = e^{-rT}E\left[(S_T - K)^+\right] = e^{-rT}\int_K^\infty (S_T - K)g(S_T)dS_T. \qquad (8.11)$$

Here, the function $g(S_T)$ is the probability density of the price S_T conditional on S_0 and other information given at time zero. The formula (8.11) should hold for all strike prices K with the same probability density. However, in the Black-Scholes model, the density function will be lognormal with the volatility depending on the strike price. Hence, if there is a volatility smile then the lognormal density function also depends on the strike price, and the model is not consistent. Let us relax the Black-Scholes model assumptions and assume that $g(\cdot)$ is a general density function. Given the call option market premiums for a range of strike prices, the formula (8.11) can be differentiated twice in order to obtain the single *implied density function* $g(S_T)$:

$$\frac{\partial c}{\partial K} = -e^{-rT}(K-K)g - e^{-rT}\int_K^\infty g(S_T)dS_T = e^{-rT}\int_K^\infty g(S_T)dS_T,$$

$$\frac{\partial^2 c}{\partial K^2} = e^{-rT}g(K), \text{ and so}$$

$$g(K) = e^{rT}\frac{\partial^2 c}{\partial K^2}.$$

In practice, the call option market premiums are given only on a grid of strike prices, and so the second-order derivative must be estimated numerically with the result depending on the interpolation and extrapolation methods employed. Figure 8.4 shows an example of the implied volatility derived from EUR/USD volatility smile quotes. The option premiums have been calculated from quoted volatilities using the Black-Scholes formula. The lognormal density in Fig. 8.4 corresponds to at-the-money volatility. The implied volatility is more peaked and has fatter tails than the lognormal distribution. It appears that the market indeed expects two possible regimes: a normal one, when the exchange rates do not move too much, and a crisis regime, when the exchange rates move significantly in one direction or another.

The Black-Scholes model assumes that a single volatility σ should work for all possible strike prices K and maturities T. We have already noticed that different volatilities are quoted for different maturities (Fig. 4.2), and the existence of a volatility smile means that the constant volatility assumption does not hold for different strike prices. Generally, the market (implied or quoted) volatility depends

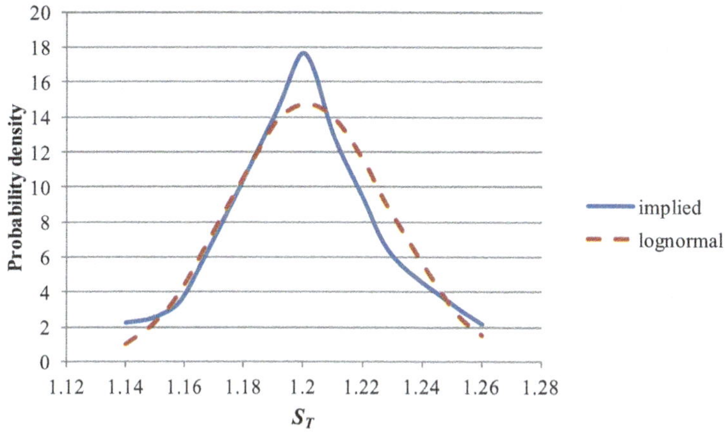

Fig. 8.4 Implied versus lognormal density of future prices

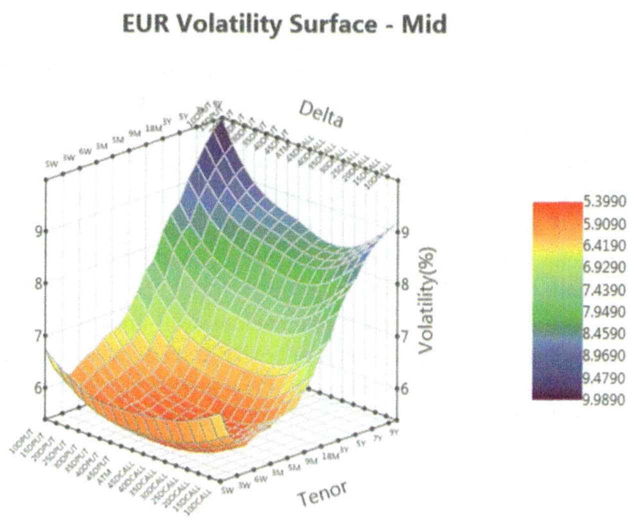

Fig. 8.5 3M EUR/USD volatility surface (Source: Thomson Reuters, 9.9.2019)

on the maturity T and the strike price K. The function of the two variables (or its chart—see Fig. 8.5) is called *a volatility surface*.

The volatility smile also changes the reality of delta hedging. Let us have a call option with maturity T and strike price K. The price of the option depends on the spot price S and on the volatility σ. But now the volatility $\sigma = \sigma(S)$ also depends on the spot price (the delta influencing σ through the volatility smile is a function of K/S). Hence, in order to calculate the true sensitivity of the option price with respect to

S (the true or adjusted delta) we have to differentiate $c(S, \sigma)$ with respect to both variables and use the derivative chain rule:

$$\Delta_{\text{adj}} = \frac{\partial}{\partial S} c(S, \sigma) = \frac{\partial c}{\partial S} + \frac{\partial c}{\partial \sigma} \frac{\partial \sigma}{\partial S} = \Delta_{BS} + \mathcal{V}_{BS} \frac{\partial \sigma}{\partial S},$$

where $\Delta_{BS} = \frac{\partial c}{\partial S}$ denotes the ordinary Black-Scholes Delta and $\mathcal{V}_{BS} = \frac{\partial c}{\partial \sigma}$ denotes the Vega. For example, in the case of a volatility skew like GBP/USD in Fig. 8.3, the implied volatility increases as the spot price increases. Therefore $\frac{\partial \sigma}{\partial S} > 0$, and the adjusted delta is higher than the BS delta.

8.3 Alternative Stochastic Models

There is a lot of empirical evidence that asset returns are not exactly lognormally distributed. For example, market moves exceeding six-standard deviations can be commonly observed in different markets, while the Gaussian probability of such a value is less than 10^{-8}. Therefore, such a move should occur less than once in a million of years, and the empirical reality contradicts the normality assumption. Generally, observed empirical returns are more peaked and have fatter tails compared to the normal distribution with the same first two moments. The same conclusion follows from the existence of volatility smiles (see Fig. 8.4). Moreover, the empirical volatility smile or skew implied distribution is often asymmetric, i.e., skewed. These empirical facts, and the need to value precisely various exotic derivatives, lead to a search for better stochastic models. From the large variety of models available in the literature, we will look at the models with jumps, with stochastic volatility, and at a few miscellaneous well-known models like the local volatility, Variance-Gamma, Heston's or SABR model. All these stochastic models (as well as the Wiener process) belong to the general class of Levy process models (see Cont and Tankov 2004).

8.3.1 Models with Jumps

The financial markets, from time to time, display jumps, unexpected falls or crashes related to bad news, or sometimes sudden positive increases based on unexpected good news. Consequently, let us allow a jump component to be added to our stochastic model. The key building block is the (counting) *Poisson process* $N(t)$ taking nonnegative integer values, so that $dN(t) \in \{0, 1\}$ and $\Pr[dN(t) = 1] = \lambda dt$, i.e., the probability of a jump over an infinitesimal time interval of the length dt is λdt, where λ, the *jump intensity*, is a positive constant. Note that although the probability of a jump over an infinitesimal time interval is infinitesimal, over an interval of length T the expected number of jumps λT is finite. The Poisson process can be represented in the context of a (hyperfinite) binomial tree by the elementary

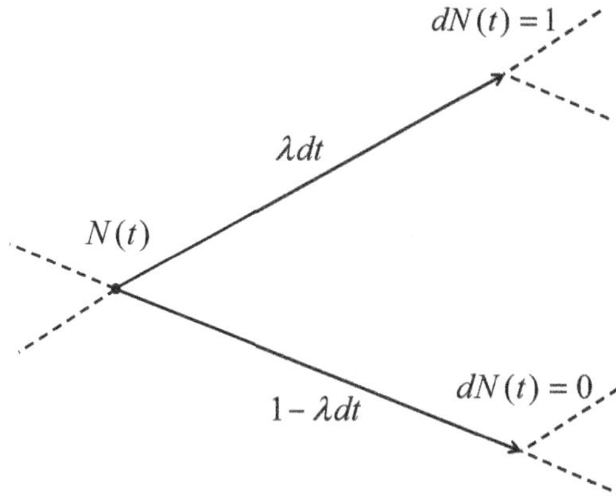

Fig. 8.6 Poisson process binomial tree branching

one-step branching with values zero (probability $1 - \lambda\delta t$) and one (probability $\lambda\delta t$), see Fig. 8.6.

The Poisson process can be used to model constant jumps. Let us assume that an asset price $S(t)$ follows the geometric Brownian motion most of the time, but from time to time the instantaneous log-return $d\ln S(t)$ suddenly jumps to m_J, i.e., $S(t)$ jumps to $S(t+dt) = e^{m_J}S(t)$. If $S(t)$ follows the pure jump process, then the increment of $S(t)$ will be

$$dS(t) = (e^{m_J} - 1)S(t)dN(t).$$

By combining the geometric Brownian motion and the pure diffusion model components, the jump–diffusion process proposed by Merton (1976) can be described by the stochastic differential equation

$$dS = \mu Sdt + \sigma Sdz + (e^{m_J} - 1)SdN. \tag{8.12}$$

The Wiener process and the Poisson process are assumed to be independent. The argument above also shows that the equivalent log-return stochastic differential equation is

$$d\ln S = \left(\mu - \sigma^2/2\right)dt + \sigma dz + m_J dN. \tag{8.13}$$

Note that the Poisson process is not a martingale as $E[dN(t)] = \lambda dt$, and so the drift must be compensated by a factor with respect to the traditional risk-neutral measure

$$dS = (r - \lambda m_J)Sdt + \sigma Sdz + (e^{m_J} - 1)SdN. \tag{8.14}$$

The constant jumps can be easily generalized to random size jumps better reflecting the market behavior. Let $Z(t)$ be a family of independent identically distributed variables, where t ranges over an infinitesimal grid from 0 to T. If there is a jump, then let the random $Z(t)$ be the jump component of $d\ln S(t)$, i.e.,

$$dS = \mu S dt + \sigma S dz + \left(e^{Z(t)} - 1\right) S dN \tag{8.15}$$

or equivalently

$$d(\ln S) = (\mu - \sigma^2/2)dt + \sigma dz + Z(t)dN.$$

A typical choice for the jump size distribution would be the normal distribution $N\left(m_J, s_J^2\right)$. The sign of the mean m_J depends on the market. For stocks, the mean would tend to be negative, since the market falls down rather than jumps up. For foreign currencies, the mean would be around zero, or it could even be positive, if the quoted currency is more stable than the quoting currency.

To compensate the effect of the jump component we need to take into account that the average jump size k as a percentage of S is not exactly m_J but using the lognormal distribution property it is

$$k = E\left[e^{Z(t)} - 1\right] = e^{m_J + s_J^2/2} - 1.$$

Therefore, with respect to the risk-neutral measure, the risk-free interest rate drift must be compensated by the factor λk, i.e.,

$$d\ln S = \left(r - \sigma^2/2 - \lambda k\right)dt + \sigma dz + Z(t)dN. \tag{8.16}$$

The introduction of jumps improves our ability to capture the dynamics of asset returns, but there are several drawbacks related to delta-hedging, analytical pricing, and calibration. The first issue is that the sudden jumps cannot be delta-hedged—the jump–diffusion models are generally incomplete. However, Merton (1976) argues that jumps present an idiosyncratic component that can be diversified away in a large portfolio, and so the price of risk of jumps equals zero. Consequently, with respect to the traditional risk-neutral measure, where S follows (8.16), we can still value derivatives based on the discounted expected payoff value, i.e., $f(0) = E[e^{-rT}f(T)]$. The expected value must usually be evaluated by a Monte Carlo simulation, where independent jumps and diffusion increments can be easily sampled. Merton (1976) has shown that there is a semi-analytic formula for European options, provided the log-return jump size is normally distributed:

$$f = \sum_{n=0}^{\infty} \frac{e^{-\tilde{\lambda}T}\left(\tilde{\lambda}T\right)^n}{n!} f_n, \tag{8.17}$$

where $\tilde{\lambda} = \lambda + \lambda k$ and f_n is the Black-Scholes option price with the modified volatility

$$\sigma_n = \sqrt{\sigma^2 + \frac{ns_J^2}{T}}, \tag{8.18}$$

and adjusted risk-free rate $r_n = r - \lambda k + n \ln(1 + k)/T$ (i.e., $\ln(1+k) = m_J + s_J^2/2$).

The number of jumps follows the *Poisson distribution*, i.e., the probability of exactly n jumps over a time period of length T, is $e^{-\lambda T}(\lambda T)^n/n!$ if the jump intensity is λ. The idea of formula (8.17) is to decompose the expected payoff $E[f(T)]$ into an infinite sum of discounted expected payoffs conditional on n jumps. Let $S_n(T)$ denote the asset price at time T conditional on n jumps taking place. If the variance of the diffusion component is $\sigma^2 T$, then the total variance of $\ln S_n(T)$ will be $\sigma^2 T + ns_J^2$ including the n independent jumps, and so the annualized volatility is indeed given by (8.18). In addition, the expected log-return will be

$$E[\ln S_n(T)/S_0] = \left(r - \lambda k - \sigma^2/2\right)T + nm_J = \left(r_n - \sigma_n^2/2\right)T.$$

Therefore, we can express the expected option payoff conditional on n jumps using the Black-Scholes model option price f_n as.

$$E[f(T)|n\,\text{jumps}] = \mathrm{e}^{r_n T} f_n = e^{rT} e^{-\lambda kT}(1+k)^n f_n,$$

and so indeed

$$f = e^{-rT} \sum_{n=0}^{\infty} \frac{e^{-\lambda T}(\lambda T)^n}{n!} E[f(T)|n\text{ jumps}] =$$

$$= e^{-rT} \sum_{n=0}^{\infty} \frac{e^{-\lambda T}(\lambda T)^n}{n!} e^{rT} e^{-\lambda kT}(1+k)^n f_n = \sum_{n=0}^{\infty} \frac{e^{-\tilde{\lambda} T}(\tilde{\lambda} T)^n}{n!} f_n$$

The idea above can be used to characterize the probability distribution of the log-return $X = \ln(S(\Delta T)/S(0))$ over a period of length ΔT. If S follows (8.15), the jumps have the distribution $N(m_J, s_J^2)$, and if there are exactly n jumps over the period of length ΔT, then the distribution of X is $N\left((\mu - \sigma^2/2)\Delta T + nm_J, \sigma^2 \Delta T + ns_J^2\right)$. Therefore, the density function can be expressed as a mix of normal distributions weighted by the Poisson distribution probabilities

$$g(x) = \sum_{n=0}^{\infty} \frac{e^{-\lambda \Delta T}(\lambda \Delta T)^n}{n!} \varphi\left(x; (\mu - \sigma^2/2)\Delta T + nm_J, \sigma^2 \Delta T + ns_J^2\right). \tag{8.19}$$

Given a series of historical log-returns, the density function (8.19) can be used to set up a likelihood function and estimate the parameters (i.e., λ, μ, σ, m_J, s_J) by the maximization of likelihood. In practice, the infinite series is truncated to a finite mixture of normal distributions. In fact, jumps should be rare events and there should be only one or a few jumps over a shorter period. For example, for daily data it is reasonable to assume at most one jump. Then the mixed distribution has a simple form:

$$g(x) = (1-\lambda)\varphi\left(x; \left(\mu - \sigma^2/2\right)\Delta T, \sigma^2 \Delta T\right) + \\ + \lambda\varphi\left(x; \left(\mu - \sigma^2/2\right)\Delta T + m_J, \sigma^2 \Delta T + s_J^2\right) \tag{8.20}$$

However, the likelihood maximization procedure often gets stuck in various local maxima, or the parameters are not well identified at all. For example, if the true returns were just normal $N((\mu - \sigma^2/2)\Delta T, \sigma^2 \Delta T)$ then the jump process specified by (8.20) with arbitrary $\lambda \in [0, 1]$ and $m_J = (\mu - \sigma^2/2)\Delta T$, $s_J^2 = \sigma^2 \Delta T$ would be consistent with the data. Therefore, the parameters must satisfy certain a priori assumptions stating that jumps are rare events and that they are much larger than the usual returns. Another issue is the identification of jump times that are not directly observable from the empirical data. The jump times are *latent state variables*. Note that there are no latent variables in the pure diffusion model. Given a return and the model parameters, we can calculate the Wiener process increment. In the case of the jump–diffusion model, given a large (positive or negative) return, we never know whether it has been caused by the diffusion component or by the jump component of the model. We can only say that there was probably a jump if the observed return is unusually large positive or negative. The probability can be expressed exactly in the Bayesian approach. And, in fact, it turns out that the Bayesian Markov Chain Monte Carlo (MCMC) estimation methods are efficient in the estimation of models with latent state variables like the jump–diffusion one (see, e.g., Witzany 2013a).

Another estimation approach is to calibrate the jump–diffusion model from plain-vanilla option prices, i.e., to fit the model parameters to the empirical volatility smile or skew. The model with normal jumps pricing the options according to (8.17) is able to capture fat tails and skewed distributions, i.e., the smiles and skews in both directions. However, the model cannot fit an arbitrary volatility term-structure because the parameters do not depend on time.

8.3.2 Models with Stochastic Volatility: An Introduction

The jump–diffusion models can filter out unusual jumps in returns, but they do not reflect the fact that the volatility itself changes over time. Figure 5.11 shows the historical volatility development of the index PX. The volatility does not appear to be constant at all—there is a crisis period of high volatility, but there are also large sudden changes in volatility (of course depending on the measurement method). Moreover, option (implied) volatility has become a market factor that is quoted and fluctuates up and down in a fashion like other market factors. Therefore, it is natural to introduce the concept of stochastic volatility. Here, we provide just an overview of some stochastic volatility models that will be studied in more depth in the following section.

Assume that the asset price follows the geometric Brownian motion

$$dS = \mu S dt + \sigma S dz, \tag{8.21}$$

but let $\sigma = \sigma(t)$ be allowed to be stochastic as well. A straightforward approach is to let the variance $V(t) = \sigma(t)^2$ follow the stochastic differential equation

$$dV = \alpha V dt + \sigma_V V dz_V \tag{8.22}$$

where the Wiener process increments dz and dz_V have a correlation ρ. Hull and White (1987) show that European options can be valued by a semi-analytical formula, provided dz and dz_V are independent, and in addition α, σ_V, and the price of volatility risk are constant.

The problem of volatility modeling is similar to the interest rate modeling: variance does not attain negative values and it should have the mean-reversion property. There is a normal level of volatility, to which the market tends to return after periods of unusually high or low volatility. One possible way to reflect this property is the following mean-reversion log-variance model

$$d \ln V = a(b - \ln V)dt + \sigma_V dz_V \tag{8.23}$$

where the Wiener process increments dz and dz_V are independent, or may possibly exhibit a certain correlation. In terms of hyperfinite binomial trees or a Monte Carlo simulation, we have to update in turn the asset price S according to (8.21), and the volatility $\sigma = \sqrt{V}$ according to (8.22) or (8.23). The calibration of stochastic volatility models from historical data is even more difficult than the calibration of jump–diffusion models, since the stochastic volatility is a latent variable not directly observable from historical returns. Besides ML (Maximum Likelihood) or GMM (Generalized Method of Moments), the Bayesian Markov Chain Monte Carlo (MCMC) or Particle Filter (PF) algorithms appear to be the most universal and efficient (see Jacquier et al. 1994; Shephard 2004, or Fičura and Witzany 2016). The log-variance stochastic model (8.23) does not have an analytical solution for option prices, the options must be valued by a Monte Carlo simulation, and so the calibration of the model to a given volatility surface is also numerically difficult.

However, a closed-form option pricing solution has been found for the Heston (1993) stochastic volatility CIR-like model

$$dV = a(b - V)dt + \sigma_V \sqrt{V} dz_V. \tag{8.24}$$

The stochastic volatility models can certainly be combined with jump–diffusion models (8.15). Moreover, not only the asset returns but also the volatility itself may exhibit jumps (see Fig. 5.11), and so there are also models, where jumps are incorporated into the stochastic volatility equations (Cont and Tankov 2004 or Witzany and Fičura 2018). The stochastic volatility models with jumps in returns, or with jumps in stochastic volatility, become even more difficult to calibrate, but are considered by many authors to be the best in terms of their ability to capture the true market dynamics.

8.3.3 Miscellaneous Stochastic Models

The jump–diffusion and stochastic volatility models outlined above can fit a given volatility surface only approximately since they have only a limited degree of freedom (a few constant parameters). Full flexibility to fit a given volatility surface is given by the *local volatility model* of the form

$$dS = \mu S dt + \sigma(S,t) S dz, \tag{8.25}$$

where the volatility $\sigma = \sigma(S, t)$ is a general deterministic function of S and t. In terms of a (recombining) binomial tree, the volatility must be specified at each branching node (corresponding to a price S and a time t) of the tree. It has been shown by Derman and Kani (1994), Dupire (1994), and Rubinstein (1994) that the model can provide an exact fit to any given volatility surface. The function $\sigma(S, t)$ is called the *implied volatility function* (IVF) and (8.25) is also called the *implied tree* model. Andersen and Brotherton-Ratcliffe (1998) provide an analytical formula for $\sigma(S, t)$ in terms of observed European call option market prices $c_{\mathrm{mkt}}(K, T)$ for different maturities and strike prices:

$$\sigma(K,T) = \left[2 \frac{\partial c_{\mathrm{mkt}}/\partial T + q(T) c_{\mathrm{mkt}} + K(r(T) - q(T)) \partial c_{\mathrm{mkt}}/\partial K}{K^2 \left(\partial^2 c_{\mathrm{mkt}}/\partial K^2 \right)} \right]. \tag{8.26}$$

The interest rate $r(T)$ and the income rate $q(T)$ may depend on time, and the first variable of σ in (8.26) is denoted by K, since c_{mkt} is expressed in terms of the strike price K and the maturity T. The formula includes the first and the second-order derivatives of c_{mkt} with respect to the time to maturity and to the strike price, and thus the results are very sensitive to the smoothness of the volatility surface. The asset prices can then be simulated according to (8.25) in a Monte Carlo simulation. Options can be also valued by solving the Black-Scholes partial differential equation and applying the finite difference method (Andersen and Brotherton-Ratcliffe 1998).

Although the model can fit a given volatility surface exactly, it does not necessarily give a correct joint distribution of the asset prices at different times. Consequently, exotic options, like the barrier or Asian options, may be priced incorrectly by the model (Hull and Suo 2002).

Finally, let us mention the quite popular *Variance Gamma* (VG) or *Normal Inverse Gaussian* (NIG) models belonging to the general class of Levy models (Cont and Tankov 2004). A stochastic process $X(t)$ is called *a Levy process* if the increments are independent and stationary (i.e., the distribution of $X(t + h) - X(t)$ does not depend on t), and if it is stochastically continuous, meaning that jumps can exist only as rare random events. Levy processes cannot, in general, be described by stochastic differential equations, but many important Levy processes can be built from the Wiener process or general jump–diffusion processes transforming the time by another stochastic (non-decreasing) process called the *subordinator*. This is the case of the VG (where the increments follow a Variance-gamma distribution generalizing the Laplace distribution) and NIG models (where the increments have the Normal-inverse Gaussian distribution). If $S(t)$ follows the geometric Brownian motion and if $g(t)$ follows the Gamma (or the inverse Gaussian) process, then $S(g(t))$ defines the VG (or NIG) models. The VG and NIG processes are both pure jump processes with infinitely many small jumps over any time interval (see Appendix A). The transformation of time into a pure jump process can be explained by the arrival of information. The function $g(t)$ can be interpreted as the *business, information or economic time*. The probability distribution of $g(t)$ in the case of the VG (NIG) model is the Gamma distribution (Inverse Gaussian) conditional on the time t, and so the prices at a fixed time T can be relatively easily sampled. Moreover, Madan et al. (1998) provide semi-analytic formulas for the pricing of European options under the VG model, thus the model parameters can be calibrated relatively well.

8.4 Stochastic Volatility Modeling and Estimation

In this section, we will start with a more detailed presentation of the Hull and White (1987) stochastic volatility (SV) model, which provides a useful insight into the pricing of derivatives with SV. Then we will derive the Heston formulas and, finally, focus on the issue of SV models' parameters and SV values estimation.

8.4.1 Hull–White Stochastic Volatility Model

Consider a derivative security with price f depending on an underlying asset price S and its instantaneous variance V following the processes:

$$\begin{aligned} dS &= \mu S dt + \sigma S dz, \\ dV &= \alpha V dt + \sigma_V V dz_V \end{aligned} \tag{8.27}$$

where $\sigma = \sqrt{V}$, the Wiener process increments dz and dz_V have a correlation ρ, the parameter μ may generally depend on S, V, and t, and similarly the parameters α and σ_V might depend on V and t. Let us assume that the instantaneous risk-free rate r is constant or deterministic and try to repeat the argument deriving the Black-Scholes partial differential equation. Since $f = f(S, V, t)$ now depends on two sources of uncertainty, we have to use Ito's two-dimensional formula (6.26)

$$df = \left(\frac{\partial f}{\partial t} + \frac{\partial f}{\partial S}\mu S + \frac{\partial f}{\partial V}\alpha V + \frac{1}{2}\frac{\partial^2 f}{\partial S^2}VS^2 + \frac{1}{2}\frac{\partial^2 f}{\partial V^2}\sigma_V^2 V^2 + \frac{\partial^2 f}{\partial S \partial V}\sigma_V S V^{3/2}\rho\right)dt + \\ + \frac{\partial f}{\partial S}V^{1/2}Sdz + \frac{\partial f}{\partial V}\sigma_V dz_V.$$

The problem is, that by combining a (long) position in the derivative and the delta-hedging position of $-\frac{\partial f}{\partial S}$ in the asset we eliminate only the source of uncertainty dz, but, unfortunately, we cannot hedge the volatility source of uncertainty dz_V unless the variable V is itself a traded asset price. However, according to Hull and White, the premium related to this residual volatility risk can be expressed in line with the CAPM as $RP_V = V\frac{\partial f}{\partial V}\beta_V(\mu^* - r)$, where $V\frac{\partial f}{\partial V}$ is the exposure with respect to V, β_V is the systematic risk measure defined as the sensitivity of returns of V with respect to the returns of the market portfolio, and μ^* is the expected return of the market portfolio. Since the expected return of the delta-hedged portfolio $\Pi = f - S\frac{\partial f}{\partial S}$ must be equal to the risk-free rate plus the volatility risk premium,

$$E[d\Pi] = r\Pi dt + RP_V dt, \text{ i.e.}$$
$$E\left[df - \frac{\partial f}{\partial S}dS\right] = r\left(f - S\frac{\partial f}{\partial S}\right)dt + RP_V dt,$$

we get the following partial differential equation for $f = f(S, V, t)$:

$$\frac{\partial f}{\partial t} + \frac{1}{2}\frac{\partial^2 f}{\partial S^2}VS^2 + \frac{1}{2}\frac{\partial^2 f}{\partial V^2}\sigma_V^2 V^2 + \frac{\partial^2 f}{\partial S \partial V}\sigma_V S V^{3/2}\rho \\ = rf - \frac{\partial f}{\partial S}rS - V\frac{\partial f}{\partial V}(\alpha - \beta_V(\mu^* - r)). \tag{8.28}$$

Hull and White argue that it is reasonable to assume that the volatility price of risk term $\beta_V(\mu^* - r)$ is zero (although a number of empirical studies have since shown that the volatility price of risk tends to be negative). Under this assumption, in order to price a derivative, we need to solve the partial differential equation

$$\frac{\partial f}{\partial t} + \frac{1}{2}\frac{\partial^2 f}{\partial S^2}VS^2 + \frac{1}{2}\frac{\partial^2 f}{\partial V^2}\sigma_V^2 V^2 + \frac{\partial^2 f}{\partial S \partial V}\sigma_V S V^{3/2}\rho \\ = rf - \frac{\partial f}{\partial S}rS - \frac{\partial f}{\partial V}\alpha V \tag{8.29}$$

with boundary conditions given by the derivative payoff function and the initial asset and volatility values. It can be noted that (8.29) also applies if the factor $\beta_V(\mu^* - r)$ is a constant replacing α by $\tilde{\alpha} = \alpha - \beta_V(\mu^* - r)$.

When valuing a European call option maturing at time T, analogously to the classical Black-Scholes formula, more insight is provided if we apply the risk-neutral valuation approach. Since neither (8.29) nor the boundary conditions depend on the risk preferences, we may apply the risk-neutral principle. Therefore,

$$f(S_0, V_0, t) = e^{-rT} \int f(S_T, V_T, T) p(S_T|S_0, V_0) dS_T \tag{8.30}$$

where $p(S_T| S_0, V_0)$ is the conditional probability density of the terminal asset price S_T and $f(S_T, V_T, T) = (S_T - K)^+$ is the payoff function. In order to simplify (8.30) we are going to assume that the Wiener increments dz and dz_V are independent, in particular $\rho = 0$. The key step is to decompose the conditional density function $p(S_T| S_0, V_0)$ that depends both on the process driving S, and on the process driving the variance V.

Let us first assume that $V = \sigma^2$ is a deterministic function of time. Since S follows the equation:

$$dS = rSdt + \sigma Sdz, \text{ or equivalently}$$

$$d(\ln S) = \left(r - \sigma^2/2\right)dt + \sigma dz,$$

we are simply summing up normally distributed independent increments with variance $V(t)dt$, and so the distribution of $\ln S_T/S_0$ is normal with the mean $rT - \overline{V}T/2$ and the integrated variance $\overline{V}T$ where $\overline{V} = \frac{1}{T}\int_0^T V(t)dt$. Consequently, the (conditional) distribution of S_T is lognormal; it depends only on S_0 and on the average variance $\overline{V}$. Therefore,

$$p(S_T|S_0, V_0) = \int g\left(S_T|S_0, \overline{V}\right) h\left(\overline{V}|V_0\right) d\overline{V},$$

where $g\left(S_T|S_0, \overline{V}\right)$ is the conditional lognormal density and $h\left(\overline{V}|V_0\right)$ is the density function of the average variance $\overline{V}$ conditional on the initial instantaneous variance V_0.

Substituting this integral into (8.30), we get

$$\begin{aligned} f(S_0, V_0, t) &= \int \left(\int e^{-rT} f(S_T) g(S_T|S_0, \overline{V}) dS_T\right) h(\overline{V}|V_0) d\overline{V} = \\ &= \int C(\overline{V}) h(\overline{V}|V_0) d\overline{V} = E[C(\overline{V})|S_0, V_0] \end{aligned} \tag{8.31}$$

where the inner term is just the Black-Scholes price of a call option conditional on the average variance $\overline{V}$ given by

$$C(\overline{V}) = \int e^{-rT} f(S_T) g(S_T|S_0, \overline{V}) dS_T.$$

Therefore, in the stochastic volatility model, an option price can be expressed as the expected value of the Black-Scholes prices conditional on the average variance $\overline{V}$ that depends on the initial asset price S_0 and on the initial instantaneous variance V_0. It is important to note that this would not be true if S and V were instantaneously correlated—in this case, the distribution of S_T generally depends on the path of V and cannot be integrated only with respect to the distribution of the average variance $\overline{V}$.

The Eq. (8.31) can be used to derive a semi-analytical formula, if we know the moments of $\overline{V}$, by expanding the option value in a Taylor series around the mean value $\mu_{\overline{V}} = E[\overline{V}]$ of $\overline{V}$:

$$C(\overline{V}) = \sum_{n=0}^{\infty} \frac{1}{n!} C^{(n)}(\mu_{\overline{V}})(\overline{V} - \mu_{\overline{V}})^n.$$

Substituting this into (8.31) we obtain

$$f(S_0, V_0, t) = \sum_{n=0}^{\infty} \frac{1}{n!} C^{(n)}(\mu_{\overline{V}}) E[(\overline{V} - \mu_{\overline{V}})^n].$$

The distribution of $\overline{V}$ is known, for example, if the parameters α and σ_V are constant. In this case, V simply follows the geometric Brownian motion and all the moments of the average variance $\overline{V}$ can be analytically expressed (Hull and White 1987). Since the derivatives of $C^{(n)}(\mu_{\overline{V}})$ are also analytically known, we get a semi-analytical formula in the form of an infinite series. According to Hull and White, the series converges fast and just a few first terms are enough in order to get a good approximation.

The Eq. (8.31) can be used to price options relatively efficiently by Monte Carlo simulation, even if the moments of $\overline{V}$ are not known: simply simulate the paths of V and the values of $\overline{V}$ in order to estimate $E[C(\overline{V})]$. This is the case, for example, if we incorporate a mean reversion into the stochastic variance formula, i.e.,

$$dV = \alpha(V_L - V)dt + \sigma_V V dz_V,$$

or in the more popular log-variance form:

$$dh = a(h_L - h)dt + \sigma_V dz_V \quad \text{with} \quad h = \ln V. \tag{8.32}$$

The Eq. (8.31) also provides a useful insight into the relationship between the Black-Scholes price $C(\mu_{\overline{V}})$ evaluated at the expected mean variance and the stochastic volatility price $E[C(\overline{V})]$. Generally, if C is a convex function of $\overline{V}$ then $C(E[\overline{V}]) < E[C(\overline{V})]$, and vice versa, if C is concave. By an analysis of the second-order derivative of C we can see that the function is convex for low values of $\overline{V}$ and concave for high values of $\overline{V}$ depending on the option moneyness. In fact, it is always concave for at-the-money options and almost always convex for options that are deeply out or in-the-money. Therefore, the Black-Scholes formula surprisingly overprices at-the-money options and underprices deeply out-of-the-money or in-the-money options.

8.4.2 Heston's Model

The model of Heston (1993) is known in the form

$$\begin{aligned} dS &= \mu S dt + \sigma S dz, \\ dV &= \kappa(\theta - V)dt + \sigma_V \sqrt{V} dz_V, \end{aligned} \tag{8.33}$$

allowing for a non-zero correlation ρ between dz and dz_V. The CIR stochastic differential equation is closely related to the Ornstein-Uhlenbeck (Vasicek-like) process for $\sigma = \sqrt{V}$:

$$d\sigma = -\beta\sigma dt + \delta dz_V. \tag{8.34}$$

Note that by applying the Ito's lemma to $V = \sigma^2$ we get

$$dV = 2\beta\left(\frac{\delta^2}{2\beta} - V\right)dt + 2\delta\sqrt{V}dz_V,$$

corresponding to the second equation in (8.33) with $\kappa = 2\beta$, $\theta = \frac{\delta^2}{2\beta}$ and $\sigma_V = 2\delta$. The Eq. (8.34) has an analytical solution (see Sect. 7.1), and it turns out that σ has a normal distribution allowing for negative values. In order to ensure the positiveness of the process V in (8.33) we need to require $2\kappa\theta > \sigma_V^2$.

The main advantage of the Heston's model is that European options can be valued by a (semi)analytical formula even for a general correlation ρ and a non-zero volatility risk premium assumed to be proportional to V, i.e., $\lambda(S, V, t) = \lambda V$. Note that the Hull–White decomposition (8.31) does not apply under these relatively general assumptions. The partial differential equation set up analogously to (8.28) for a derivative security price $f = f(S, V, t)$ depending on S and V then takes the following form:

$$\begin{aligned} &\frac{\partial f}{\partial t} + \frac{1}{2}\frac{\partial^2 f}{\partial S^2}VS^2 + \frac{1}{2}\frac{\partial^2 f}{\partial V^2}V\sigma_V^2 + \frac{\partial^2 f}{\partial S \partial V}\sigma_V VS\rho = \\ &= rf - \frac{\partial f}{\partial S}rS - \frac{\partial f}{\partial V}(\kappa(\theta - V) - \lambda V) \end{aligned} \tag{8.35}$$

An arbitrage argument (Wilmott 2006) can be used to show that the volatility price of risk λ must be the same for all derivatives depending on S and V. Given two derivative securities with the same sources of uncertainty, set up a portfolio with the first derivative security $+f_1$, offset the volatility risk by the position $-\Delta_V f_2$ where $\Delta_V = \frac{\partial f_1/\partial V}{\partial f_2/\partial V}$ in the second derivative security, and finally offset the asset risk by the position in $-\Delta = -\frac{\partial f_1}{\partial S} + \Delta_V \frac{\partial f}{\partial S}$ units of the underlying asset. The resulting portfolio Π is fully risk-free and so $d\Pi = r\Pi dt$. Expressing Π and $d\Pi$ using the partial derivatives of f_1 and f_2 and collecting the f_1 terms on the left hand side and f_2

terms on the right-hand side, we prove that λ defined from the Eq. (8.35) must be the same for both derivatives.

Let us consider a European call option with the payoff function $C(S, V, T) = (S - K)^+$ and let us show the key steps on how to get to the Heston's valuation formula. The derivation is based on the concept of the characteristic function of a probability distribution and on the concept of Fourier transformation. These techniques are useful also in other financial engineering problems.

Given a real valued random variable, its cumulative distribution function fully characterizing its probabilistic properties can be written in the form

$$F_X(x) = E\left[1_{\{X \le x\}}\right].$$

Similarly, the characteristic function is defined as

$$\varphi_X(t) = E\left[e^{itX}\right],$$

where i is the imaginary unit. Again, the characteristic function completely determines the behavior of the random variable X. In particular, if $f_X(x)$ is the probability density function, then the characteristic function equals its (inverse) Fourier transformation

$$\varphi_X(t) = \int_R e^{itx} f_X(x) dx.$$

In the opposite direction, the density function can be recovered from the characteristic function by the Fourier transformation of the characteristic function:

$$f_X(x) = \frac{1}{2\pi} \int_R e^{-itx} \varphi_X(t) dt.$$

Similarly, we can recover the cumulative distribution function given the characteristic function

$$F_X(x) = \frac{1}{2} - \frac{1}{\pi} \int_0^{+\infty} \frac{\mathrm{Im}\left[e^{-itX} \varphi_X(t)\right]}{t} dt. \tag{8.36}$$

The complex valued characteristic function is obviously less intuitive than the ordinary cumulative distribution function, but it has many useful properties related to the fact that it is defined as an expectation of the complex exponential function which has derivatives of all orders, which is bounded, and so on.

Let us return to the problem of valuing a European option. It is sufficient to find formulas for asset-or-nothing and cash-or-nothing binary call options. First, let $B = f(S, V, t)$ denote the price of a binary call option that pays $1_{\{S_T \ge K\}}$ at the maturity

T. The function f is a solution of the partial differential Eq. (8.35) satisfying the boundary condition $f(S, V, T) = 1_{\{S \geq K\}}$. It is not at all obvious how to guess the solution in an analytical form. However, we can use the classical risk-neutral valuation argument as follows: let us adjust the coefficients κ and θ so that (8.35) remains unchanged, but also so that the volatility price of risk becomes zero, i.e., $\kappa^* = \kappa + \lambda$ and $\theta^* = \kappa\theta/(\kappa + \lambda)$. Therefore, we can assume, without loss of generality, that the asset price and variance follow the risk-adjusted stochastic differential equations

$$\begin{aligned} dS &= rSdt + \sqrt{V}Sdz, \\ dV &= \kappa^*(\theta^* - V)dt + \sigma_V\sqrt{V}dz_V, \end{aligned} \tag{8.37}$$

and that there is no price of asset and volatility risk. It means that for any payoff function the derivative value at time t can be expressed as the discounted expected value

$$f(s, v, t) = e^{-r(T-t)}E[f(S_T, V_T, T)|S(t) = s, V(t) = v].$$

To see this more formally, one can apply the Ito's lemma to the forward option price

$$e^{r(T-t)}f(s, v, t) = E[f(S_T, V_T, T)|S(t) = s, V(t) = v]$$

and use the fact that it is a martingale (i.e., the coefficient of dt is zero) in order to verify that f satisfies the adjusted PDE (8.35).

The ingenious idea is to solve first the equation for the boundary condition $g(X, V, T; u) = e^{iuX}$ where $X = \ln S$ and u is a parameter. Since the solution in the form

$$g(x, v, t; u) = E\left[e^{iuX(T)}|X(t) = x, V(t) = v\right]$$

is also the characteristic function of $X(T)$ conditional on $X(t)$ and $V(t)$, it can be transformed using the Fourier formula (8.36) to get

$$p(x, v, t; \ln K) = \Pr[X(T) \geq \ln K|X(t) = x, V(t) = v] = 1 - F_{X(T)}(\ln K).$$

After the substitution $X = \ln S$ the transformed stochastic differential Eq. (8.37) are

$$\begin{aligned} dX &= (r - V/2)dt + \sqrt{V}dz \\ dV &= \kappa^*(\theta^* - V)dt + \sigma_V\sqrt{V}dz_V \end{aligned}$$

and the martingale equation for g takes the simplified form

$$\frac{\partial g}{\partial t}+\frac{\partial g}{\partial X}(r-V/2)+\frac{\partial g}{\partial V}\kappa^{*}(\theta^{*}-V)+ \\ +\frac{1}{2}\frac{\partial^{2}g}{\partial X^{2}}V+\frac{1}{2}\frac{\partial^{2}g}{\partial V^{2}}V\sigma_V^2+\frac{\partial^{2}g}{\partial S\partial X}\sigma_V V\rho=0. \quad (8.38)$$

This PDE is linear in V and its solution with the boundary condition $g(X,V,T;u)=e^{iuX}$ can be guessed in the analytical form

$$g(X,V,t;u)=\exp\left(C(T-t)+D(T-t)V+iuX\right)$$

where $C(\tau)$ and $D(\tau)$ are functions of one variable subject to the boundary conditions $C(0)=0$ and $D(0)=0$. It turns out that (8.38) then splits into two ordinary differential equations for C and D, which have an analytical solution (Heston 1993). Hence, using the Fourier transformation (8.36) we finally obtain the cash-or-nothing binary options price in the form

$$\begin{aligned}B(S,V,t)&=e^{-r(T-t)}p(\ln S,V,t;\ln K)=\\&=e^{-r(T-t)}\left(1-F_{X(T)}(\ln K)\right)\\&=e^{-r(T-t)}\left(\frac{1}{2}+\frac{1}{\pi}\int_0^{+\infty}\frac{\operatorname{Im}\left[e^{-iu\ln K}g(\ln S,V,t;u)\right]}{t}du\right).\end{aligned}$$

We can proceed similarly for an asset-or-nothing option paying $S_T\times 1_{\{S_T\geq K\}}$ at time T. Heston proposes to look for the price in the form $\widetilde{B}(S,V,t)=S\widetilde{p}(S,V,t)$ satisfying (8.35) and appropriate boundary conditions. It turns out that after the substitution $X=\ln S$ the function $\widetilde{p}$ satisfies the PDE

$$\frac{\partial\widetilde{p}}{\partial t}+\frac{\partial\widetilde{p}}{\partial X}(r+V/2)+\frac{\partial\widetilde{p}}{\partial V}\widetilde{\kappa}\left(\widetilde{\theta}-V\right)+\frac{1}{2}\frac{\partial^{2}\widetilde{p}}{\partial X^{2}}V+\frac{1}{2}\frac{\partial^{2}\widetilde{p}}{\partial V^{2}}\sigma_V^2V+\frac{\partial^{2}\widetilde{p}}{\partial S\partial X}\sigma_V V\rho \\ =0.$$

where $\widetilde{\kappa}=\kappa+\lambda-\rho\sigma_V$ and $\widetilde{\theta}=\kappa\theta/\widetilde{\kappa}$. Therefore, we can again "risk-neutralize" the stochastic differential Eq. (8.33) as

$$\begin{aligned}dX&=(r+V/2)dt+\sqrt{V}dz\\dV&=\widetilde{\kappa}\left(\widetilde{\theta}-V\right)dt+\sigma_V\sqrt{V}dz_V\end{aligned}$$

and find $\widetilde{p}(x,v,t;u)=E\left[1_{\{X(T)\geq\ln K\}}|X(t)=x,V(t)=v\right]$ using the characteristic function of $X(T)$ and the Fourier transformation as above.

Finally, the call option value can be expressed by the formula:

$$C(S,V,t)=\widetilde{B}(S,V,t)-KB(S,V,t)=S\widetilde{p}(S,V,t)-Ke^{-r(T-t)}p(S,V,t).$$

8.4.3 SABR Model

A popular stochastic volatility model for valuing interest rate options is denoted by the acronym SABR meaning: Stochastic, Alpha, Beta, and Rho. It was proposed by Hagan et al. (2002) as a two-factor model for a maturity T forward price (or rate) $F(t)$ and stochastic volatility $\alpha(t)$ of the following form under the forward risk-neutral measure:

$$\begin{aligned} dF(t)&=\alpha(t)F(t)^{\beta}dz_1, \quad F(0)=f\\ d\alpha(t)&=\nu\alpha(t)dz_2, \quad \alpha(0)=\alpha \end{aligned}$$

where α, β, ν are constants, f is the current forward price and $dz_1dz_2=\rho dt$. The authors argue that this model captures well the volatility smile and its dynamics—unlike the local volatility model of Derman and Kani (1994) and Dupire (1994). They demonstrate that under the local volatility model, the asset prices and the volatility smile move in a direction that is opposite to typical market dynamics where smiles move in the same direction as the underlying price. One of the negative consequences of this phenomenon is an inefficient hedging performance when the local volatility model is used.

Hagan et al. (2002) use singular perturbation techniques to obtain semi-analytic formulas for the prices of European options under the SABR model and the (Black's model) implied volatilities as functions of the initial forward price $F(0)=f$ and the strike price K:

$$\begin{aligned} \sigma_B(f,K)&=U\cdot\left(\frac{z}{x(z)}\right)\cdot V,\\ U&=\alpha\left\{(fK)^{(1-\beta)/2}\left(1+\frac{(1-\beta)^2}{24}\ln^2 f/K+\frac{(1-\beta)^4}{1920}\ln^4 f/K\right)\right\}, \\ V&=1+\left\{\frac{(1-\beta)^2}{24}\frac{\alpha^2}{(fK)^{1-\beta}}+\frac{\rho\beta\nu\alpha}{4(fK)^{(1-\beta)/2}}+\frac{2-3\rho^2}{24}\nu^2\right\}T+\ldots, \end{aligned} \tag{8.39}$$

where

$$z=\frac{\nu}{\alpha}(fK)^{(1-\beta)/2}\ln f/K \quad \text{and}$$

$$x(z)=\ln\left(\frac{\sqrt{1-2\rho z+z^2}+z-\rho}{1-\rho}\right).$$

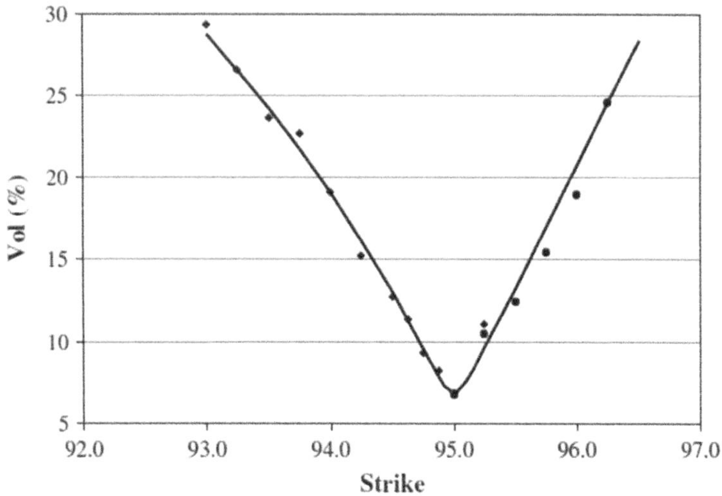

Fig. 8.7 Implied volatility for the Eurodollar options together with the volatilities predicted by the SABR model (Source: Hagan et al. 2002)

Although the formulas look formidable, they involve only elementary functions and can be easily implemented, e.g., in Excel or other calculation tools. The model implied volatility $\sigma_B(f, K)$ can then be used as an input of Black's standard formula to price European options. There is one drawback to the solution; when implementing the formula for V, the higher-order terms " $+ \cdots$ " need to be omitted. However, Hagan et al. (2002) argue that the omitted terms have a negligible effect on the precision of the result.

The implemented pricing model can be used to fit the parameters α, β, ρ, v given the market values of options with different strike prices in a standard way, minimizing the sum of squared differences between the market and model prices. Hagan et al. (2002) analyze the fit precision in a different market situation (see also Molodkina 2018), and demonstrate that it works well, even for very unusual volatility smile shapes (see Fig. 8.7).

The parameter β can be set based separately, or estimated from the data together with other parameters. Note that $\beta = 0$ transforms the SABR to a normal model, $\beta = 1$ to a lognormal type model, and $\beta = 0.5$ to a CIR-like model. The parameter β also significantly influences the dependence of at-the-money volatility $\sigma_B(f,f)$ on the underlying price f. The SABR formula (8.39) for $\sigma_B(f,f)$ simplifies, since $\ln f/K = 0$ for $K = f$, and to get an insight into the dependence one can use the following simple approximation:

$$\sigma_B(f,f) \approx \frac{\alpha}{f^{1-\beta}}.$$

Therefore, for $\beta = 0$, the ATM volatility decreases with increasing f, while for $\beta = 0$ it stays constant, and the parameter can also be used to fit the observed market

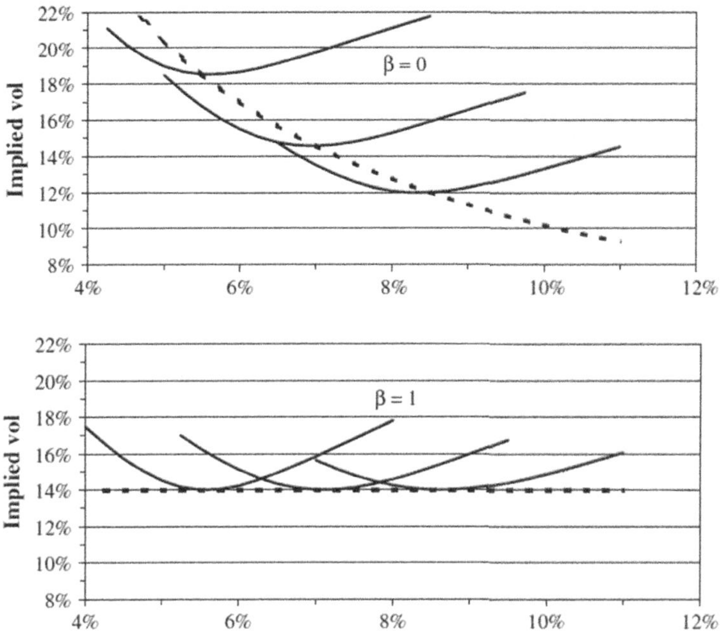

Fig. 8.8 Backbone and smiles for varying f with $\beta = 0$ and $\beta = 1$ (Source: Hagan et al. 2002)

dynamics. Figure 8.8 illustrates the curve where $\sigma_B(f,f)$ varies with f, known as the *backbone* for these two cases. The other parameters have different effects on the volatility smile. We can see from the approximation above that α controls the level of the ATM volatility. On the other hand, the correlation ρ can be shown to control the smile skewness, while the volatility of volatility ν impacts the curvature. The analytical formulas also allow the calculation of the sensitivities of option prices known as *vanna* and *volga* (vol gamma) with respect to the correlation ρ and the volatility of volatility ν that can also be viewed as market factors (implied by the market volatility smile). The sensitivities can be used to hedge an option portfolio with respect to volatility smile skewness and curvature changes.

8.4.4 Stochastic Volatility Estimation

In order to apply a stochastic volatility model, for example, to value exotic options, one needs to estimate its parameters. A possible approach is to use the existing quotations of plain vanilla options and calibrate the parameters of the model, so that the model prices are as close as possible to the quoted prices, e.g., minimizing the sum of squared differences

$$\sum \left(P_i^{\text{model}} - p_i^{\text{quoted}}\right)^2.$$

However, if there are no option quotations, we need to estimate the parameters from a series of historical returns. An analysis of the time series in the context of stochastic volatility models is also in itself of econometric interest.

Compared to the geometrical Brownian motion (GBM), it is much more difficult to estimate the parameters of a stochastic volatility model given a series of daily returns.

Let us assume that the log-returns $r_i = \ln (S_i/S_{i+1})$ over a regular time interval of length Δt follow the discretized stochastic differential equation specified as

$$r_i = \tilde{\mu} + \tilde{\sigma}\varepsilon_i, \varepsilon_i \sim N(0,1) \tag{8.40}$$

where $\widetilde{\mu} = (\mu - \sigma^2/2)\Delta t$ and $\widetilde{\sigma} = \sigma\sqrt{\Delta t}$. Recall that in this case, it is straightforward to set up the likelihood function directly expressing $\varepsilon_i = (r_i - \widetilde{\mu})/\widetilde{\sigma}$ and $L = \prod \varphi(\varepsilon_i)$. The parameters $\widetilde{\mu}$ and $\widetilde{\sigma}$ are then estimated maximizing the log-likelihood ln$L$.

Consider now, for example, the discretized log-variance model (8.32) in the form

$$\begin{aligned} r_i &= \tilde{\mu} + e^{h_i/2}\varepsilon_i, \varepsilon_i \sim N(0,1), \\ h_i &= \alpha + \beta h_{i-1} + \gamma\varepsilon_i^V, \varepsilon_i^V \sim N(0,1), \end{aligned} \tag{8.41}$$

where $\alpha = \alpha h_L \Delta t$, $\beta = 1 - \alpha\Delta t$, $\gamma = \sigma_V\sqrt{\Delta t}$, the annualized variance is $V_i = e^{h_i}/\Delta t$, and the innovations $\varepsilon_i, \varepsilon_i^V$ are independent. Given a time series of log-returns $r_i = \ln (S_i/S_{i+1})$, there is no obvious way how to express the innovations ε_i, ε_i^V, and the corresponding likelihood function that would enable us to estimate the parameters $\widetilde{\mu}, \alpha, \beta$, and γ. The problem is that the log-variances h_i are latent, i.e., not directly observable. There are certainly some naïve approaches, e.g., estimate first the volatilities using EWMA or GARCH from the return time series, and then estimate the constant parameters. However, both EWMA and GARCH implicitly assume that the variance innovations directly depend on the realized squared returns, and so are not consistent with (8.41) where ε_i and ε_i^V are assumed to be independent. In fact, we need to estimate not only the model parameters $\Theta = \langle\widetilde{\mu}, \alpha, \beta, \gamma\rangle$ but also the series of the latent state variables $X = \langle h_i; =1, \ldots, T\rangle$. The known estimation approaches include the method of moments (efficient, simulated, etc.), quasi-maximum likelihood, or Bayesian methods such as Markov Chain Monte Carlo (MCMC)—see, e.g., Shephard (2004). Among these, the MCMC approach, first applied by Jacquier et al. (1994), is considered to be the most efficient.

8.4.5 MCMC Estimations

The Bayesian MCMC sampling algorithm has become a strong and frequently used tool to estimate complex models with multidimensional parameter vectors, including latent state variables. Examples include financial stochastic models with jumps, stochastic volatility processes, models with complex correlation structure, or switching-regime processes. For a more complete treatment of MCMC methods and applications we refer the reader, for example, to Johannes and Polson (2009), Rachev et al. (2008), or Lynch (2007).

MCMC provides a method of sampling from multivariate densities that are not easy to sample from directly, by breaking these densities down into more manageable univariate or lower dimensional multivariate densities. To estimate a vector of unknown parameters $\Theta = (\theta_1, \ldots, \theta_k)$ from a given dataset, where we are able to write down the Bayesian marginal densities $p(\theta_j | \theta_i, i \neq j, \text{data})$ but not the multivariate density $p(\Theta | \text{data})$, the MCMC *Gibbs sampler* works according to the following generic procedure:

0. Assign a vector of initial values to $\Theta^0 = \left(\theta_1^0, \ldots, \theta_k^0\right)$ and set $j = 0$.
1. Set $j = j + 1$.
2. Sample $\theta_1^j \sim p(\theta_1 | \theta_2^{j-1}, \ldots, \theta_k^{j-1}, \text{data})$.
3. Sample $\theta_2^j \sim p(\theta_2 | \theta_1^j, \theta_3^{j-1}, \ldots \theta_k^{j-1}, \text{data})$.

⋮

$k + 1$. Sample $\theta_k^j \sim p(\theta_k | \theta_1^j, \theta_2^j, \ldots, \theta_{k-1}^j, \text{data})$ and return to step 1.

According to the Clifford-Hammersley theorem, the conditional distributions $p(\theta_j | \theta_i, i \neq j, \text{data})$ fully characterize the joint distribution $p(\Theta | \text{data})$, and, moreover, under certain mild conditions, the Gibbs sampler distribution converges to the target joint distribution (Johannes and Polson 2009).

The conditional probabilities are typically obtained by applying the Bayes theorem[1] to the likelihood function and a prior density, e.g.,

$$p\left(\theta_1 | \theta_2^{j-1}, \ldots, \theta_k^{j-1}, \text{data}\right) \propto L\left(\text{data} | \theta_1, \theta_2^{j-1}, \ldots, \theta_k^{j-1}\right) \text{prior}\left(\theta_1 | \theta_2^{j-1}, \ldots, \theta_k^{j-1}\right). \tag{8.42}$$

The symbol $\propto$ means that the density on the left-hand side equals to the expression on the right-hand side multiplied by an unknown constant independent of the density function variable, i.e., θ_1 in the case of (8.42). We will often use uninformative priors, i.e., $\text{prior}(\theta_i) \propto 1$, and assume that the parameters are independent. In order to apply the Gibbs sampler, the right-hand side of the proportional

[1] The Bayes theorem can, in this context, be stated as $\frac{p(\theta|\text{data})}{p(\theta)} = \frac{p(\text{data}|\theta)}{p(\text{data})}$ and so $p(\theta | \text{data}) \propto p(\text{data} | \theta) p(\theta)$ where $\propto$ reads as "proportionate".

relationship needs to be normalized, i.e., we need to be able to integrate the right-hand side with respect to θ_1 conditional on $\theta_2^{j-1}, \ldots, \theta_k^{j-1}$.

Useful Gibbs sampling distributions are univariate or multivariate normal, Inverse Gamma[2] or inverse Wishart, and the Beta distributions. If $\mathbf{y} = \langle y_1, \ldots, y_T \rangle$ is an observed series and assuming that $y_i \sim N(\mu, \sigma^2)$ are independent identically distributed (iid) with unknown parameters μ and σ, we have

$$p(\mu|\mathbf{y},\sigma) \propto L(\mathbf{y}|\mu,\sigma^2)p(\mu) = \prod_{i=1}^{T} \varphi(y_1;\mu,\sigma) \propto \left(\frac{1}{\sqrt{2\pi\sigma}}\right) \exp\left(-\frac{\sum (y_i-\mu)^2}{2\sigma^2}\right)$$
$$\propto \exp\left(-\frac{T\mu^2 - 2\mu\sum y_i}{2\sigma^2}\right) \propto \varphi\left(\mu; \frac{\sum y_i}{T}, \frac{\sigma}{\sqrt{T}}\right) \tag{8.43}$$

using the uninformative prior $p(\mu) \propto 1$. Moreover,

$$p(\sigma^2|\mathbf{y},\mu) \propto L(\mathbf{y}|\mu,\sigma^2) \cdot p(\sigma^2) = \frac{1}{\sigma^2} \prod_{i=1} \varphi(y_i;\mu,\sigma)$$
$$\propto (\sigma^2)^{-\frac{T}{2}-1} \exp\left(-\frac{\sum (y_i-\mu)^2}{2\sigma^2}\right) \propto IG\left(\sigma^2; \frac{T}{2} \frac{\sum (y_i-\mu)^2}{2}\right) \tag{8.44}$$

using the prior $p(\sigma^2) \propto 1/\sigma^2$ equivalent to the uninformative log-variance prior p $(\log \sigma^2) \propto 1$. Hence, the Bayesian distributions for μ and σ can be obtained by the Gibbs sampler iterating (8.43) and (8.44). The prior distributions are often specified in order to improve convergence, but not to influence (significantly) the final results. Typically, a conjugate prior wide normal distribution for μ and a flat inverse gamma distribution for σ^2 are used.

If the series is multivariate normal then the distributions are generalized to multivariate normal and inverse Wishart (Lynch 2007). A multivariate discrete-time diffusion process (8.40) is, in fact, equivalent to a multivariate normal return series model with iid $\mathbf{r}_i \sim N(\boldsymbol{\mu}, \Sigma)$, where $\mathbf{r}_i = (r_{i,1}, \ldots, r_{i,m})'$ is a vector of returns on m assets observed at time i, $\boldsymbol{\mu}$ is a vector of means, and Σ a covariance matrix. The marginal distributions are

$$p(\boldsymbol{\mu}|\mathbf{r},\Sigma) = \varphi\left(\boldsymbol{\mu}; \frac{1}{T}\sum_{i=1}^{T} \mathbf{r}_i, \frac{1}{T}\Sigma\right) \quad \text{and} \tag{8.45}$$

[2]The inverse gamma probability distribution density function with the shape parameter α and scale parameter β is given by $IG(x;\alpha,\beta) = \frac{\beta^\alpha}{\Gamma(\alpha)} x^{-\alpha-1} \exp(-\beta/x)$ where Γ is the Gamma function. The mean of x is $\mu = \frac{\beta}{\alpha-1}$ for $\alpha > 1$ and the variance is $\sigma^2 = \frac{\beta^2}{(\alpha-1)^2(\alpha-2)}$ for $\alpha > 2$. Alternatively, given μ and σ^2 we get $\alpha = \frac{\mu^2}{\sigma^2} + 2$ and $\beta = \mu\left(\frac{\mu^2}{\sigma^2} + 1\right)$.

$$p(\Sigma|\mathbf{r},\boldsymbol{\mu}) = IW(\Sigma;T,S) \propto |\Sigma|^{-(T+m+1)/2}\exp\left(-\frac{1}{2}\mathrm{tr}\left(S\Sigma^{-1}\right)\right) \tag{8.46}$$

where $IW(\Sigma; T, S)$ denotes the inverse Wishart distribution, $S = \sum_{i=1}^{T}(\mathbf{r}_i - \boldsymbol{\mu})'(\mathbf{r}_i - \boldsymbol{\mu})$ is the scale matrix, and the improper prior $p(\Sigma) \propto |\Sigma|^{-\frac{m+1}{2}}$ analogous to the univariate case is used.

If $\mathbf{b} = \langle b_1, \ldots, b_T \rangle$ is a binary series where b_i~Bern(λ) iid, then λ can be sampled using the beta[3] distribution:

$$\begin{aligned} p(\lambda|\mathbf{b}) &\propto L(\mathbf{b}|\lambda)p(\lambda) \propto \prod_{i=1}^{T}\lambda^{b_i}(1-\lambda)^{1-b_i} = \lambda^n(1-\lambda)^{T-n} \propto \\ &\propto \mathrm{Beta}(\lambda; n+1, T-n+1) \end{aligned} \tag{8.47}$$

with the uninformative prior $p(\lambda) \propto 1$. Generally, the beta distribution Beta($x; \alpha, \beta$) would be a conjugate prior where α and β can be interpreted as prior "successes" and "failures."

If the integration on the right-hand side of (8.42) is not analytically feasible (which is usually the case for volatility models), then the *Metropolis-Hastings algorithm* can be used. It is based on a rejection sampling algorithm. For example, in step 2, the idea is first to sample a new proposal value of θ_1^j, and then accept it or reject it (i.e., reset $\theta_1^j := \theta_1^{j-1}$) with appropriate probability so that, intuitively speaking, we move rather to the parameter estimates with higher corresponding likelihood probabilities.

Specifically, step 2 is replaced with the following two-step procedure:

A. Draw θ_1^j from a proposal density $q\left(\theta_1|\theta_1^{j-1}, \theta_2^{j-1}, \ldots \theta_k^{j-1}, \mathrm{data}\right)$

B. Accept θ_1^j with the probability $\alpha = \min(R, 1)$, where

$$R = \frac{p\left(\theta_1^j|\theta_2^{j-1}, \ldots, \theta_k^{j-1}, \mathrm{data}\right) q\left(\theta_1^{j-1}|\theta_1^j, \theta_2^{j-1}, \ldots, \theta_k^{j-1}, \mathrm{data}\right)}{p\left(\theta_1^{j-1}|\theta_2^{j-1}, \ldots, \theta_k^{j-1}, \mathrm{data}\right) q\left(\theta_1^j|\theta_1^{j-1}, \theta_2^{j-1}, \ldots, \theta_k^{j-1}, \mathrm{data}\right)}. \tag{8.48}$$

In practice, the step 2B is implemented by sampling a value u~$U(0, 1)$ from the uniform distribution and accepting θ_1^j if, and only if, $u < R$.

[3]The pdf of the beta distribution is $\mathrm{Beta}(x; \alpha, \beta) = \frac{\Gamma(\alpha+\beta)}{\Gamma(\alpha)\Gamma(\beta)}x^{\alpha-1}(1-x)^{\beta-1}$ for $0 < x < 1$. The mean can be expressed as $\mu \frac{\alpha}{\alpha+B}$ and the variance $\sigma^2 = \frac{\alpha\beta}{(\alpha+\beta)^2(\alpha+\beta+1)}$.

It is again shown (see Johannes and Polson 2009) that under certain mild conditions the limiting distribution is the joint distribution $p(\Theta|\,\text{data})$ of the parameter vector. Note that the limiting distribution does not depend on the proposal density, or on the starting parameter values. The proposal density and the initial estimates only make the algorithm more or less numerically efficient.

A popular simple proposal density is the random walk, i.e., sampling by

$$\theta_1^j \sim \theta_1^{j-1} + N(0,c). \tag{8.49}$$

The algorithm is then called *Random Walk Metropolis-Hastings*. The scaling constant c can be based on the expected parameter estimation error. The proposal density is in this case symmetric, i.e., the probability of going from θ_1^{j-1} to θ_1^j is the same as the probability of going from θ_1^j to θ_1^{j-1} (conditional on the other parameters), and so the second part of the fraction in the formula (8.48) for α in step 1B cancels out. Consequently, assuming a non-informative prior, the acceptance or rejection is driven only by the likelihood ratio

$$R = \frac{L\left(\text{data}|\theta_1^j, \theta_2^{j-1}, \ldots, \theta_k^{j-1}\right)}{L\left(\text{data}|\theta_1^{j-1}, \theta_2^{j-1}, \ldots, \theta_k^{j-1}\right)}.$$

Another popular approach we shall use is the *Independence Sampling Metropolis-Hastings algorithm* where the proposal density $q(\theta_1^j)$ does not depend on θ_1^{j-1} (given the other parameters). The acceptance probability ratio (8.48) is slightly simplified, but the proposal densities do not cancel out. In order to achieve efficiency, the shape of the proposal density q should be close to the shape of the target density p which is known only up to a normalizing constant.

Typically, when estimating complex stochastic models, we need to estimate the parameter vector with a few model parameters Θ, and a vector with a large number of state variables X (proportional to the number of observations). We know that

$$p(\Theta, X|\text{data}) \propto p(\text{data}|\Theta, X) \cdot p(X, \Theta),$$

and so we may estimate iteratively the parameters and the state variables:

$$p(\Theta|X, \text{data}) \propto p(\text{data}|\Theta, X) \cdot p(X|\Theta) \cdot p(\Theta),$$
$$p(X|\Theta, \text{data}) \propto p(\text{data}|\Theta, X) \cdot p(X|\Theta) \cdot p(X).$$

The parameters and state variables are sampled step by step, or in blocks, often combining the Gibbs and Metropolis–Hastings sampling steps.

8.4.6 Stochastic Volatility and Jumps Estimations: An Empirical Study

Jacquier et al. (1994), in their seminal paper, estimated the stochastic volatilities and parameters of the model (8.41) using an MCMC algorithm based on the NYSE stock index returns. Their contribution lies, especially in their efficient choice of the Metropolis–Hastings proposal density to sample the stochastic volatilities. To illustrate the method, we will present a generalization of the algorithm to the stochastic volatility model with jumps. We follow Witzany (2013a), where the main result is an estimation of a more general bivariate stochastic volatility model with correlated jumps, allowing us to study the transfer of volatility and jumps between different markets.

In order to present the model and its estimation algorithm, let us first consider the discrete-time jump–diffusion model

$$r_i = \mu + \sigma\varepsilon_i + Z_i J_i,$$
$$\varepsilon_i \sim N(0,1), Z_i \sim N(\mu_J, \sigma_J), J_i \sim \text{Bern}(\lambda),$$

with a constant volatility σ. Given a sequence of observed returns data $= \{r_i; i = 1,\ldots,T\}$, the parameters and latent state variables to be estimated are: μ, , λ, μ_J, σ_J, $\mathbf{Z} = \{Z_i; i = 1,\ldots,T\}$, $\mathbf{J} = \{J_i; i = 1,\ldots,T\}$.

In this case, we may use the pure Gibbs MCMC algorithm:

1. Sample reasonable initial values $\mu^{(0)}$, $\sigma^{(0)}$, $\lambda^{(0)}$, $\mu_J{}^{(0)}$, $\sigma_J{}^{(0)}$, $\mathbf{Z}^{(0)}$, $\mathbf{J}^{(0)}$
2. For $i = 1, \ldots, T$ sample $Z_i^{(g)} \propto \varphi\left(Z; \mu_J^{(g-1)}, \sigma_J^{(g-1)}\right)$ if $J_i^{(g-1)} = 0$, and $Z_i^{(g)} \propto \varphi\left(r_i; \mu^{(g-1)} + Z, gs^{(g-1)}\right)\varphi\left(Z; \mu_J^{(g-1)}, \sigma_J^{(g-1)}\right)$ *if* $J_i^{(g-1)} = 1$.
3. For $i = 1, \ldots, T$ sample $J_i^{(g)} \in \{0,1\}$ with $\Pr[J=1] = p_1/(p_0 + p_1)$, where $p_0 = \varphi\left(r_i; \mu^{(g-1)}, \sigma^{(g-1)}\right)\left(1 - \lambda^{(g-1)}\right)$, $p_1 = \varphi\left(r_i; \mu^{(g-1)} + Z, \sigma^{(g-1)}\right)\lambda^{(g-1)}$
4. Sample $\mu^{(g)}$, $\sigma^{(g)}$ based on the normally distributed time series $r_i - Z_i^{(g)} J_i^{(g)}$ according to (8.43) and (8.44).
5. Sample $\lambda^{(0)}$ based on Bernoulli distributed $J_i^{(g)}$ binary time series according to (8.47).
6. Sample $\mu_J^{(g)}, \sigma_J^{(g)}$ based on the normally distributed time series $Z_i^{(g)}$ again according to (8.43) and (8.44).

Next, let us consider a jump–diffusion model with stochastic volatility following the discrete time specification:

$$\begin{aligned} r_i &= \mu + \sqrt{V_i}\varepsilon_i + Z_i J_i, \\ \ln V_i &= \alpha + \beta \ln V_{i-1} + \gamma \varepsilon_i^V, \\ \varepsilon_i, \varepsilon_i^V &\sim N(0,1), Z_i \sim N(\mu_J, \sigma_J), J_i \sim \text{Bern}(\lambda), \text{iid}. \end{aligned} \tag{8.50}$$

In this case we need to estimate not only the latent state variable vectors **Z**, **J**, but also the vector of latent stochastic variances **V**. The MCMC estimation unfortunately requires an application of the Metropolis–Hastings approach since the conditional distribution for the variance V_i (conditional on the other variances and parameters) is not known. It follows from (8.50) and the Bayes Theorem that:

$$p\left(V_i | \mathbf{V}_{(-i)}, \Theta, \mathbf{r}, \mathbf{Z}, \mathbf{J}\right) \propto p(r_i | V_i, \Theta, Z_i, J_i) p(V_i | V_{i-1}, \Theta) p(V_{i+1} | V_i, \Theta). \tag{8.51}$$

Here, the first part of the right-hand side of (8.51) is inverse gamma in V_i:

$$\begin{aligned} p(r_i | V_i, \Theta, Z_i, J_i) &= \varphi(r_i; \mu + Z_i J_i, V_i) \\ &\propto V_i^{-0.5} \exp\left(-0.5(r_i - \mu - Z_i J_i)^2 / V_i\right). \end{aligned} \tag{8.52}$$

But the remaining two factors are lognormal[4] in V_i:

$$\ln V_i \propto \varphi(\ln V_i; \alpha + \beta \ln V_{i=1}, \gamma), \text{i.e.,}$$

$$p(V_i | V_{i-1}, \Theta) \propto V_i^{-1} \exp\left(-(\ln V_i - \alpha - \beta \ln V_{i-1})^2 / (2\gamma^2)\right),$$

and similarly

$$\ln V_{i+1} \sim \varphi(\ln V_{i+1}; \alpha + \beta \ln V_i, \gamma), \text{i.e.,}$$

$$p(V_{i+1} | V_i, \Theta) \propto \exp\left(-(\ln V_{i+1} - \alpha - \beta \ln V_i)^2 / (2\gamma^2)\right) \text{ in terms of } V_i.$$

It is easy to verify that the product of the two lognormal distributions is proportional to the lognormal distribution with the corresponding normal distribution mean and standard deviation:

[4]The lognormal probability density function with parameters μ and σ is given by $LN(x;\mu,\sigma) = \frac{1}{x\sigma\sqrt{2\pi}} \exp\left(-0.5(\ln x - \mu)^2/\sigma^2\right)$. It is useful to note that the mean of x is $\exp(\mu + \sigma^2/2)$ and the variance is $(\exp(\sigma^2) - 1)\exp(2\mu + \sigma^2)$.

$$\mu_i = (\alpha(1-\beta) + \beta(\log V_{i+1} + \log V_{i-1}))/(1-\beta^2),$$
$$\sigma = \gamma/\sqrt{1+\beta^2}. \tag{8.53}$$

In order to obtain a proposal distribution, Jacquier et al. (1994) suggest replacing the lognormal distribution with an inverse gamma distribution fitting the first two moments. It is confirmed empirically that the choice of a proposal distribution with a shape closely mimicking the original distribution is of key importance, since high dimensionality of the variance state variable vector makes convergence of the MCMC algorithm difficult.

The product of two inverse gamma distribution density functions is an inverse gamma distribution density function, hence combining the inverse gamma distribution (8.52) and the fitted inverse gamma distribution, we finally obtain the proposal density function:

$$q(V_i|\mathbf{V}_{(-i)}.\Theta,\mathbf{r},\mathbf{Z},\mathbf{J}) =$$
$$= IG\big(V_i;\phi+0,5,(\phi-1)\exp\left(\mu_i+0.5\sigma^2\right)$$
$$+0.5(r_i-\mu-Z_iJ_i)^2\big), \text{where}\,\phi = \frac{1-2\exp\left(\sigma^2\right)}{1-\exp\left(\sigma^2\right)}. \tag{8.54}$$

The proposal density is used in the Metropolis–Hasting algorithm within a new block, e.g., following the step 3 in the MCMC procedure for the jump–diffusion processes. This block updates all the variances V_i, $i = 1, \ldots, T$. For V_1 and V_T the formula (8.53) needs to be modified slightly, since V_0 and V_{T+1} are not known. The diffusion volatility σ is obviously replaced by the square root of the latest estimate of the variance V_i and we also need to add a new MCMC step for the AR(1) coefficients α, β and γ (for example, following the step that updates $\mathbf{V}$). The coefficients α, β can be sampled with a bivariate normal distribution, and γ with the inverse gamma distribution. The extended MCMC algorithm can be described in detail as follows:

1. Sample reasonable initial values $\mu^{(0)}, \lambda^{(0)}, \mu_J{}^{(0)}, \sigma_J{}^{(0)}, \alpha^{(0)}, \beta^{(0)}, \gamma^{(0)}, \mathbf{V}^{(0)}, \mathbf{Z}^{(0)}, \mathbf{J}^{(0)}$
2. For $i = 1, \ldots, T$ sample $Z_i^{(g)} \propto \varphi\left(Z;\mu_J^{(g-1)},\sigma_J^{(g-1)}\right)$ if $J_i^{(g-1)} = 0$, and
$$Z_I^{(g)} \propto \varphi\left(r_i;\mu^{(g-1)}+Z,\sqrt{V_i^{(g-1)}}\right)\varphi\left(Z;\mu_J^{(g-1)},\sigma_J^{(g-1)}\right)$$
if $J_i^{(g-1)} = 1$.
3. For $i = 1, \ldots, T$ sample $J_i^{(g)} \in \{0,1\}$ with $\Pr[J=1] = p_1/(p_0+p_1)$, where
$$p_0 = \varphi\left(r_i;\mu^{(g-1)},\sqrt{V_i^{(g-1)}}\right)\left(1-\lambda^{(g-1)}\right),$$
$$p_1 = \varphi\left(r_i;\mu^{(g-1)}+Z,\sqrt{V_i^{(g-1)}}\right)\lambda^{(g-1)}.$$
4. Sample new stochastic variances $V_i^{(g)}$ for $i = 1, \ldots, T$ using Metropolis–Hastings (8.48) with the proposal density given by (8.54).

5. Sample new stochastic volatility autoregression coefficients $\alpha^{(g)}$, $\beta^{(g)}$, $\gamma^{(g)}$ from $h_i = \log V_i^{(g)}$ for $i = 1, \ldots, T$ using the Bayesian linear regression model (Lynch 2007):

$$\widehat{\boldsymbol{\beta}} = (\mathbf{X}'\mathbf{X})^{-1}\mathbf{X}\mathbf{y}, \widehat{\mathbf{e}} = \mathbf{y} - \mathbf{X}\widehat{\boldsymbol{\beta}}, \text{where } \mathbf{X} = \begin{pmatrix} 1 \ldots 1 \\ h_1 \ldots h_{T-1} \end{pmatrix}', \mathbf{y} = (h_2 \ldots h_T)',$$

$$\left(\gamma^{(g)}\right)^2 \propto IG\left(\frac{n-2}{2}, \frac{\widehat{\mathbf{e}'\mathbf{e}}}{2}\right),$$

$$\left(\alpha^{(g)}, \beta^{(g)}\right)' \propto \varphi\left((\alpha, \beta)'; \widehat{\boldsymbol{\beta}}, \left(\gamma^{(g)}\right)^2 (\mathbf{X}'\mathbf{X})^{-1}\right)$$

6. Sample $\mu^{(g)}$ based on the normally distributed time series $r_i - \mathbf{Z}_i^{(g)} J_i^{(g)}$ with variances $V_i^{(g)}$, i.e.,

$$p\left(\mu^{(g)} | \mathbf{r}, \mathbf{Z}^{(g)}, \mathbf{J}^{(g)}, \mathbf{V}^{(g)}\right) \propto \varphi\left(\mu; \sum_{i=1}^{T} \frac{r_i - J_i^{(g)} Z_i^{(g)}}{V_i^{(g)}} / \sum_{i=1}^{T} \frac{1}{V_i^{(g)}}, 1/\sum_{i=1}^{T} \frac{1}{V_i^{(g)}}\right).$$

7. Sample $\lambda^{(0)}$ based on Bernoulli distributed $J_i^{(g)}$ binary time series according to (8.47).
8. Sample $\mu_J^{(g)}, \sigma_J^{(g)}$ based on the normally distributed time series $Z_i^{(g)}$ according to (8.43) and (8.44).

It can be argued that the algorithm is becoming more and more complicated, but it should be noted that it is built from individual, not so complex estimation blocks. If the blocks are well defined and put together, then the MCMC "machine" is quite comprehensible and works well.

Witzany (2013a) applied the MCMC algorithm to a data set consisting of CZK/EUR exchange rates and the Czech stock index PX values from Sep 1, 2004, to Feb 2, 2011. The time series of daily returns (Fig. 8.9) visually indicate many jumps and overlapping periods of relatively high volatility.

The estimated parameters of the stochastic volatility model and their standard deviations based on 3000 MCMC simulations are shown in Table 8.1. Figure 8.10 shows a relatively fast convergence of the coefficient β in the case of the PX return process. The parameter means and standard deviations in Table 8.1 are based on the last 2500 simulations, discarding the first 500. The model with constant volatility is also estimated in Witzany (2013a), and it turns out that by introducing the stochastic volatility into the model, the probabilities of jumps were significantly reduced, and the jump size standard deviation went up. The high value of the stochastic volatility (log-variance) autocorrelation coefficient β, almost 99% for CZK/EUR and almost 98% for PX, shows a high persistence of stochastic volatilities, which is in line with other empirical studies on US stock market data (e.g., Jacquier et al. 1994; Eraker et al. 2003). The volatility of the stochastic volatility, i.e., the coefficient γ, around 13% for CZK/EUR and over 21% for PX, is also in the range estimated on the US data.

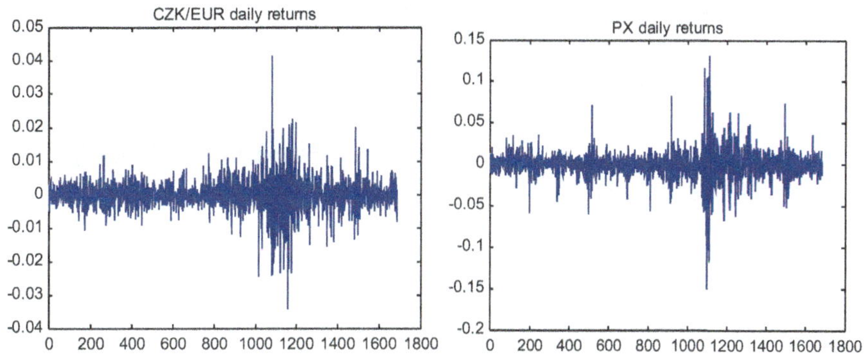

Fig. 8.9 CZK/EUR exchange rate and PX stock index daily returns

Table 8.1 Estimated parameters for the CZK/EUR and PX univariate jump–diffusion models with stochastic volatility

CZK/EUR univariate jump–diffusion process with stochastic volatility						
μ_1	λ_1	$\mu_{1,J}$	$\sigma_{1,J}$	α_1	β_1	γ_1
1.8506e-004 (6.374e-005)	0.0284 (0.0083)	−2.2616e-004 (0.0024)	0.0117 (0.0018)	−0.1205 (0.0545)	0.9893 (0.0048)	0.1313 (0.0193)
PX univariate jump–diffusion process with stochastic volatility						
μ_2	λ_2	$\mu_{2,J}$	$\sigma_{2,J}$	α_2	β_2	γ_2
0.0012 (1.921e-004)	0.0237 (0.0068)	0.0011 (0.0079)	0.0427 (0.0066)	−0.1957 (0.0613)	0.9781 (0.0069)	0.2119 (0.0247)

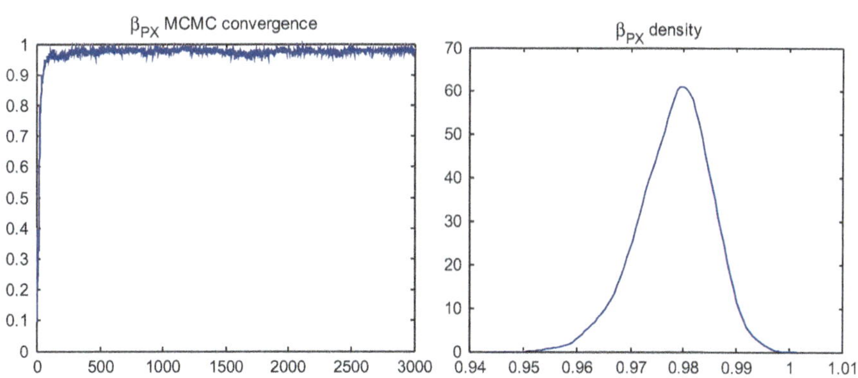

Fig. 8.10 Convergence and MCMC simulated density of the parameter β for PX returns

The latent stochastic volatilities are sampled at each MCMC simulation run and we get distribution for each particular day. In order to investigate the relationship between the CZK/EUR and PX volatilities, we can use the mean estimates, specifically given by the equation $\widehat{\sigma}_i = \exp\left(\overline{h}_i/2\right)$, where $\overline{h}_1$ is the MCMC mean of

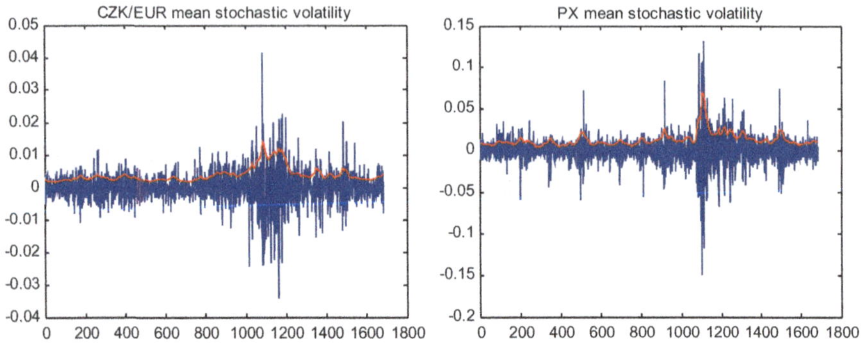

Fig. 8.11 CZK/EUR and PX returns (bars) and mean stochastic volatilities (lines)

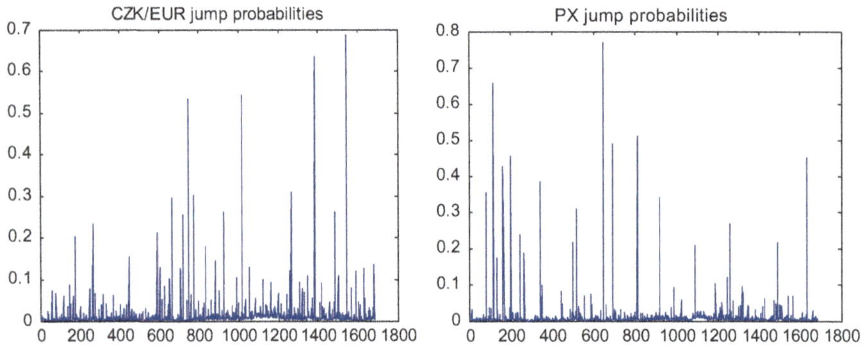

Fig. 8.12 CZK/EUR and PX returns jump probabilities assuming stochastic volatility

normally distributed $\ln V_i$ sampled values. Figure 8.11 shows that the mean stochastic volatility for both series copies well the pattern of the observed returns. The figures also explain why many jumps that can be identified in the constant volatility model have been filtered out in the stochastic volatility model.

Similarly, we can look at sampled jump probabilities. Note that in each run, the sampled jump indicator is 0 or 1. Figure 8.12 shows their average (discarding the first 500 iterations), which can be interpreted as Bayesian probabilities of jumps over individual days. As noted above, the intensity of jumps is much smaller in the model with stochastic volatility than in the model with constant volatility, as indicated by Fig. 8.13. The intensity of jumps went up dramatically during the crisis period of high volatility, which is not sufficiently reflected by the constant volatility estimated by the model.

The mean stochastic volatility series and mean estimated coefficients α, β, and γ may be used to obtain the residuals of the two series $e^V_{FX,i}$ and $e^V_{PX,i}$. The correlation of the residuals is a relatively low 5.7%, and not significant at the 1% confidence level. However, the correlation of $\overline{h}_{FX,i}$ and $\overline{h}_{PX,i}$ comes out at a highly significant 61.66%,

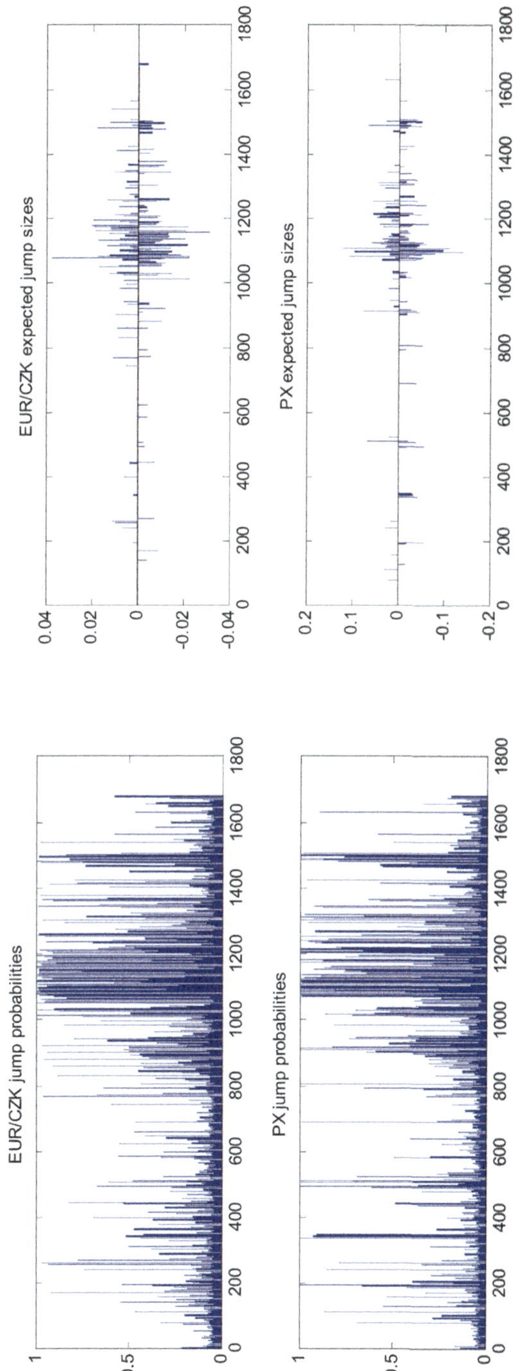

Fig. 8.13 CZK/EUR and PX returns jump probabilities and mean jump sizes assuming constant volatility

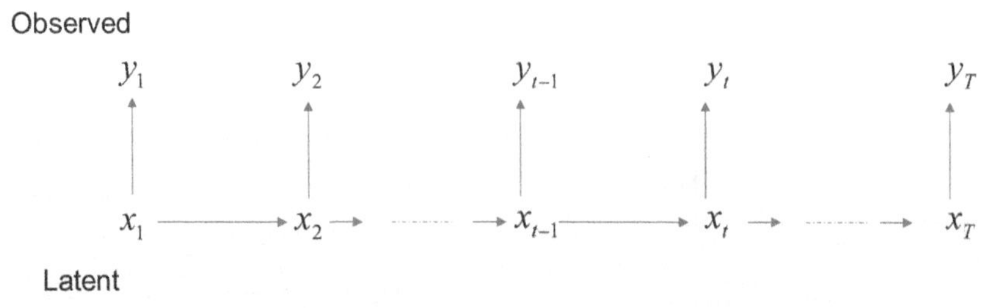

Fig. 8.14 A general state-space model

as indicated visually by Fig. 8.11. The study (Witzany 2013a) then goes on to formulate and estimate a bivariate stochastic volatility model with correlated volatilities and jumps confirming these preliminary results and, in particular, showing that it is the FX volatility which influences the PX volatility rather than vice versa.

8.4.7 Particle Filters

The MCMC algorithm is not efficient when we need to estimate latent state variables and model parameters in a real-time regime, when new observations arrive step by step. Figure 8.14 shows schematically a general latent state-space model where we are observing a time series $y_{t:T} = \langle y_1, \ldots, y_T \rangle$ that is related to a latent state process $x_{t:T} = \langle x_1, \ldots, x_T \rangle$. The model can be specified by the following two equations:

$$\begin{aligned} y_t &= H(x_t, w_t, \theta), \\ x_t &= F(x_{t-1}, v_t, \theta) \end{aligned} \tag{8.55}$$

where w_t and v_t are mutually independent noise variables and θ is a vector of unknown model parameters. The density $p(y_t | x_t, \theta)$ determined by the first equation is called the *observation density*, while the density $p(x_t | x_{t-1}, \theta)$ given by the second equation is called the *transition density* of the Markov process of the hidden state with an initial distribution $p(x_0 | \theta)$. The observation variable may differ from the latent variable only by a measurement error, or the relationship may be more complicated, as in the case of an observed asset return with an underlying latent volatility.

The goal of *smoothing* is to estimate $x_{1:T}$ and θ given the time series $y_{1:T}$ of observations over the full-time window, or, more precisely, to estimate the conditional density $p(x_{1:T}, \theta | y_{1:T})$, while the goal of *filtering* is to estimate sequentially $p(x_{1:t}, \theta | y_{1:t})$ with $t = 1, \ldots, T$. The MCMC algorithm is appropriate for smoothing, but not very efficient for filtering where it is not able to use the already estimated density $p(x_{1:t-1}, \theta | y_{1:t-1})$ when estimating $p(x_{1:t}, \theta | y_{1:t})$.

The well-known Kalman filter assumes that the relationships (8.55) are linear and the innovations Gaussian. The conditional distribution of x_t, given the observations $y_{1:t}$ and the parameters θ, is then normal, and its mean and variance can be easily updated to the normal distribution of x_{t+1} given a new observation y_{t+1} This follows from the Bayesian decomposition

$$p(x_{t+1}|y_{1:t+1}) = p(x_{t+1}|x_t, y_{t+1}) \propto p(y_{t+1}|x_{t+1})p(x_{t+1}|x_t). \tag{8.56}$$

Since the densities on the right-hand side are normal, the density of the left-hand side must be also normal. The Kalman filter technique may be applied both to filtering and smoothing. It can also be used to express analytically the marginal multivariate normal density $p(y_{1:T}|\theta)$, and so, if the parameters are unknown we obtain their Bayesian density $p(\theta|y_{1:T}) \propto p(y_{t:T}|\theta)p(\theta)$ given a prior distribution $p(\theta)$.

The method of particle filters (or sequential Monte Carlo) firstly proposed by Gordon et al. (1993) and Del Moral (1996) is more general—the idea is to represent the probability distribution $p(x_{1:t}|y_{1:t})$, given the model parameters θ, by an empirical approximation $\left\{\left(x_{1:t}^i, w_t^i\right)\right\}_{i=1}^N$ where w_t^i are weights of sampled values (paths) $x_{1:t}^i$ which are also called the particles. Note that w_t^i must be nonnegative and sum up to 1, but should not be interpreted as probabilities, since $x_{1:t}^i$ do not have to be (and usually are not) uniformly distributed. Generally, if f is a function (e.g., to calculate moments of the empirical distribution of the latent variable) then we can use the approximation

$$E[f(x_{1:t})] \approx \sum_{i=1}^N f\left(x_{1:t}^i\right)w_t^i.$$

The particle filter algorithm sequentially samples x_{t+1}^i given $x_{1:t}^i$ and the new observation y_{t+1}. The weights are updated according to (8.56) and the elementary decomposition

$$p(x_{1:t+1}|y_{1:t+1}) = p(x_{t+1}|x_t, y_{t+1})p(x_{1:t}|y_{1:t})$$

that follows from the model assumption (see Fig. 8.14). A common approach is to sample x_{t+1}^i from a proposal density $q\left(x_{t+1}|x_{t+1}^i, y_{t+1}\right)$, e.g., from $p\left(x_{t+1}|x_{t+1}^i\right)$, and update the weight w_t^i based on (8.56), i.e.,

$$\widehat{w}_{t+1}^i = \frac{p\left(y_{t+1}|x_{t+1}^i\right)p\left(x_{t+1}^i|x_t^i\right)}{q\left(x_{t+1}^i|x_t^i, y_{t+1}\right)}w_t^i.$$

The updated weights do not have to sum up to one and need to be normalized, i.e., by setting $w_{t+1}^i = \widetilde{w}_{t+1}^i / \sum_j \widetilde{w}_{t+1}^j$.

It would be optimal to use directly $p(x_{t+1}|x_t, y_{t+1})$ as the proposal density. But since this is only rarely possible and the proposal density $q\left(x_{t+1}|x_{t+1}^i, y_{t+1}\right)$ only approximates the target density, the weights tend to degenerate, i.e., some become relatively large and other very small. For example, if this were the case of $\left\{\left(x_{1:t}^i, w_t^i\right)\right\}_{i=1}^N$ we could resample new N particles with repetition from these empirical distributions. The weights, after resampling, are then uniform, i.e., $1/N$. Some of the paths after resampling might be identical up to time t, but will be "jittered" after the sampling of the $t+1$ values (see Fig. 8.15 and Speekenbrink 2016 for a detailed tutorial on Particle filters).

So far, we have assumed that the vector of parameters θ is known. For example, if the parameters of a stochastic volatility model are known, then the sequential algorithm quite efficiently estimates the latent volatilities (posterior distribution), given the time series of observed returns. The application of the sequential Monte Carlo becomes more complicated. One possibility is to combine the MCMC and Particle filter techniques (see, e.g., the seminal stochastic volatility article by Kim et al. 1998). In a naïve approach (see Speekenbrink 2016), one may sequentially update also the parameters based on the likelihood $p(\theta| y_{1:t+1})$, which can be recursively approximated during the Particle filter algorithm based on the following decomposition:

$$p\left(\theta|y_{1:t+1}\right) = p(\theta|y_{1:t}) \int p\left(y_{t+1}|x_{t+1}, \theta\right) p(x_{t+1}|x_t, \theta) p(x_{1:t}|y_{1:t}, \theta) dx_{1:t+1}.$$

The parameters θ_{t+1}^i can be sampled starting from θ_t^i, for example, in random walk fashion with weights updated using the formula above and the latent state particle weights. This approach may work well in practice, but the estimation approach, which treats the parameters as time-varying stochastic variables, introduces an inconsistency into the model, which assumes that the parameters are

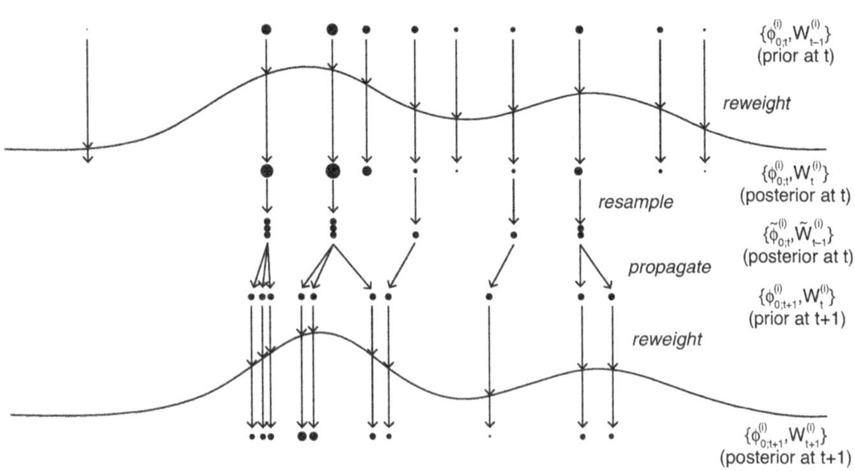

Fig. 8.15 A schematic representation of a resample-move particle filter (Speekenbrink 2016)

time invariant. Consistent particle filter algorithms estimating stochastic volatilities and the unknown model parameters have been proposed for example by Gilks and Berzuini (2001), Andrieu et al. (2004), Fulop and Li (2013), or Witzany and Fičura (2018).

8.4.8 Realized Volatility

While methods such as EWMA, GARCH, or MCMC work typically with daily returns, the concept of *realized volatility*, first defined by Andersen et al. (1999), depends on the availability of high-frequency data. The idea is simple—define the daily realized volatility by summing up intraday squared returns. By sampling the intraday returns sufficiently frequently, the realized volatility should approximate the *integrated volatility* arbitrarily well. To see this theoretically, let us consider a pure diffusion price process

$$dr = \mu dt + \sigma dz,$$

where $r = \ln S$ and the volatility itself is generally stochastic. Since

$$dr^2 = \mu^2 dt^2 + 2\mu\sigma dt dz + \sigma^2 dt, \quad \text{and}$$

$$\int_{t-1}^{t} \mu^2 ds^2 = 0, \quad \int_{t-1}^{t} 2\mu\sigma ds dz = 0,$$

the integrated volatility

$$IV(t) = \int_{t-1}^{t} \sigma^2(s) ds = \int_{t-1}^{t} dr^2$$

can be formally expressed as an integral of the squared instantaneous returns. The path-dependent integral of squared returns can be approximated by the realized volatility

$$RV(t,\Delta) = \sum_{j=1}^{n} r^2(t-1+j\Delta,\Delta) \tag{8.57}$$

with $\Delta = 1/n$ dividing the daily interval (from $t - 1$ to t) into n subintervals, and $r(s,\Delta)$ denoting the log-return between $s - \Delta$ and s. Assuming no jumps, the realized volatility (8.57) theoretically converges to the integrated volatility, i.e., $\lim_{n \to \infty} RV(t, 1/n) = IV(t)$. However, for a diffusion process with jumps

$$dr = \mu dt + \sigma dz + \kappa dN,$$

the quadratic variance of returns $QV(t) = \int_{t-1}^{t} dr^2$ splits into the integrated volatility and into the sum squared realized jumps

$$QV(t) = IV(t) + \sum_{t-1<s\leq t,\, dN(s)=1} \kappa^2(S).$$

Since the realized volatility calculated according to (8.57) converges to the quadratic variance rather than to stochastic volatility, Barnfdorff-Nielsen and Shephard (2004) proposed an alternative estimator called the *realized bipower variation*:

$$BV(t,\Delta) = \frac{\pi}{2}\sum_{j=1}^{n} \mid r(t-1+(j-1)\Delta,\Delta) \mid \times \mid r(t-1+j\Delta,\Delta) \mid . \qquad (8.58)$$

Even in the presence of jumps, the realized bipower volatility converges to the integrated volatility, $\lim_{n\to\infty} BV(t, 1/n) = IV(t)$. As a byproduct, the difference $RV(t, \Delta) - BV(t, \Delta)$ can be used to measure the presence of jumps.

A number of subsequent empirical studies have shown that the realized volatility does not necessarily converge to the theoretical implied volatility using very short time steps (e.g., tick data) due to microstructure effects such as bid-ask spreads and the autocorrelation of returns (Andersen et al. 2005). One simple way to solve this empirical issue is to use a frequency that is sufficiently high to estimate the daily realized volatility, but not so high that the impact of the infrastructure effects remains limited.

The realized volatility estimations are used to study the properties of historical volatility and to predict future expected volatility. The volatility prediction models of the ARIMA or ARFIMA type are based on persistence, i.e., the autoregression properties of volatility. Heterogeneous autoregressive (HAR) models incorporate different time horizons, which can be typical for different groups of market participants—see, e.g., Ficura (2013) for a detailed overview. The Bayesian MCMC estimation methods can be significantly improved by incorporating the high-frequency estimators (Fičura and Witzany 2016, 2018).

8.4.9 Model-Free Implied Volatility

In Sect. 8.2, we discussed the concept of Black-Scholes implied volatility. Its advantage, compared to estimations based on historical data, is that it is directly derived from the option market prices and represents an observable market consensus. However, this consensus is ambiguous due to the existence of a volatility smile or skew, i.e., different volatilities implied not only for different maturities but also for different strike prices. The existence of the volatility smile was one of the motives for the development of alternative stochastic models with stochastic volatility and jumps. Given a stochastic volatility model of the type (8.27), possibly with jumps, it does not make sense to ask what the constant future volatility for a maturity T is. We should ask rather what the expected integrated volatility $E\left[\int_0^T \sigma_t^2 dt\right]$ is, i.e., the expected future realized variance, or, equivalently, the expected (annualized) volatility realized over the time horizon from now to T:

$$\widehat{\sigma} = \sqrt{\frac{1}{T} E\left[\int_0^T \sigma_t^2 dt\right]}. \tag{8.59}$$

The concept of the model-free implied volatility aims to give an answer precisely to this question. According to the result of Britten-Jones and Neuberger (2000), given an asset following a diffusion process with stochastic volatility and given European call option prices $C(T, K)$ for all strike prices $K \geq 0$, the expected integrated volatility under the forward risk-neutral measure (making the maturity T forward price F_t a martingale) can be calculated as follows:

$$\begin{aligned} E_0^F\left[\int_0^T \sigma_t^2 dt\right] &= E_0^F\left[\int_0^T \left(\frac{dF_t}{F_t}\right)^2 dt\right] \\ &= 2\int_0^\infty 0 \frac{C^F(T, K) - \max(0, F_0 - K)}{K^2} dK, \end{aligned} \tag{8.60}$$

where $C^F(T, K) = C(T, K)/P(0, T)$ denotes the option price rediscounted forward to the time T.

This result was later generalized by Jiang and Tian (2005) to stochastic volatility models with jumps. Hence, literally speaking, the concept is not completely model-free, but it is applicable to a wide class of diffusion stochastic volatility models with jumps.

The main issue of practical implementation is the scarcity of data. There is usually a limited scale of strike prices (or deltas) with reliable quoted option prices. For other strike prices, the option values must be interpolated or extrapolated, and the result then certainly depends on the applied interpolation/extrapolation approach. In spite of these problems, on a liquid option market, the model-free volatility provides relatively good volatility predictions compared to other approaches (see, e.g., Ficura 2013).

Although the result (8.60) looks mysterious, its proof is relatively simple. We are going to outline the simplified proof of Jiang and Tian (2005) for a pure diffusion process. Let V_t be the stochastic variance and let F_t be the forward price following the martingale process $dF_t = \sqrt{V_t}F_t dz$ under the forward risk-neutral measure. Then, according to Ito's lemma,

$$d\ln F_t = -\frac{1}{2}V_t dt + \sqrt{V_t}dz. \tag{8.61}$$

Therefore, integrating the left-hand side and the right-hand side of (8.61) from 0 to T and taking the expectation, we have

$$E_0^F\left[\int_0^T V_t dt\right] = 2\left(\ln F_0 - E_0^F[\ln F_T]\right). \tag{8.62}$$

In order to show that the right-hand side of (8.62) equals to the right-hand side of (8.60), it is useful to note that $\int \frac{1}{K}dK = \ln K$ and $C^F(T,K) = E_0^F[\max(0, F_T - K)]$. Let us integrate the right-hand side of (8.60) by parts[5] in order to replace the term $u' = \frac{1}{K^2}$ by $u = \frac{-1}{K}$, i.e.,

$$\int_0^\infty \frac{C^F(T,K) - \max(0, F_0 - K)}{K^2}dK = \left[-\frac{C^F(T,K) - \max(0, F_0 - K)}{K}\right]_0^\infty + $$
$$+\int_0^\infty \frac{C_K^F(T,K) + 1_{\{F_0>K\}}}{K}dK. \tag{8.63}$$

Since

$$C_K^F(T,K) = \frac{\partial}{\partial K}E_0^F[\max(0, F_T - K)] = -E_0^F\left[1_{\{F_T>K\}}\right]$$

the integral on the right-hand side of (8.63) can be expressed as

$$\int_0^\infty \frac{C_K^F(T,K) + 1_{\{F_0>K\}}}{K}dK = E_0^F\left[\int_0^\infty \frac{-1_{\{F_T>K\}} + 1_{\{F_0>K\}}}{K}dK\right] = E_0^F\left[\int_{F_T}^{F_0}\frac{1}{K}dK\right] =$$
$$= E_0^F[\ln F_0 - \ln F_T] = \ln F_0 - E_0^F[\ln F_T].$$

It remains to be shown that the first term on the right-hand side of (8.63) disappears. First, for K large ($K = \infty$), the numerator of the fraction approaches zero and the fraction approaches zero as well. On the other hand, for K small,

[5] $\int u'v = uv - \int uv'$.

$K < F_0$, we have to work a little bit more. If $g(F_T)$ denotes the density of F_T then, since

$$\int_0^\infty g(F_T)dF_T = 1 \quad \text{and} \quad \int_0^\infty F_T g(F_T)dF_T = E[F_T] = F_0$$

the numerator can be written as

$$C^F(T,K) - (F_0 - K) = E_0^F[\max(0, F_T - K)] - (F_0 - K) =$$
$$= \int_K^\infty (F_T - K)g(F_T)dF_T - \int_0^\infty (F_T - K)g(F_T)dF_T = -\int_0^K (F_T - K)g(F_T)dF_T.$$

Under the mild condition that the density function $g(F_T)$ is bounded by a finite constant M the integral on the right-hand side can be estimated as

$$\left| -\int_0^K (F_T - K)g(F_T)dF_T \right| \leq \int_0^K KMdF_T = K^2M.$$

Therefore, the fraction

$$\left| \frac{C^F(T,K) - \max(0, F_0 - K)}{K} \right| \leq KM$$

vanishes when K goes to zero completing the proof. For a generalization of stochastic volatility processes with jumps, see Jiang and Tian (2005).

8.4.10 VIX Index and VIX Trading

The model-free volatility formula (8.60) is not only an interesting theoretical concept, but it has also become a practical tool for defining the well-known volatility index VIX. The VIX index, also called the fear index or fear gauge, was introduced by the Chicago Board Options Exchange (CBOE) in 1993, based on a methodology proposed by Whaley (2000). The goal of VIX is to measure the volatility of one-month at-the-money S&P 500 index options. Originally it was defined as an average of implied volatilities of a benchmark set of S&P 500 put and call at-the-money options. In 2003, CBOE introduced a new methodology based on the concept of model-free volatility. Since the used European style CBOE "weekly" SPX options have listed maturities with weekly frequency (settled on Fridays), to calculate the 30-day volatility there are two maturities that can be used: the *near term* (on Friday before the target 30-day maturity) and the *next term* one week later. The time-weighted average from the two (near term and next term) expected variances calculated according to (8.64) and annualized based on (8.59) then constitutes the exact VIX definition. The VIX formula for a (near or next) maturity T utilizes SPX call and put option prices $Q(K_i)$ for a list of strike prices K_i and takes the following form (see CBOE 2019):

$$\widehat{\sigma}^2 = \frac{2}{T}\sum_i \frac{e^{RT}Q(K_i)\Delta K_i}{K_i^2} - \frac{1}{T}\left(\frac{F_0}{K_{ATM}} - 1\right)^2, \tag{8.64}$$

where R denotes the risk-free rate to expiration, F_0 the forward index price, K_{ATM} the first listed strike below F_0, and $\Delta K_i = (K_{i+1} - K_{i-1})/2$ (for the leftmost and rightmost K_i it is just the length of the adjacent interval). The price $Q(K_i)$ is the mid (out-of-the-money) put option quoted price for $K_i < K_{ATM}$, the mid (out-of-the-money) call option price for $K_i > K_{ATM}$, and the average of the put and call (at-the-money) prices for $K_i = K_{ATM}$. To understand the link between the numerical formula (8.64) and the theoretical model free volatility formula (8.60), note that it follows from its proof that the indefinite integral can be approximated by a definite integral from K_{min} to K_{max} where K_{min}, K_{max} are far out-of-the money listed strike prices, i.e., $0 < K_{min} < F_0 < K_{max} < \infty$. The definite integral can be further split into the following three parts utilizing the put-call parity $C^F(T,K) - P^F(T,K) = F_0 - K$ and the fact that $\max(0, F_0 - K) = F_0 - K$ for $K \le K_{ATM}$ and $\max(0, F_0 - K) = 0$ for $K > K_{ATM}$:

$$\int_{K_{min}}^{K_{max}} \frac{C^F(T,K) - \max(0, F_0 - K)}{K^2} dK =$$

$$\int_{K_{min}}^{K_{ATM}} \frac{P^F(T,K)}{K^2} dK + \int_{K_{ATM}}^{K_{max}} \frac{C^F(T,K)}{K^2} dK - \int_{K_{ATM}}^{F_0} \frac{F_0 - K}{K^2} dK.$$

The first two integrals are simply numerically approximated by the sum in (8.64), while the third one can be evaluated and simplified analytically:

$$\int_{K_{ATM}}^{F_0} \frac{F_0 - K}{K^2} dK = \left[-\frac{F_0}{K} - \ln K\right]_{K_{ATM}}^{F_0}$$

$$= -\left(\ln \frac{F_0}{K_{ATM}} + 1 - \frac{F_0}{K_{ATM}}\right) \approx \frac{1}{2}\left(1 - \frac{F_0}{K_{ATM}}\right)^2.$$

The last approximation follows from Taylor's approximation $\ln(1 - x) + x \approx -\frac{1}{2}x^2$ with $x = 1 - \frac{F_0}{K_{ATM}}$.

Figure 8.16 shows the development of the VIX index since 2000. The volatility clearly reverts to a long-term level around 15% after several periods of high volatility representing crises or turmoils in the financial markets. The historical data provide

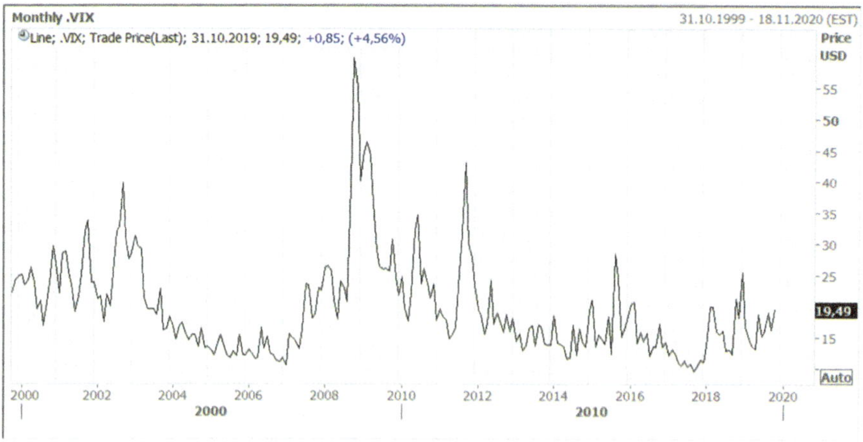

Fig. 8.16 VIX index history 2000–2019 (Source: Thomson Reuters)

strong evidence in favor of mean reverting stochastic volatility models with positive rather than negative jumps.

The introduction of VIX by CBOE was followed by many other volatility indices, e.g., VSTOXX based on the Dow Jones EURO STOXX 50 designed by EUREX, VDAX by Deutsche Börse, the Nikkei 225 volatility index JNIV, etc. In addition, in 2004 CBOE launched trading with futures, and later with options on VIX. Like stock index futures, the contracts are settled based on the difference between the settlement index value VIX_T and the futures price F_0, i.e.,

$$\text{payoff} = M(VIX_T - F_0).$$

In the case of VIX futures, the multiplier is $M = 1000$ USD, so that a one percentage point difference corresponds to 1000 USD.

The quoted VIX futures prices provide information on the market expectations about future equity index volatility. The VIX contracts can be used both for hedging and speculation. Since volatility is negatively correlated with the underlying index returns, a long position in the index can be used as a diversification tool by equity portfolio managers. Small investors, on the other hand, may enter into short positions, collecting the risk premium paid by the hedgers. Many empirical studies (see, e.g., Krol 2019) confirm that, due to negative systematic risk, the option implied volatility is in contango with respect to the objective probability measure expectation of the realized volatility. Practically, it means that the ex post realized volatility is on average lower than the volatility predicted by the option prices implied index. The difference is also called the *volatility risk premium,* and is explained by the fact that systematic risk insurers are willing to pay a premium when entering the hedging positions. The volatility risk premium is also reflected in the structure of VIX futures prices that usually go up with increasing contract expiration (Table 8.2).

Table 8.2 VX—Cboe S&P 500 Volatility Index (VIX) Futures quotes

Symbol	Expiration	Last	Change	High	Low	Settlement	Volume	Interest
VXV19	10/16/2019	19.00	0.45	19.96	18.50	18.58	16,510	94,395
VXX19	11/20/2019	19.50	0.25	20.05	19.20	19.28	16,603	172,195
VXZ19	12/18/2019	19.40	0.20	19.82	19.17	19.18	4489	52,789
VXF20	01/22/2020	19.95	0.25	20.30	19.72	19.78	2265	28,397
VXG20	02/19/2020	20.10	0.19	20.45	19.95	19.98	1293	22,686
VXH20	03/18/2020	19.96	0.11	20.31	19.85	19.88	654	14,734
VXJ20	04/15/2020	20.00	0.13	20.32	19.95	19.93	220	9339
VXK20	05/20/2020	19.85	0.07	20.15	19.85	19.83	147	5428
VXM20	06/17/2020	20.20	0.14	20.20	20.16	19.90	16	618

Source: www.cboe.com, 10.10.2019

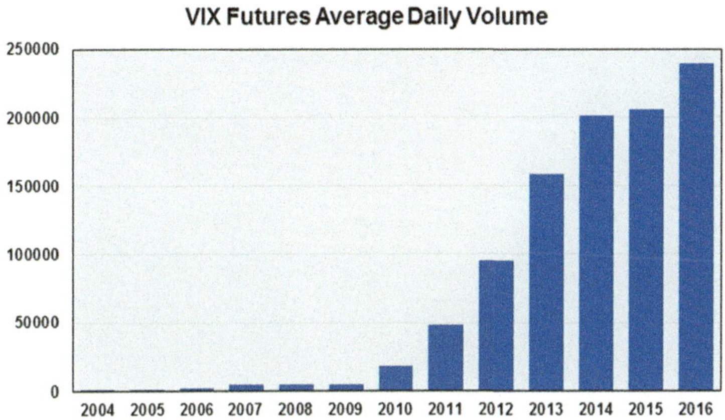

Fig. 8.17 VIX futures average daily volume—number of contracts (Source: CBOE Futures Exchange)

The VIX futures contracts have become very popular, with volumes going up almost exponentially (Fig. 8.17). The success of VIX trading has been again followed by other exchanges introducing similar contracts, such as VSTOXX futures on Euronext, Nikkei Volatility Index futures, etc.

Appendix A: Infinitesimals and the Elementary Stochastic Calculus

Throughout the text, we have liberally used the concept of infinitesimals in order to introduce and analyze the properties of various continuous-time stochastic processes built on hyperfinite binomial trees. The goal of this appendix is to provide a better foundation of infinitesimal calculus, hyperfinite probability theory, and elementary stochastic calculus with infinitesimals.

The notion of infinitesimals was discovered in the seventeenth century by Leibnitz and Newton and used to formulate classical differential and integral calculus. Later, it was unfortunately abandoned due to its weak foundations, and in 1966 it was rehabilitated by Abraham Robinson, the famous logician. Working with infinitesimals one must necessarily deal with the concept of a *horizon*. There is a "natural" perspective and an "extended" perspective, going beyond the horizon, under which numbers and objects indiscernible from the natural point of view become discernible from the extended point of view. This intuition has been consistently formalized by the results of Robinson (1966) or by other set-theoretical approaches like Vopěnka (1979).

During their education, pure mathematicians usually go through the construction of rational numbers defined as pairs of integers factored by an equivalence relation, and of real numbers defined as Cauchy sequences of rational numbers factored by another equivalence relation. Similarly, the extended real (*nonstandard*) or *hyperreal* numbers R^* are built from the standard real numbers as their sequences factored out by an equivalence relation based on a nontrivial *ultrafilter*. The construction is performed within the framework of classical set theory. Consequently, it is as well-founded and consistent as any other part of modern mathematics. Since the foundation of mathematics is not the subject of this text, we will not go into the details of this construction, but only state the key results and properties of the set of hyperreal numbers and the universe built on them.

© The Editor(s) (if applicable) and The Author(s), under exclusive license to Springer Nature Switzerland AG 2020

J. Witzany, *Derivatives*, Springer Texts in Business and Economics,
https://doi.org/10.1007/978-3-030-51751-9

Differential and Integral Calculus with Infinitesimals

The extension operator * mapping standard objects to their nonstandard counterparts can be applied not only to the set of real numbers itself, but similar to any function, relation, or operation built in one or more steps from the set of real numbers. The key *Transfer Principle* says that the (mutual) properties of the standard objects are preserved by the extension. The operation can be intuitively interpreted as an extension of objects only partially seen in front of the horizon, to the space beyond the horizon. The extended objects are marked with the * superscript, which can be dropped if there is no danger of confusion. So, for example, + denotes the addition operation on R and on R^* as well, $\leq$ the inequality relation on R or R^*, etc. The standard real numbers are identified with the corresponding elements of R^*. There is a new element of R^* that is not in R, in other words we do assume that $R \subset R^*$. Similarly, the set of natural numbers N is extended to the set of *hyper-finite natural numbers*$N \subset N^* \subset R^*$.

At this point, we can define the notion of infinitesimals and infinite numbers. For two hyperreal numbers $x, y \in R^*$ we say that

(a) x is *finite* if $|x| < n$ for some finite $n \in N$,
(b) x is *infinite* if $|x| > n$ for all finite $n \in N$,
(c) x is *infinitesimal*, $x \approx 0$, if $|x| < 1/n$ for all finite $n \in N$,
(d) x is *infinitesimally closeto* y, $x \approx y$, if $|x - y| \approx 0$.

Since R^* is a proper extension of R, there are infinite, as well as infinitesimal, numbers. It is easy to show that for any finite hyperreal number x there is a uniquely determined infinitesimally close standard real number $r \approx x$ called the standard part of x and denoted $st(x) = r$. The standard part can be interpreted as the projection of x onto the horizon between the standard and extended worlds. In other words, behind every real number r (observable in front of the horizon) there is a *monad* of all infinitesimally close hyperreal numbers

$$\mathrm{Mon}(r) = \{x \in R^*; x \approx r\} = st^{-1}(r),$$

i.e., the set of all numbers in the form $x + \epsilon$, where $\epsilon \approx 0$. For x and y infinitesimal, we also define $x \approx y$ if $|x/y| \approx 1$, i.e. if the difference between x and y is infinitesimal of a lower order with respect to x and y.

In the language of infinitesimals, we say that a function $f: R \to R$ is continuous at a point $x \in R$ if $f(x) \approx f(y)$ whenever $x \approx y$. Note that we implicitly work with the extended function $f^*: R^* \to R^*$. The Transfer Principle can be used to show that this intuitive definition is equivalent to the classical $\varepsilon - \delta$ definition of continuity.

The notions of limits, infinite series, derivatives, and integrals can be easily defined, and their properties are shown using the concept of infinitesimals. It is often useful to do some calculations in the extended world, but then to go back to the standard world where the infinitesimal "calculation errors" can be neglected and the results become much simpler.

For example, to calculate the derivative of a real function $f : R \to R$ at a point $x \in R$, we look at the slope

$$\frac{f(x+dx)-f(x)}{dx}=\frac{df}{dx} \tag{9.1}$$

calculated for a nonzero infinitesimal dx. If the result is finite and its standard part is a (standard) number y independent of our choice of dx, then we say that the derivative exists and equals to y,

$$f'(x)=\frac{df}{dx}=y.$$

Again, the "f" in (9.1) denotes, in fact, the extended function $f^* : R^* \to R^*$ while the derivative, if defined on R, is just a standard real-valued function $f' : R \to R$. To illustrate the concept, let $f(x) = x^2$. Then

$$\frac{(x+dx)^2-x^2}{dx}=\frac{2x\,dx+dx^2}{dx}=2x+dx\approx 2x$$

for any nonzero infinitesimal dx. Therefore, the derivative exists and $f'(x) = 2x$. Similarly, we can relatively easily differentiate other elementary functions and prove many key rules of the differential calculus.

Regarding integration, given a function $f : [a, b] \to R$, we look at the sums of the form

$$\sum_{i=0}^{H-1} f(x_i)dx_i \tag{9.2}$$

where $a = x_0 < x_1 < \cdots < x_H = b$ is an *infinitesimalgrid* on the interval $[a, b]$, i.e., $dx_i = x_{i+1} - x_i$ is infinitesimal for each $i = 0, \ldots H - 1$. For a nonnegative f the sum (9.2) corresponds to the area under the curve f approximated in the usual way. If the standard part of (9.2) exists and does not depend on the infinitesimal grid, then we say that the function has a definite integral on the interval $[a, b]$ and define

$$\int_a^b f(x)dx = st\left(\sum_{i=0}^{H-1} f(x_i)dx_i\right). \tag{9.3}$$

After two paragraphs defining derivatives and integrals, we are now prepared to prove the *Fundamental Theorems of Calculus*.

First, let f be a continuous function on $[a, b]$. Then we want to prove that the integral $F(z) = \int_a^z f(x)dx$ is defined for every $z \in [a, b]$ and that $F' = f$. In order to verify the existence of the integral we need to show that for any two infinitesimal grids the summations (9.2) are infinitesimally close. Refine the two grids to a common one. Then the difference between the two summations can be written as

$$\sum_{i=0}^{H-1} \left(f\left(x_i'\right) - f\left(x_i''\right)\right) dx_i \tag{9.4}$$

with $x_i' \approx x_i''$. Since f is continuous $f\left(x_i'\right) \approx f\left(x_i''\right)$, the maximum of the absolute differences $\max_i \left| f\left(x_i'\right) \approx f\left(x_i''\right) \right| = \epsilon$ is also infinitesimal. Therefore, the absolute difference between the two summations given by (9.4) is at most $\epsilon(z - a) \approx 0$ proving that the integral is well defined. Next, by the definition of F, for any standard real numbers $z_1 < z_2$,

$$u(z_2 - z_1) \leq F(z_2) - F(z_1) \leq v(z_2 - z_1),$$

where u is the minimum of f on $[z_1, z_2]$, and v is the maximum. The same statement holds for any $z_1 < z_2$ in R^*, and so $F' = f$ by the definition of the derivative and by the continuity of f. Similarly, we can use infinitesimals to prove easily the Second Fundamental Theorem of Calculus, which states that if F is an antiderivative of f, i.e., $F' = f$ on $[a, b]$ then

$$\int_a^b f(x)dx = F(b) - F(a).$$

The proof of the First Fundamental Theorem of Calculus illustrates how we can use the advantages of both worlds. Various definitions and constructions can be elaborated in the world with infinitesimals, but then it is also useful to work in the classical world without infinitesimals, where the infinitesimal errors can be neglected, i.e., the hyperreal numbers are "rounded" to the standard real numbers.

Hyperfinite Probability Theory

A set A is called *hyperfinite* if it can be enumerated by hyperfinite numbers $k = 1, \ldots, H$, where $H \in N^*$. Hyperfinite sets have the same properties as finite sets according to the Transfer Principle. At this point we must be a little bit more careful. In the standard world, any subset of a finite set is finite. However, in the extended world, it is not true that any subset of a hyperfinite set is hyperfinite. Since any finite subset of N has a maximum element, the same must hold for hyperfinite subsets of N^*. But the set $N \subset \{1, \ldots, H\}$ of all finite natural numbers is a subset of a hyperfinite set with $H \in N^* \backslash N$ elements, and yet it does not clearly have a maximum element. Therefore, the set N itself is not a hyperfinite set. This is explained by the fact that the Transfer Principle can be applied only to statements with bounded quantifiers, i.e., bounded to sets from the universe $V(R)$, which means to sets and relations that are built from R in the standard world. When the statement is transferred to the extended world, every quantifier is limited to a set A^*, where $A \in V(R)$, which does not necessarily contain all possible subsets and relations built from R^*. The elements of sets A^*, where $A \in V(R)$, are called*internal sets.* The Transfer Principle says that internal sets have the same properties as the standard sets, and the argument above simply shows that N is not an internal set, but that it is, rather, an *externalset.*

In Sect. 4.2, we introduced finite probability spaces and the basic probability concepts, such as mean, variance, and standard deviation. A *hyperfiniteprobability space* is a hyperfinite set Ω with an elementary probability function $P:\Omega \to [0,1]$ so that $\sum_{\omega\in\Omega} P(\omega) = 1$. A typical example of a hyperfinite probability space is a hyperfinite set Ω with the *counting measure*$P(\omega) = 1/\mid\Omega|$. All the finite probability concepts apply to the hyperfinite probability spaces. The elementary probability function can be easily extended to a real-valued probability measure Pr on the algebra of internal subsets of Ω (closed under the operations of intersection, complement, and union). Since every internal subset A of Ω is hyperfinite, we can simply set

$$\Pr(A) = st\left(\sum_{\omega\in A} P(\omega)\right).$$

If needed, the measure can be extended to a σ-complete measure by the Caratheodory extension theorem (Loeb 1979).

Similarly, a general hyperfinite measure $m : \Omega \to [0,1]$, without the condition $\sum_{\omega\in\Omega} m(\omega) = 1$, can be extended to a σ-complete standard measure. For example, if

$$\Omega = \{0, \delta t, \ldots, H\delta t\} \subset [0,T]^*$$

where $\delta t = T/H$, and $m(\omega) = T/(H+1)$, then the σ-completion of the measure yields the classical Lebesgue measure on the interval $[0, T]$.

Elementary Stochastic Calculus

In Sect. 4.2, we have defined the key stochastic calculus concepts on finite binomial trees, and freely extended the binomial tree definitions and results to continuous time through the *hyperfinite binomial trees.* Our goal is to make this idea a little bit more precise following Albeverio et al. (1986), Cutland et al. (1991), Witzany (2008), or Herzberg (2013).

We define a hyperfinite binomial tree as the set of sequence of "ups" and "downs"

$$\Omega_H = \{U, D\}^H$$

where H is a hyperfinite integer. For an $\omega \in \Omega_H$ it is also useful to define the corresponding $\omega_s \in \{+1, -1\}^H$ by $\omega_s(k) = \pm 1$ if $\omega(k) = U$ or D (for $k \le H$). We use the tree $\Omega_T = \Omega_H$ to model stochastic processes with elementary infinitesimal time steps of length $\delta t = T/H$ over a time interval $[0, T]$. We call the time step δt elementary, because it is the shortest infinitesimal step on the grid $\mathcal{T} = \{0, \delta t, 2\delta t, \ldots, H\delta t\}$—note that an infinite multiple of δt can still be infinitesimal, and we typically denote such a time step dt. The basic probability measure on Ω_T is the counting measure with $P(\omega) = 1/2^H$ for every $\omega \in \Omega_T$ corresponding to 50% branching probability at each one-step binomial tree. The key building block used in stochastic asset price modeling is the Wiener process formally defined as a function $z : \mathcal{T} \times \Omega_T \to R^*$. For an $\omega \in \Omega_T$ and $t = k\delta t$ the value of the Wiener process depends only on the path up to (and including) the time t, i.e.$\omega \upharpoonright k = \langle \omega(1), \ldots, \omega(k) \rangle$. Inductively, $z(0, \omega) = 0$ and

$$z(k\delta t, \omega) = z((k-1)\delta t, \omega) + \omega_s(k)\sqrt{\delta t} \quad \text{for} \quad k = 1, \ldots, H.$$

The distribution of

$$z(T) \in \left\{ -H\sqrt{\delta t}, \ldots, 0, \ldots, H\sqrt{\delta t} \right\}$$

can be expressed in terms of the hyperfinite binomial distribution. It can be shown algebraically (Cutland et al. 1991) that

$$\Pr[z(T) \leq x] \approx \Phi(x; 0, T) \tag{9.5}$$

where $x \in R^*$ and $\Phi(x; 0, T)$ is the cumulative distribution function of the normal distribution with the mean 0 and variance T.

The same conclusion follows from the nonstandard counterpart of the Central Limit Theorem (CLT): If $\langle X_k; k = 1, \ldots, H \rangle$ is a hyperfinite sequence of independent identically distributed variables with the mean 0 and variance 1, then the distribution of $Y = \frac{1}{\sqrt{H}} \sum_{k=1}^{H} X_k$ is infinitesimally close to the standard normal distribution, i.e., for every standard real x

$$\Pr[Y \leq x] \approx \Phi(x).$$

Note that $z(T) = \frac{\sqrt{T}}{\sqrt{H}} \sum_{k=1}^{H} X_k$ where $X_k(\omega) = \omega_s(k) \in \{-1, +1\}$, and so (9.5) is a direct consequence of the CLT. Analogously, the increments $z(t_2) - z(t_1)$ are $N(0, t_2 - t_1)$ distributed, whenever $t_2 - t_1 > 0$ is an infinite multiple of δt. Moreover, by the construction, the increments are mutually independent, and so z (its standard part) has the classical properties of a Wiener process.

However, since we often change the measure on Ω_T, we need to be a little bit more general. The change of measure we consider is always based on a change in the splitting probabilities to $0.5 \mp \frac{\lambda}{2}\sqrt{\delta t}$ where λ is finite. Then the values of $X_i(\omega)$ must be appropriately adjusted to $X_i(\omega) = \omega_s(k) - \lambda\sqrt{\delta t}$ to keep $E[X_i] = 0$ and

$\mathrm{var}[X_i] \approx 1$. To prove that $z(T) = \frac{\sqrt{T}}{\sqrt{H}} \sum_{i=1}^{H} X_i$ is still $N(0, T)$ normally distributed, we need to apply a nonstandard counterpart of a more general CLT, e.g., in the Lindberg form. Again, it follows that z has the classical distributional properties of the Wiener process. Therefore, we may consider the wider class of probability measures and Wiener process representations.

For the purpose of most of our applications, we use stochastic differential equations (SDE) of the form

$$dx = m(x,t)dt + s(x,t)dz. \tag{9.6}$$

If $\mathcal{T} = \{0, \delta t, 2\delta t, \ldots, H\delta t\} \subset [0,1]$ is an infinitesimal grid and if z is a Wiener process representation then (9.6) can be solved recursively calculating

$$\begin{aligned} \delta x(t) &= m(x(t),t)\delta t + s(x(t),t)\delta z(t), \quad \text{and} \\ x(t+\delta t) &= x(t) + \delta x(t). \end{aligned} \tag{9.7}$$

The process (9.7) is always well defined on Ω_T, but the question is (similarly to the definite integral definition) whether its distributional properties depend on the grid and on the Wiener process representation. Only if the answer is yes, can we define the standard part of (9.7) as a well-defined solution of the stochastic differential Eq. (9.6). Moreover, we often need to neglect the terms of order lower than dt, and again the question is whether the distributional properties of the solution of (9.7) are preserved. It is shown in Albeverio (1986), Witzany (2008), or Herzberg (2013) that the answer is yes, provided the functions $m(x,t)$ and $s(x,t)$ satisfy certain regularity conditions.

The second property is needed for the proof of the *Ito'slemma*. Let $G(x,t)$ be a differentiable function, then the Taylor's expansion theorem states that

$$\begin{aligned} \delta G(x,t) &= G(x+\delta x, t+\delta t) - G(x,t) \approx \\ &\approx \delta x \frac{\partial G}{\partial x} + \delta t \frac{\partial G}{\partial t} + \frac{1}{2}\delta x^2 \frac{\partial^2 G}{\partial x^2} + \frac{1}{2}\delta t^2 \frac{\partial^2 G}{\partial t^2} + \delta x \delta t \frac{\partial^2 G}{\partial x \partial t} + \cdots \end{aligned} \tag{9.8}$$

If δx from (9.7) is substituted into (9.8), then we can use the fact $\delta z^2 = \delta t$ and neglect all the terms of an order lower than δt (considering δz to be of the order $\sqrt{\delta t}$). Therefore, we get

$$\delta G(x,t) \approx \left(\frac{\partial G}{\partial t} + m(x,t)\frac{\partial G}{\partial x} + \frac{1}{2} s\left(x,t\right)^2 \frac{\partial^2 G}{\partial x^2} \right) \delta t + s(x,t)\frac{\partial G}{\partial x}\delta z. \tag{9.9}$$

Because the difference between the left-hand side and the right-hand side of (9.9) is of an order lower than δt and because the coefficients of δt and δz satisfy the regularity conditions, any solution of the Ito's stochastic differential equation

$$dG = \left(\frac{\partial G}{\partial t} + m\frac{\partial G}{\partial x} + \frac{1}{2}s^2\frac{\partial^2 G}{\partial x^2}\right)dt + s\frac{\partial G}{\partial x}dz. \qquad (9.10)$$

is infinitesimally close to the transformed process $G(x(t), t)$. That is, the transformed process is a unique (in terms of the standard distributional properties) and well-defined solution of (9.10) with the initial condition $G(0) = G(x(0), 0)$.

Jump Processes

The concept of hyperfinite binomial trees can also be used to define a pure Poisson jump process J with a probability of jump λ and constant jump size m_J. In this case

$$J(k\delta t, \omega) = J((k-1)\delta t, \omega) + m_J I(\omega(k) = U) \quad \text{for} \quad k = 1, \ldots, H, \ \ \omega \in \Omega_T,$$

and where I is the indicator function. Since a jump should happen only with the probability $\lambda\delta t$ over a time interval of length δt, we have to define a specific Poisson jump process probability measure on Ω_T by setting

$$P(\omega) = \prod_k (\lambda\delta t)^{I(\omega(k)=U)}(1-\lambda\delta t)^{I(\omega(k)=D)}.$$

If both λ and m_J are finite, then the expected number of jumps is finite and the expected value of J at time T is finite, $E[J(T)] = \lambda m_J T$. Notice that the construction works also for an infinite λ and an infinitesimal m_J such that $\lambda\delta t$ is still infinitesimal and λm_J finite. In this case, we get a Levy type process with infinitely many (infinitesimal) jumps over $[0, T]$ and still with a finite expected value $\lambda m_J T$.

A jump diffusion process can then be built on the product of the diffusion process space $\Omega_{D,T}$ and the jump process space $\Omega_{J,T}$. The jump process hyperfinite tree becomes slightly more complicated if want to include random jump size with certain given distribution. One possibility is to introduce a third space $\Omega_{Z,T}$ representing a family of independent jump size random variables $Z(t)$. In this case, $\Omega_{Z,T}$ would be just a product of H copies of a hyperfinite space representing the desired jump size distribution (not a hyperfinite tree). For a finite λ and normally distributed jump size we get only the Merton model pure jump process (which can be further combined with the diffusion process to obtain the Merton model jump diffusion process). With infinitesimal λ, and by applying the principle of superposition (see Sect. 8.3), again on a product of two or more trees, we may get hyperfinite representations of various Levy stochastic processes (such as the NIG or VG processes).

Appendix B: A Primer of Classical Stochastic Calculus

Although we believe that infinitesimals allow us to explain many key ideas in an efficient way, and use this approach throughout the book, we must admit that the mainstream stochastic finance literature is based on the concepts of classical calculus. The goal of this appendix is to provide the reader with an overview of the key classical concepts and compare them briefly with our (hyper)finite space and infinitesimal calculus-based approach.

Infinite Probability Spaces and Filtrations

The first issue that must be handled by the classical probability theory is that on an infinite probability space Ω the probability measure cannot be defined just by assigning probabilities to individual elementary events $\omega \in \Omega$, since the probabilities would have to be zero, except at finitely many elements of Ω. The solution is to assign probabilities $P(A)$ to subsets $A \subseteq \Omega$ so that the *probability measure* satisfies the properties analogous to finite probability spaces, in particular: $P(\varnothing) = 0$, $P(\Omega) = 1$, and $P(A \cup B) = P(A) + P(B)$ for disjoint $A.\ B \subseteq \Omega$. The (hyper)finite additivity in fact corresponds to the *σ-additivity* property:

$$P\left(\bigcup_{n=1}^{\infty} A_n\right) = \sum_{n=1}^{\infty} P(A_n) \quad \text{for disjoint subsets} \quad A_1, A_2, \ldots \quad \text{of} \quad \Omega$$

It turns out that, unfortunately, a probability measure satisfying all these properties cannot, in general, be defined on all subsets of an infinite space Ω but only on a collection $\mathcal{F}$ of subsets of Ω having the properties of a *σ-algebra*:

(i) $\varnothing, \Omega \in \mathcal{F}$,
(ii) whenever $A \in \mathcal{F}$ then $A^c \in \mathcal{F}$, and
(iii) whenever $A_1, A_2, \ldots \in \mathcal{F}$ then $\bigcup_{n=1}^{\infty} A_n \in \mathcal{F}$.

© The Editor(s) (if applicable) and The Author(s), under exclusive license to Springer Nature Switzerland AG 2020

J. Witzany, *Derivatives*, Springer Texts in Business and Economics,
https://doi.org/10.1007/978-3-030-51751-9

The most well-known example is the *uniform (Lebesgue) measure* on the unit interval $[0, 1]$ satisfying the property that $P([a, b]) = b - a$ for any $0 \leq a \leq b \leq 1$. The probability measure is defined on all *Borel* subsets of $[0, 1]$ and it can be extended to all Lebesgue subsets satisfying the completeness property, i.e., if $A \subseteq [0, 1]$ is a null set, $P(A) = 0$, then all subsets of A are also measurable (belong to $\mathcal{F}$ and their probability measure is necessarily zero).

Things become even more complicated when we try to generalize the finite binomial trees to infinite ones in the classical context. A possible attempt (see Appendix A and Shreve 2004) is to define an infinite binomial tree as a set of infinite sequences of ups and downs, $\Omega_\infty = \{U, D\}^{\{1,2,3,\ldots\}}$. The space is infinite and the probability of each path $\omega \in \Omega_\infty$ must be zero if the probability of an upmove is $p \in (0, 1)$, and $1 - p$ for a downmove. In order to define the probability measure corresponding to the finite case concept, we may define simple collections $\mathcal{F}_n$ of subsets of Ω_∞ representing the information until step n. That is, given any $\widetilde{\omega} = \{\widetilde{\omega}(1), \ldots, \widetilde{\omega}(n)\} \in \Omega_n$ let us firstly define (an atom) $A(\widetilde{\omega}) = \{\omega \in \Omega_\infty; \omega \uparrow n = \widetilde{\omega}\}$ to be the set of all infinite sequences starting with $\widetilde{\omega}$ using the notation from Appendix A. For any such set $A = A(\widetilde{\omega}) \in \mathcal{F}_n$, it is straightforward to define the probability $P_n(A) = p^{n_{\mathrm{up}}}(1 - p)^{n - n_{\mathrm{up}}}$ where n_{up} is the number of ups in $\widetilde{\omega}$. Next, given a finite set $\widetilde{\mathrm{A}} \subseteq \Omega_n$ define $A = \left\{\omega \in \Omega_\infty; \omega \restriction \mathrm{n} \in \widetilde{\mathrm{A}}\right\}$ also as a member of $\mathcal{F}_n$ and $P_n(A)$ to be the finite sum of probabilities corresponding to individual $\widetilde{\omega} \in \widetilde{A}$. Therefore, $\mathcal{F}_n$ naturally extends the collection of all subsets of Ω_n to subsets of Ω_∞ and the finitely additive measure defined on the sets of finite sequences to the sets of infinite sequences that belong to the collection $\mathcal{F}_n$. Note that the collections $\mathcal{F}_n$ get finer and finer with increasing n as there is more and more information, formally $\mathcal{F}_0 \subseteq \mathcal{F}_1 \subseteq \mathcal{F}_2 \subseteq \cdots$ where $\mathcal{F}_0 = \{\varnothing, \Omega_\infty\}$ represents no information algebra. Every $\mathcal{F}_n$ is a σ-algebra on Ω_n since it is finite, but this property does not hold for the union of all $\mathcal{F}_n$. It can be shown that the algebra $\cup\mathcal{F}_n$ of subsets Ω_∞ can be extended to a σ-algebra $\mathcal{F}_\infty$ just by putting in all other sets in order to satisfy the σ-algebra properties. The collection of increasing σ-algebras $\mathcal{F}_0 \subseteq \mathcal{F}_1 \subseteq \mathcal{F}_2 \subseteq \cdots \subseteq \mathcal{F}_\infty$ is an example of a *filtration*. Since the probability measures P_n defined on $\mathcal{F}_n$ are mutually consistent in the sense that $P_n(A) = P_m(A)$ for $n < m$ and $A \in \mathcal{F}_n$, the measure $P(A) = P_n(A)$ is well defined on the algebra $\cup\mathcal{F}_n$. It can, with some effort, be uniquely extended to a σ-additive probability measure defined on $\mathcal{F}_\infty$.

The theoretical construction above unfortunately is still not enough for the purpose of continuous time stochastic modeling where the time is indexed by $t \in [0, T]$. A *filtration* on a nonempty set Ω is generally defined as a system of σ-algebras $\mathcal{F}(t)$, $t \in [0, T]$ such that $\mathcal{F}(s) \subseteq \mathcal{F}(t)$ for any $s < t$.

Note that this abstract definition does not involve any time structure of the individual elementary events $\omega \in \Omega$. The information until time t is now encoded in the σ-algebra $\mathcal{F}(t)$. On the other hand, many standard probability spaces used in stochastic calculus do have some time structure embedded into the elementary events.

Example B.1 The space $\Omega = C_0[0, T]$ of continuous functions $\omega : [0, T] \to R$ such that $\omega(0) = 0$ provides the simplest setting in which one may construct the Brownian motion. If we start with a σ-algebra $\mathcal{F}(T)$ containing all reasonable subsets of $C_0[0, T]$ then $\mathcal{F}(t)$ for $t < T$ can be obtained by filtering $\mathcal{F}(T)$ to keep only the sets A that are homogenous with respect to the information available at time t. That is for any $\widetilde{\omega} \in C_0[0, t]$ the set (atom) $\{\omega \in C_0[0, T]; \omega\restriction[0, t] = \widetilde{\omega}\}$ must be either a subset of A or disjoint with A. It remains to define a probability measure on $\mathcal{F}(T)$ so that the functions $\omega \in C_0[0, T]$ represent the paths of the Brownian motion.

Other key concepts that need to be defined in the context of stochastic calculus are the expectation and conditional expectation of a random variable.

Random Variables and Expectations

For a finite space Ω, any real-valued function $X : \Omega \to R$ represents a *random variable*. For an infinite probability space $(\Omega, \mathcal{F}, P)$, we have to be more careful and limit ourselves only to measurable functions $X : \Omega \to R$, i.e., to functions such that the set $X^{-1}(B) = \{\omega \in \Omega; X(\omega) \in B\}$ is $\mathcal{F}$-measurable for any Borel[1] measurable $B \in \mathcal{B}(R)$. This condition is necessary if we want to speak about the probability $P[X \in B] = P(X^{-1}(B))$ for reasonable subsets B of R. In this case, we define $\mu_X(B) = P[X \in B]$ as a new probability measure on all Borel subsets od R. We can also introduce the standard cumulative distribution function (cdf) $F_X(x) = P[X \leq x]$ for $x \in R$.

For a finite probability space, the expected value of a random variable X is defined as the probability-weighted average and can be written in integral form as

$$E[X] = \sum_{\omega \in \Omega} X(\omega)P(\omega) = \int_{\omega} XdP.$$

In the case of infinite probability spaces, the summation is not well defined, and we need to work hard to extend properly the notion of the integral. The idea is to define firstly the integral for a random variable X that takes only finitely many values $x_1, \ldots, x_n$ using the finite summation:

$$\int_{\Omega} XdP = \sum_{i=1}^{n} x_i P[X = x_i].$$

[1]The collection $\mathcal{B}(R)$ of Borel measurable subsets of R contains all intervals and all the other sets generated by the σ-algebra operations. It represents the collection of all "reasonable" subsets of R. The Lebesgue measure, which we have defined on [0,1], can be analogously extended to all Borel subsets of R.

For a general nonnegative random variable X, the (Lebesgue) integral $\int_\Omega XdP$ is defined as the limit of approximations of X with finitely many values from above and from below. Of course, one has to prove that the definition is correct. Note that the limit might be infinite; in this case we set $\int_\Omega XdP = \infty$. Finally, for a general $X : \Omega \to R$ that takes positive and negative values, we set

$$\int_\Omega XdP = \int_\Omega X^+ dP - \int_\Omega X^- dP$$

where $X^+ = \max(X^+, 0)$ and $X^- = \max(-X^-, 0)$ so that $X = X^+ - X^-$. The integral on the left-hand side is defined only if at least one of the integrals on the right-hand side is finite (and so it takes values in $[-\infty, +\infty]$). Otherwise, it is not assigned any value since the expression $\infty - \infty$ is not well defined. The integral is additive, and so

$$E[aX + bY] = \int_\Omega (aX + bY)dP = a\int_\Omega XdP + b\int_\Omega YdP = aE[X] + bE[Y],$$

provided the expectations on the right-hand side exist and a, b are constants. The Lebesgue integral also satisfies the Monotone Convergence Theorem, which can be formulated in terms of random variables as follows:

Let $0 \le X_1 \le X_2 \le \cdots$ be an (almost surely[2]) increasing sequence of random variables converging (almost surely) to X. Then $\lim_{n\to\infty} E[X_n] = E[X]$.

This convergence theorem is important for the transfer of various properties that can be easily shown for piecewise constant random variables taking only finitely many values to general random variables. Having the concept of integral, we can formally define the notion of a probability density function of a random variable X. We say that a nonnegative measurable function $f : R \to R_0^+$ is a probability density function (pdf) of X if

$$F_X(x) = \int_{-\infty}^{x} f(u) \quad \text{for any} \quad x \in R.$$

It follows that

---

[2]Generally, we say that a property of $\omega \in \Omega$ holds *almost surely* if the set of ω satisfying the property has probability 1, and so the set of ω not satisfying the property (exceptions) is null (has probability 0 but is not necessarily empty).

$$P[X \in B] = \int_B f(u)du = \int_{-\infty}^{\infty} I(u \in B)f(u)du$$

where $I(u \in B)$ is the indicator function. The cumulative distribution function is well defined for all random variables, but the density function does not always have to exist. Note that if X takes only finitely many values $x_1, \ldots, x_n$ then it is tempting to set $f(x_i) = P[X = x_i]$ and zero otherwise. But if f is positive only at finitely many points, it is zero almost surely, and so the integral $\int_B f(u)du$ is always zero. The density function does exist for random variables with a continuous differentiable cumulative distribution function where it follows that it is obtained by differentiating the cdf, $f(x) = F'(x)$. In this case $P[X = x] = 0$ necessarily for all $x \in R$ but $f(x) > 0$ on a non-null set. In this case, $f(x)\Delta x$ only approximates the probability that X takes a value in the neighborhood of $x \in R$,

$$P[x - \Delta x/2 \leq X < x + \Delta x/2].$$

Change of Measure

For a finite probability space Ω, the change of measure P to another measure Q can be expressed by a random variable Z satisfying the property $Q(\omega) = Z(\omega)P(\omega)$. The variable is well defined only if $Q(\omega) = 0$ whenever $P(\omega) = 0$. In that case, we can simply set $Z(\omega) = \frac{Q(\omega)}{P(\omega)}$ for $P(\omega) > 0$ and $Z(\omega) = 0$ otherwise. The variable Z tells us where to revise the probability upward ($Z > 1$) and downward ($Z < 1$). The expectation of Z (with respect to P) must be 1 since

$$E_P[Z] = \sum_{\Omega} Z(\omega)P(\omega) = \sum_{\Omega} Q(\omega) = 1.$$

In financial modeling, we want to keep the space of possible scenarios and simply adjust the probabilities, so that whatever scenario was possible before the change of measure ($P(\omega) > 0$) will be possible after the change ($Q(\omega) > 0$) and vice versa. In this case, we say that the probability measures are *equivalent.*

In the case of infinite probability spaces, the concept again becomes more difficult to handle, since we cannot work directly with elementary events, but rather with measurable sets. However, if $(\Omega, \mathcal{F}, P)$ is a probability space and Z an almost surely nonnegative random variable such that $E_P[Z] = 1$ then it is relatively easy to show (using the Monotone Convergence Theorem) that $Q(A) = \int_A Z\,dP$ defines a new probability measure. It also follows from the definition that $E_Q[X] = E_P[ZX]$ for any nonnegative random variable X and equivalently $E_Q[X/Z] = E_P[X]$ provided Z is almost surely positive.

We say that two probability measures P and Q on $(\Omega, \mathcal{F})$ are *equivalent* if $P(A) > 0$ whenever $Q(A) > 0$ for any $A \in \mathcal{F}$, in other words if the two measures agree which sets have probability zero. If the variable Z is almost surely positive, then the measures P and Q are equivalent. The Radon–Nikodým Theorem proves the opposite: If P and Q on $(\Omega, \mathcal{F})$ are equivalent, then there exists an almost surely positive variable Z such that $E[Z] = 1$ and $Q(A) = \int_A Z\,dP$ for any $A \in \mathcal{F}$. The variable Z is called the Radon–Nikodým derivative of Q and it is formally written as

$$Z = \frac{dQ}{dP}.$$

Example B.2 We say that a random variable X is standard normal if it has the probability density function

$$\varphi\left(x; \mu, \sigma^2\right) = \frac{1}{\sigma\sqrt{2\pi}} e^{-\frac{(x-\mu)^2}{2\sigma^2}}.$$

It follows that its mean is $E[X] = \mu$ and the variance $\mathrm{Var}[X] = E[(X - EX)^2] = \sigma^2$. A normal variable is called standard if $\mu = 0$, $\sigma = 1$. Let us assume that we have a normal variable X on a probability space $(\Omega, \mathcal{F}, P)$ having a positive mean θ and variance 1 and we would like to change it to a standard variable. One possibility is to simply subtract θ from X and get standardized $\widetilde{X} = X - \theta$. But let us say that we do not want to change the variable itself but just the probability measure P to a measure Q so that it becomes standard normal, i.e., will have mean zero instead of θ. Intuitively, we need, somehow, to revise the probability downward when $X > 0$ and upward it when $X < 0$. Let us assume that the Radon–Nikodým derivative that determines Q is in the form $Z(\omega) = z(X(\omega))$ where z is a real value function. Generally, if f is the density function of X under the original measure P then zf is the density function under Q since

$$Q[X \le b] = E_Q[I\{X \le b\}] = E_P[z(X)I\{X \le b\}] = \int_{-\infty}^{b} z(x)f(x)dx.$$

Therefore, if g is our target density then it is enough to set $z = g/f$ provided f is (almost surely) positive. In the case of the normal distributions we get the following function:

$$z(x) = \frac{\varphi(x; 0, 1)}{\varphi(x; \theta, 1)} = \exp\left(-\frac{x^2}{2} + \frac{(x-\theta)^2}{2}\right) = \exp\left(-x\theta + \frac{\theta^2}{2}\right),$$

Thus, the Radon–Nikodým derivative is $Z(\omega) = \exp\left(-X(\omega)\theta + \theta^2/2\right)$.[3] Notice that $Z < 1$ if $X > \theta/2$ and $Z > 1$ if $Z < \theta/2$. By construction, the mean of the variable

[3]The variable Z can be expressed as well in terms of the standard normal (under P) variable $\widetilde{X} = X - \theta$ as $Z(\omega) = \exp\left(-\widetilde{X}(\omega)\theta - \theta^2/2\right)$.

X becomes zero under the new measure Q and, in addition, the variable remains normal with unit variance as desired.

The change of probability measure technique outlined above lies at the core of the equivalent martingale measure theorem, according to which a process with positive drift (expectation over a period) can be changed to a martingale with zero drift (expectation) just by changing the underlying probability measure and not the process itself.

Conditional Expectations and Stochastic Processes

In the context of finite binomial trees an *adapted stochastic process* is defined by a sequence of random variables $X_0, \ldots, X_n$ on $\Omega_n = \{U, D\}^n$ such that $X_k(\omega)$ depends only on the information until time k, i.e., on $\omega \upharpoonright k$ for $\omega \in \Omega_n$, $k \leq n$. This condition can be equivalently formulated by requiring that X_k is $\mathcal{F}_k$-measurable where $\mathcal{F}_k$ is the (σ-)algebra on Ω_n corresponding to the collection of all subsets of Ω_k. For any $\widetilde{\omega} \in \Omega_k$ and $A \in \mathcal{F}_k$, the atom $\{\omega \in \Omega_n; \omega \upharpoonright \mathrm{k} = \widetilde{\omega}\}$ either belongs to A or is disjoint with A, and so the $\mathcal{F}_k$-measurability condition is indeed only about respecting the information until time k.

Let us assume that $\mathcal{F}(t), t \in [0, T]$ is a filtration on a nonempty space. Let $X(t)$ be a collection of random variables indexed by $t \in [0, T]$ such that $X(t)$ is $\mathcal{F}(t)$-measurable. The collection is then called an *adapted stochastic process*.

Given a probability measure P defined on $\mathcal{F}_T$ we can directly calculate the expected values $E_p[X_t]$ for any $t \in [0, T]$, since P is also defined on $\mathcal{F}_t \subseteq \mathcal{F}_T$. What we need to define is the concept of *conditional expectation* $E_P[X_T | \mathcal{F}_t]$ of X_T given the information until time t.

Going back to finite binomial trees, we were able to define directly the conditional expectation of X_n for any $\widetilde{\omega} \in \Omega_n$ and $k < n$ as the probability weighted average of values of X_n over all paths $\omega \in \Omega_n$ that start with $\widetilde{\omega} \upharpoonright \mathrm{k}$, i.e.,

$$E[X_n | \widetilde{\omega} \upharpoonright \mathrm{k}] = \frac{1}{P(\widetilde{\omega} \upharpoonright \mathrm{k})} \sum_{\omega \upharpoonright \mathrm{k} = \widetilde{\omega} \upharpoonright \mathrm{k}} X(\omega) P(\omega), \quad \text{where} \quad P(\widetilde{\omega} \upharpoonright \mathrm{k}) = \sum_{\omega \upharpoonright \mathrm{k} = \widetilde{\omega} \upharpoonright \mathrm{k}} P(\omega).$$

Note that the conditional expectation $Y(\widetilde{\omega}) = E[X_n | \widetilde{\omega} \upharpoonright \mathrm{k}]$ is a function of $\widetilde{\omega} \in \Omega_n$ that depends only on the information until time n, and so it is $\mathcal{F}_k$-measurable. In addition, its partial average over any $A \in \mathcal{F}_k$ will be the same as for X_n, i.e., $\int_A Y dP = \int_A X_n dP$.

Thus, let us use these observations and define generally the conditional expectation $E[X | \mathcal{G}]$ for an $\mathcal{F}$-measurable variable X and $\mathcal{G} \subseteq \mathcal{F}$. We say that a random variable Y is the conditional expectation denoted $E[X | \mathcal{G}]$ if

(i) Y is $\mathcal{G}$-measurable, and
(ii) the partial average $\int_A Y dP = \int_A X dP$ for any $A \in \mathcal{G}$.

That looks like a sound definition, but one needs to work hard to show that the conditional expectation exists and is in a sense unique. Let us assume (without loss of generality) that X is nonnegative. The uniqueness of the conditional expectation follows from the property (ii). If Y_1, Y_2 were two random variables satisfying (i) and (ii) then $\int_A (Y_2 - Y_1)dP = 0$ for any $A \in \mathcal{G}$, and so $Y_1 = Y_2$ almost surely. The existence can be proved by a trick using the Radon–Nikodým Theorem defining the new measure $Q(A) = \int_A \frac{X+1}{E[X+1]} dP$ for any $A \in \mathcal{F}$. Restricting the two measures P and Q to $\mathcal{G}$ we can apply the Radon–Nikodým Theorem and get the derivative $Z = dQ/dP$, i.e., a $\mathcal{G}$-measurable random variable such that

$$\int_A \frac{X+1}{E[X+1]} dP = \int_A Z\, dP, \text{or equivalently}$$

$$\int_A XdP = \int_A (E[X+1]Z - 1)\, dP \quad \text{for any} \quad A \in \mathcal{G}.$$

Therefore $Y = E[X + 1]Z - 1$ will be the desired random variable satisfying the properties (i) and (ii), i.e., the conditional expectation.

We could certainly continue in the development of the classical stochastic calculus, for example, by introducing the Markov property, the Brownian motion, Itô's integral, Itô's Lemma, etc. However, the goal of this appendix was simply to provide an overview of the key concepts, and so we shall stop at this point. The aim was also to highlight the difference between the intuitive approach based on infinitesimals and the theoretically robust, but sometimes counter-intuitive classical approach. We have seen that the concepts and properties of finite probability spaces that are more-or-less automatically transferred to hyperfinite spaces must, in the classical approach, be generalized to infinite spaces in an abstract, ingenious, and often technically more complicated way with much more effort. By saying that the classical approach is theoretically robust, we do not mean that the infinitesimal approach is not. It must be reiterated that the infinitesimal approach is based on robust mathematical foundations. But it is true that the classical stochastic calculus is the mainstream and well developed approach in financial mathematics literature, and so it might be useful to understand its foundations.

References

Acerbi, C., & Szekely, B. (2014). Back-testing expected shortfall. *Risk, 27*(11), 76–81.

Albeverio, S., Fenstad, J. E., Hoegh-Krohn, R., & Lindstrom, T. (1986). *Nonstandard methods in stochastic analysis and mathematical physics*. New York: Dover Publications – Mineola.

Andersen, T. G., Bollerslev, T., Christoffersen, P. F., & Diebold, F. X. (2005). *Volatility forecasting*. National Bureau of Economic Research, March.

Andersen, T. G., Bollerslev, T., Diebold, F. X., & Labys, P. (1999). *Understanding, optimizing, using and forecasting realized volatility and correlation*. Working paper. http://public.econ.duke.edu/~boller/Published_Papers/risk_00.pdf

Andersen, L. B. G., & Brotherton-Ratcliffe, R. (1998). The equity option volatility smile: An implicit finite difference approach. *Journal of Computational Finance, 1*(2), 5–37.

Andrieu, C., Doucet, A., Singh, S. S., & Tadic, V. B. (2004). Particle methods for change detection, system identification, and control. *Proceedings of the IEEE, 92*(3), 423–438.

Andrieu, C., Doucet, A., & Holenstein, R. (2010). Particle Markov chain Monte Carlo methods. *Journal of the Royal Statistical Society: Series B (Statistical Methodology), 72*(3), 269–342.

Arlt, J., & Arltová, M. (2007). *Ekonomické časové řady*. Prague: Grada Publishing.

Artzner, P., Delbaen, F., Eber, J. M., & Heath, D. (1999). Coherent measures of risk. *Mathematical Finance, 9*, 203–228.

Bachelier, L. (1900). Theory of speculation (translation of 1900 French edition). In P. H. Cootner (Ed.), *The random character of stock market prices* (Vol. 1964, pp. 17–78). Cambridge: MIT Press.

Baran, J. (2016). *Post-crisis valuation of derivatives*. Doctoral thesis, Faculty of Finance and Accounting – University of Economics, Prague.

Baran, J., & Voříšek, J. (2011). A note on multivariate EVT VaR and ES estimation. In Jiří Málek (Ed.), *Modely řízení finančních rizik* (pp. 45–50). Prague: Oeconomica.

Baran, J., & Witzany, J. (2012). A comparison of EVT and standard VaR estimations. *Bulletin of the Czech Econometric Society, 19*(29).

Baran, J., & Witzany, J. (2013). Konstrukce výnosových křivek v pokrizovém období. In Jiří Málek (Ed.), *Risk Management 2012* (pp. 69–100). Prague: Oeconomica.

Barndorff-Nielsen, O. E., & Shephard, N. (2004). Power and bipower variation with stochastic volatility and jumps. *Journal of Financial Econometrics, 2*(1), 1–37.

Bauwens, L., Laurent, S., & Rombouts, J. (2006). Multivariate GARCH models: A survey. *Journal of Applied Econometrics, 21*, 79–109.

BCBS. (1988). *Basel committee on banking supervision*. "Basel Capital Accord", BIS, July.

BCBS. (1996). *Amendment to the capital accord to incorporate the market risk*. BIS, January.

BCBS. (2004). *International convergence of capital measurement and capital standards: A revised framework*. BIS, June.

BCBS. (2006). *International convergence of capital measurement and capital standards – A revised framework comprehensive version*. BIS, June.

© The Editor(s) (if applicable) and The Author(s), under exclusive license to Springer Nature Switzerland AG 2020

J. Witzany, *Derivatives*, Springer Texts in Business and Economics,
https://doi.org/10.1007/978-3-030-51751-9

BCBS. (2009a). *Revisions to the Basel II market risk framework*. BIS, July.

BCBS. (2009b). *Enhancements to the Basel II framework*. BIS, July.

BCBS. (2010a). *Basel III: International framework for liquidity risk measurement, standards and monitoring*. BIS, December.

BCBS. (2010b). *Basel III: A global regulatory framework for more resilient banks and banking systems*. BIS, December.

BCBS. (2016a). *Minimum capital requirements for market risk*. Standards, BIS, January.

BCBS. (2016b). *Interest rate risk in the banking book*. Standards, BIS, April.

BCBS. (2018). *Basel III monitoring report*. BIS, October.

BCBS. (2019). *Minimum capital requirements for market risk*. BIS, January (rev. February 2019).

Black, F. (1976, March). The pricing of commodity contracts. *Journal of Financial Economics, 3*, 167–179.

Black, F., & Karasinski, P. (1991). Bond and option pricing when short rates are lognormal. *Financial Analysts Journal, 47*, 52–59.

Black, F., & Scholes, M. (1973). The pricing of options and corporate liabilities. *Journal of Political Economy, 81*, 637–659.

Bollerslev, T. (1986). Generalized autoregressive conditional heteroscedasticity. *Journal of Econometrics, 31*, 307–327.

Brace, A., Gatarek, D., & Musiela, M. (1997). The market model of interest rate dynamics. *Mathematical Finance, 7/2*, 127–155.

Breeden, D. T., & Litzenberger, R. H. (1978). Price of state-contingent claims implicit in option prices. *Journal of Business, 51*, 621–651.

Brigo, D., & Mercurio, F. (2006). *Interest rate models – Theory and practice with smile, inflation and credit*. Heidelberg: Springer.

Brigo, D., & Pallavicini, A. (2008). *Counterparty risk and contingent CDS valuation under correlation between interest-rates and default*. SSRN worling paper (pp. 1–19). Available at SSRN: http://ssrn.com/abstract=926067

Britten-Jones, M., & Neuberger, A. (2000). Option prices, implied price processes and stochastic volatility. *Journal of Finance, 55*, 839–866.

Brotherton-Ratcliffe, R., & Iben, B. (1993). *Yield curve applications of swap products, advanced strategies in financial risk management* (pp. 400–450). New York: Institute of Finance.

CBOE. (2019). *CBOE volatility index – VIX Whitepaper, Chicago Board Options Exchange*. http://www.cboe.com/micro/vix/vixwhite.pdf

Cerny, J. (2011). *Stochastic interest rate modeling*. Diploma thesis, Charles University, Faculty of Mathematics and Physics.

Cerny, J., & Witzany, J. (2013). *Interest rate swap credit value adjustment* (pp. 1–15). SSRN working paper. Available at SSRN: http://ssrn.com/abstract=2302519 (July 29, 2013).

Chernobai, A. S., Rachev, S. T., & Fabozzi, F. J. (2008). *Operational risk: A guide to Basel II capital requirements, models, and analysis* (vol. 180). Wiley.

Christoffersen, P. F. (1998). Evaluating interval forecasts. *International Economic Review, 39*, 841–862.

Cipra, T. (2008). *Finanční ekonometrie*. Praha: Ekopress.

Cipra, T. (2010). *Financial and insurance formulas*. New York: Springer.

Cont, R., & Tankov, P. (2004). *Financial modelling with jump processes*. Chapman & Hall/CRC Financial Mathematics Series.

Cox, J. C., Ingersoll, J. E., & Ross, S. A. (1981). The relation between forward prices and futures prices. *Journal of Financial Economics, 9*, 321–346.

Cox, J. C., Ingersoll, J. E., & Ross, S. A. (1985). A theory of the term structure of interest rates. *Econometrica, 53*(2), 385–407.

Cox, J. C., Ross, S. A., & Rubinstein, M. (1979). Option pricing: A simplified approach. *Journal of Financial Economics, 7*, 229–264.

Crépey, S., Bielecki, T. R., & Brigo, D. (2014). *Counterparty risk and funding: A tale of two puzzles*. New York: Chapman and Hall/CRC.

Cutland, N. J., Kopp, P. E., & Willinger, W. (1991). A nonstandard approach to option pricing. *Mathematical Finance, 1*(4), 1–38.

Danielsson, J., Jorgensen, B. N., Mandira, S., Samorodnitsky, G., & de Vries, C. G. (2005). *Subadditivity re-examined: The case for value-at-risk* (No. 24668). London School of Economics and Political Science, LSE Library.

Del Moral, P. (1996). Non-linear filtering: Interacting particle resolution. *Markov Processes and Related Fields, 2*(4), 555–581.

Derman, E., & Kani, I. (1994, February). Riding on a smile. *Risk, 7*, 32–39.

Dothan, L. U. (1978). On the term structure of interest rates. *Journal of Financial Economics, 6*, 59–69.

Du, Z., & Escanciano, J. C. (2017). Backtesting expected shortfall: Accounting for tail risk. *Management Science, 63*(4), 940–958.

Dupire, B. (1994). Pricing with a smile. *Risk*, February, 18–20.

Dvořák, Petr (2010). *Deriváty*, 2nd ed. Oeconomica – University of Economics in Prague, 2006.

EBA. (2018a). *Guidelines on the management of interest rate risk arising from non-trading book activities*. European Banking Authority, EBA/GL/2018/02, 19 July.

EBA. (2018b). *Guidelines on the revised common procedures and methodologies for the supervisory review and evaluation process (SREP) and supervisory stress testing*. Final Report, European Banking Authority, EBA/GL/2018/03, 19 July.

EBA. (2019). *2020 EU-wide stress test, methodological note*. European Banking Authority, 25 June.

ECB. (2018). *ECB Guide to the internal capital adequacy assessment process (ICAAP)*. European Banking Authority, November.

Embrechts, P., & Wang, R. (2015). Seven proofs for the subadditivity of expected shortfall. *Dependence Modeling, 3*(1).

Engle, R. F. (2002). Dynamic conditional correlation – A simple class of multivariate GARCH models. *Journal of Business and Economic Statistics, 20*, 339–350.

Engle, R., & Kroner, F. K. (1995). Multivariate simultaneous generalized ARCH. *Econometric Theory, 11*, 122–150.

Eraker, B., Johannes, M., & Polson, N. (2003). The impact of jumps in volatility and returns. *The Journal of Finance, LVIII*(3), 1269–1300.

Feller, W. (1968). *Introduction to probability theory and its applications*. New York: Wiley.

Fičura, M. (2013). *Metody předvídání volatility*. Diploma thesis, University of Economics in Prague.

Ficura, M., & Witzany, J. (2016). Estimating stochastic volatility and jumps using high-frequency data and Bayesian methods. *Czech Journal of Economics and Finance (Finance a uver), 66*(4), 278–301.

Fičura, M., & Witzany, J. (2018). *Estimation of SVJD models with Bayesian methods and power-variation estimators*. ssrn.com/abstract=3212975

Fulop, A., & Li, J. (2013). Efficient learning via simulation: A marginalized resample-move approach. *Journal of Econometrics, 176*(2), 146–161.

Garman, M. B., & Kohlhagen, S. W. (1983). Foreign currency option values. *Journal of International Money and Finance, 2*(3), 231–237.

Geske, R. (1979). The valuation of compound options. *Journal of Financial Economics, 7*, 63–81.

Gilks, W. R., & Berzuini, C. (2001). Following a moving target—Monte Carlo inference for dynamic Bayesian models. *Journal of the Royal Statistical Society: Series B (Statistical Methodology), 63*(1), 127–146.

Gnedenko, B. (1943). Sur la distribution limite du terme maximum d'une serie aleatoire. *Annals of Mathematics*, 423–453.

Gordon, N. J., Salmond, D. J., & Smith, A. F. (1993). Novel approach to nonlinear/non-Gaussian Bayesian state estimation. In *IEE proceedigngs F (Radar and signal processing)* (vol. 140, no. 2, pp. 107–113). IET Digital Library.

Gregory, J. (2010). *Counterparty credit risk*. Wiley Finance.

Gregory, J. (2015). *The XVA challenge: Counterparty credit risk, funding, collateral, and capital*. New York: Wiley.

Hagan, P. S., Kumar, D., Lesniewski, A. S., & Woodward, D. E. (2002). Managing smile risk. *The Best of Wilmott, 1*, 249–296.

Heath, D., Jarrow, R., & Morton, A. (1992). Bond pricing and the term structure of interest rates: A new methodology for contingent claims valuation. *Econometrica: Journal of the Econometric Society*, 77–105.

Herzberg, F. (2013). *Stochastic calculus with infinitesimals*. New York: Springer.

Heston, S. (1993). Closed-form solution of options with stochastic volatility with application to bond and currency options. *Review of Financial Studies, 6*, 327–343.

Ho, T. S. Y., & Lee, S. B. (1986). Term structure movements and pricing interest rate contingent claims. *Journal of Finance, 41*, 1011–1029.

Hull, J. (2010). *Risk management and Financial Institutions* (2nd ed.). Upper Saddle River, NJ: Pearson.

Hull, J. (2018). *Options, futures, and other derivatives* (10th ed.). Upper Saddle River, NJ: Prentice Hall.

Hull, J., & Suo, W. (2002, June). A methodology for the assessment of model risk and its application to the implied volatility function model. *Journal of Financial and Quantitative Analysis, 37*(2), 297–318.

Hull, J., & White, A. (1987). The pricing of options on assets with stochastic volatilities. *The Journal of Finance, 42*, 281–300.

Hull, J., & White, A. (1990a). Pricing interest rate derivative securities. *Review of Financial Studies, 3*, 573–592.

Hull, J., & White, A. (1990b). Valuing derivative securities using the explicit finite difference method. *Journal of Financial and Quantitative Analysis, 25*(1), 87–100.

Hull, J., & White, A. (2012a). *LIBOR vs. OIS: The derivatives discounting dilemma*. Working paper, University of Toronto.

Hull, J., & White, A. (2012b). The FVA debate. *Risk, 25*(8), 83–85.

Hull, J., & White, A. (2012c). The FVA debate continues: Hull and White respond to their critics. *Risk, 52*, 18–22.

Hull, J., & White, A. (2014). Valuing derivatives: Funding value adjustments and fair value. *Financial Analysts Journal, 70*(3), 46–56.

Hurd, A. E., & Loeb, P. A. (1985). *An introduction to nonstandard real analysis*. New York: Academic Press.

IASB. (2011). *IFRS 13 fair value measurement*. https://www.ifrs.org/issued-standards/list-of-standards/ifrs-13-fair-value-measurement/, International Accounting Standard Board, May.

IASB. (2014). *IFRS 9 financial instruments*. https://www.ifrs.org/issued-standards/list-of-standards/ifrs-9-financial-instruments/, International Accounting Standard Board, July.

Jacquier, E., Polson, N., & Rossi, P. (1994, October). Bayesian analysis of stochastic volatility models. *Journal of Business & Economic Statistics, 12*(4), 69–87.

Jamshidian, F. (1989). An exact bond option pricing formula. *The Journal of Finance, 44*, 205–209.

Jamshidian, F. (1995). A simple class of square-root interest rate models. *Applied Mathematical Finance, 2*, 61–72.

Jamshidian, F. (1997). Libor and swap market models and measures. *Finance and Stochastics, 1*, 293–330.

Jiang, G. J., & Tian, Y. S. (2005). The model-free implied volatility and its information content. *Review of Financial Studies, 18*, 1305–1342.

Johannes, M., & Polson, N. (2009). MCMC methods for financial econometrics. In Ait-Sahalia & L. P. Hansen (Eds.), *Handbook of financial econometrics* (pp. 1–72). Amsterdam: North Holland.

Jolliffe, I. T. (2002). *Principal component analysis, series: Springer series in statistics* (2nd ed.). New York: Springer.

Keisler, H. J. (1976). *Foundations of infinitesimal calculus* (Vol. 20). Boston: Prindle, Weber & Schmidt.

Keisler, H. J. (2013). *Elementary calculus: An infinitesimal approach*. Courier Corporation.

Kim, S., Shephard, N., & Chib, S. (1998). Stochastic volatility: Likelihood inference and comparison with ARCH models. *The Review of Economic Studies, 65*(3), 361–393.

Kopp, E., Malczak, J., & Zastawniak, T. (2013). *Probability for finance*. Cambridge: Cambridge University Press.

Krol, A. (2019). *CBOE Volatility index and volatility premium structure analysis*. Master Thesis, University of Economics, Prague.

Kupiec, P. (1995). Techniques for verifying the accuracy of risk management models. *Journal of Derivatives, 3*, 73–84.

Loeb, P. A. (1979). An introduction to nonstandard analysis and hyperfinite probability theory. In Bharucha-Reid (Ed.), *Probabilistic analysis and related topics II*. New York: Academic Press.

Lynch, S. M. (2007). *Introduction to applied bayesian statistics and estimation for social scientists*. Springer.

Madan, D. B., Carr, P. P., & Chang, E. C. (1998). The variance-gamma process and option pricing. *European Finance Review, 2*, 79–105.

Málek, J. (2005). *Dynamika úrokových měr a úrokové deriváty*. Praha: Ekopress.

Markowitz, H. (1952). Portfolio selection. *Journal of Finance, 7*(1), 77–91.

Merton, R. C. (1973). Theory of rational option pricing. *Bell Journal of Economics and Management Science, 4*, 141–183.

Merton, R. C. (1976). Option pricing when underlying stock returns are discontinuous. *Journal of Financial Economics, 3*, 125–144.

Molodkina, K. (2018). *SABR model a jeho využití při modelování volatility*. Diploma thesis, University of Economics, Prague.

Nelson, E. (1977). Internal set theory: A new approach to nonstandard analysis. *Bulletin of the American Mathematical Society, 83*(6), 1165–1198.

Pelsser, A. (2003). Mathematical foundation of convexity correction. *Quantitative Finance, 3*, 59–65.

Pykhtin, M. (2012). Model foundations of the Basel III standardised CVA charge. *Risk, 25*(7), 60.

Rachev, S. T., Hsu, J. S, Bagasheva, B. S., & Fabozzi, F. J. (2008). *Bayesian methods in finance*. The Frank J. Fabozzi Series. Wiley.

Rendleman, R. J., & Bartter, B. J. (1980). The pricing of options on debt securities. *Journal of Financial and Quantitative Analysis, 15*, 11–24.

Repullo, R., Saurina, J., & Trucharte, C. (2009). *Mitigating the procyclicality of basel II*. CEPR discussion paper 7382.

Robinson, A. (1966). *Nonstandard analysis*. Amsterdam: North-Holland.

Rubinstein, M. (1994). Implied binomial trees. *Journal of Finance, LXIX*, 771–818.

Shephard, N. (2004). *Stochastic volatility: Selected readings*. Oxford: Oxford University Press.

Shreve, S. (2004). *Stochastic calculus for finance II – Continuous time models*. New York: Springer.

Shreve, S. (2005). *Stochastic calculus for finance I – The binomial asset pricing model*. New York: Springer.

Speekenbrink, M. (2016). A tutorial on particle filters. *Journal of Mathematical Psychology, 73*, 140–152.

Tasche, D. (2013). *Expected Shortfall is not elicitable. So what*? Presentation at Imperial College, London. https://workspace.imperial.ac.uk/mathfin/Public/Seminars

Turnbull, S. M., & Wakeman, L. M. (1991). A quick algorithm for pricing European average options. *Journal of Financial and Quantitative Analysis*, 377–389.

Vasicek, O. (1977). An equilibrium characterization of the term structure. *Journal of Financial Economics, 5*, 177–188.

Večeř, J. (2001). A new PDE approach for pricing arithmetic average Asian options. *Journal of Computational Finance, 4*(4), 105–113.

Vopěnka, P. (1979). *Mathematics in the alternative set theory*. Teubner-Leipzig.

Vopěnka, P. (2010). *Calculus infinitesimalis – Pars Prima*. OPS-Nymburk.

Vopěnka, P. (2011). *Calculus infinitesimalis – Pars Secunda*. OPS-Nymburk.
Whaley, R. E. (2000). The investor fear gauge. *The Journal of Portfolio Management, 26*(3), 12–17.
Wilmott, P. (2006). *Paul Wilmott on quantitative finance* (2nd ed.). New York: Wiley.
Witzany, J. (2007). *International financial markets* (1st ed.). Prague: Oeconomia.
Witzany, J. (2008). Construction of equivalent martingale measures with infinitesimals. *Charles University, KPMS Preprint, 60*, 1–16.
Witzany, J. (2009). Valuation of convexity related derivatives. *Prague Economic Papers, 4*, 309–326.
Witzany, J. (2010). *Credit risk management and modeling*. Prague: Oeconomica.
Witzany, J. (2011). *Financial derivatives and market risk management – Parts I and II*. Praha: Nakladatelství Oeconomica.
Witzany, J. (2013a). *Estimating correlated jumps and stochastic volatilities* (pp. 251–283). Prague Economic Papers, 2/2013.
Witzany, J. (2013b). *Financial derivatives – Valuation, hedging and risk managements*. Praha: Nakladatelství Oeconomica.
Witzany, J. (2017). *Credit risk management – Pricing, measurement, and modeling*. Cham: Springer.
Witzany, J., & Fičura, M. (2018). *Sequential Gibbs particle filter algorithm with an application to stochastic volatility and jumps estimation*. ssrn.com/abstract=3194544

Index

© The Editor(s) (if applicable) and The Author(s), under exclusive license to Springer Nature Switzerland AG 2020
J. Witzany, *Derivatives*, Springer Texts in Business and Economics,
https://doi.org/10.1007/978-3-030-51751-9

S

T

U

V

The manufacturer's authorised representative in the EU is Springer Nature Customer Service Centre GmbH, Europaplatz 3, 69115 Heidelberg, Germany. If you have any concerns regarding our products, please contact ProductSafety@springernature.com

Printed and bound by CPI Group (UK) Ltd, Croydon, CR0 4YY
22/09/2026
02221048-0002